Process Technology Troubleshooting

Charles E. Thomas

DELMAR
CENGAGE Learning™

Australia • Brazil • Japan • Korea • Mexico • Singapore • Spain • United Kingdom • United States

Process Technology Troubleshooting
Charles E. Thomas

Vice President, Technology and Trades ABU: David Garza

Director of Learning Solutions: Sandy Clark

Managing Editor: Larry Main

Acquisitions Editor: David Boelio

Product Manager: Sharon Chambliss

Marketing Manager: Kevin Rivenburg

Marketing Coordinator: Mark Pierro

Director of Production: Patty Stephan

Production Manager: Stacy Masucci

Content Project Manager: Michael Tubbert

Art Director: Benj Gleeksman

Technology Director: Joe Pliss

Technology Project Manager: Christopher Catalina

Editorial Assistant: Lauren Stone

Cover Image: Getty Images, Inc.

For product information and technology assistance, contact us at
Cengage Learning Customer & Sales Support, 1-800-354-9706
For permission to use material from this text or product,
submit all requests online **www.cengage.com/permissions**
Further permissions questions can be emailed to
permissionrequest@cengage.com

Library of Congress Control Number: 2007942513

ISBN-13: 978-1-4283-1100-8

ISBN-10: 1-4283-1100-9

Delmar
Executive Woods
5 Maxwell Drive
Clifton Park, NY 12065
USA

Cengage Learning is a leading provider of customized learning solutions with office locations around the globe, including Singapore, the United Kingdom, Australia, Mexico, Brazil, and Japan. Locate your local office at **www.cengage.com/global**

Cengage Learning products are represented in Canada by Nelson Education, Ltd.

To learn more about Delmar, visit **www.cengage.com/delmar**

Purchase any of our products at your local college store or at our preferred online store **www.cengagebrain.com**

Notice to the Reader
Publisher does not warrant or guarantee any of the products described herein or perform any independent analysis in connection with any of the product information contained herein. Publisher does not assume, and expressly disclaims, any obligation to obtain and include information other than that provided to it by the manufacturer. The reader is expressly warned to consider and adopt all safety precautions that might be indicated by the activities described herein and to avoid all potential hazards. By following the instructions contained herein, the reader willingly assumes all risks in connection with such instructions. The publisher makes no representations or warranties of any kind, including but not limited to, the warranties of fitness for particular purpose or merchantability, nor are any such representations implied with respect to the material set forth herein, and the publisher takes no responsibility with respect to such material. The publisher shall not be liable for any special, consequential, or exemplary damages resulting, in whole or part, from the readers' use of, or reliance upon, this material.

Printed in the United States of America
3 4 5 6 7   26 25 24 23 22

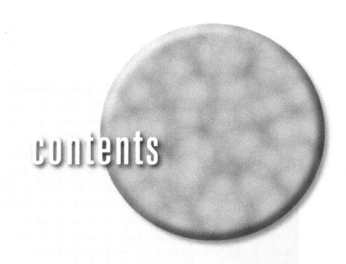

contents

v

Appendix 2 Principles of Total Quality Management 437

Index 447

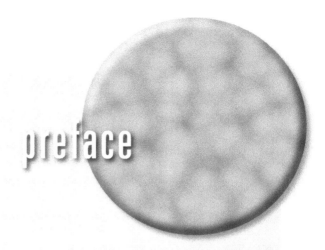

preface

Process technology is defined as "the study and application of the scientific principles (math, physics, chemistry) associated with the operation (instruments, equipment, systems, troubleshooting) and maintenance (safety and quality) of the chemical processing industry . . ." (as defined in the regionally accredited process curriculum).

This book, *Process Technology Troubleshooting,* is the product of years of design, development, and implementation within the community college system. It is the first text to use the model of the original educators for teaching process troubleshooting. This model applies concepts from advanced instrumentation, control loop, process equipment, and systems, as well as a series of what-if scenarios supported by a complex array of nine standardized processes. These processes include simple pump-around systems, compressor model, heat transfer model, cooling tower model, boiler model, furnace model, distillation, reactions, and separations. This text is presented from a process technician's point of view and has numerous illustrations and schematics. It is the only textbook to use modern troubleshooting methods and models presently being taught in the community college system.

Unique features include:
- simple to complex ideals
- line-art schematics and illustrations
- usability in several classes
- customizable systems for college professors
- simulation of real life problems
- simulation of typical equipment, systems, and operation processes
- illustration of the application of basic instrumentation concepts
- conveying upper-level process instrumentation principles
- demonstration of correct instrument systems and operator responses
- preparation of students for console operations
- simulation of reaction–response scenarios
- simulation of equipment–instrument failures
- preparation of students to solve a variety of analytical and qualitative variables
- demonstration of basic equipment and system relationships
- presentation of information from a process technician's viewpoint, and
- building upon the foundation of previous textbooks authored by the writer.

The future process technician will have a specialized degree in process technology that will include instruction in engineering principles, math, physics, chemistry, maintaining unit operations: safety, checking equipment, sampling, taking readings, making rounds, troubleshooting, using quality tools, and operating new computer systems. These new apprentice technicians will need strong technical and problem-solving skills, as well as the ability to assimilate cutting edge technologies quickly and apply innovative ideas. In addition to these skills, a process technician will need to be able to handle conflict, look at a complex situation, see the overall picture, and communicate effectively.

The author would like to express his thanks to the many educators who have been involved in the success of the process technology program and to the thousands of technicians who have been successfully placed in the chemical processing industry.

Charles E. Thomas, Ph.D.
2008

chapter 1

Introduction to Process Instrumentation and Troubleshooting

LEARNING OBJECTIVES

After studying this chapter, the student will be able to:

- Describe the basic instruments used in the process industry.
- Describe the key terms associated with process control and troubleshooting.
- List the various types of control loops.
- Explain how symbols and diagrams are used by process technicians in the chemical processing industry.
- Describe the various troubleshooting models.
- Describe the methods used in troubleshooting process problems.

Key Terms

Analytical variable—process variable associated with the chemical or physical composition of a material.

Analyzer—a device used to measure physical and chemical compositions of materials. Analyzers are devices designed to identify the presence of a substance or the concentration (%) of the substance in a process stream.

Alarm—a device designed to alert operators when a process variable exceeds established parameters.

Analog signal—transmitted continuously as air pressure, electric current, or voltage.

Bellows pressure element—a corrugated metal tube that contracts and expands in response to pressure changes.

Bourdon tube—a hook-shaped, thin-walled tube attached to a mechanical linkage that expands and contracts in response to pressure changes.

Control loop—a collection of instruments that work together to automatically control process variables such as analytical, pressure, temperature, flow, or level. A control loop typically has five parts: primary elements or sensors are coupled to a transmitter that sends a signal to a controller, which compares the process variable to a fixed set point and sends a signal to a transducer that converts the signal and sends it to a final control element that is typically a valve.

Controller—an instrument used to compare a process variable to a desired value and initiate a change to return the process to set point if a variance exists.

Control station—contains display consoles that can be used to control the process.

Demultiplexer—the signal separator for a complex data highway in a distributed control system.

Differential pressure (DP) cell—measures the difference between two pressure points.

Distributed control system—a computer-based system that controls and monitors process variables.

Engineer's console—a computer workstation used by engineers to change control schemes and modify control systems.

Final control element—the device in a control loop that actually adjusts the process, typically a control valve.

Graph display—historical and current information is displayed in bar and line graphs.

Graphics display—an electronic process flow diagram used to simplify console operations.

Indicator—an instrument used to indicate process variables.

Instrument—a device used to measure or control flow, temperature, level, pressure, or composition.

List display—an electronic list that displays information on alarms, equipment status, process averages, and process variables.

Multiplexer—a device used to translate signals as they enter a data highway.

Pressure transmitter—an instrument used to measure pressure. It sends a signal to a controller, recorder, or indicator.

Process instrumentation—transmitters, controllers, transducers, primary elements, and sensors that control and monitor process variables.

Recorder—an instrument used to record process variables.

Resistive temperature detector (RTD)—a device used to measure temperature changes by changes in electrical resistance.

Rotameter—a flow meter that allows fluid to move through a clear tube that has a ball or float in it. Numbers are usually displayed on the side of the tube to indicate flow rate.

Sight glass gauge—a level measurement device attached to a vessel that allows an operator to see the corresponding liquid level.

Set point—the desired value of process variable fed to a controller.

Supervisory control system—a digital control system used to control an entire plant.

Thermocouple—a wire composed of dissimilar metals that are connected at one end. When exposed to heat, a proportional charge is generated that corresponds to the temperature change.

Thermowell—a chamber installed in vessels or piping to hold thermocouples and RTDs.

Transducer—a device used to transmit one form of energy to another; typically, electrical to pneumatic.

Transmitter—a device used to sense a process variable and produce a signal that is sent to a controller, recorder, or indicator.

Introduction to Process Instrumentation

Automatic control and modern instrumentation has made it possible for the chemical processing industry (CPI) to operate vast networks of pipes and equipment with a much smaller group of technicians than was required just 10 years ago. The application of instrumentation and electronics is the foundation for efficient continuous flow processes, controlling **analytical variables,** pressure, temperature, and level. To a process technician, this means that an **instrument** or computer has the ability to control the opening, closing, and positioning of valves; start and stop equipment; measure process variables; and respond automatically. This automation enhances the ability of a technician to control larger and more complex process networks and to troubleshoot a much wider variety of process problems.

Figure 1-1
Basic Instruments

Process instrumentation provides a transparent window to the process and enables the technician to detect potential problems earlier and troubleshoot complex process problems with more detailed information. Process technicians use a variety of instruments to control complex industrial processes. Not long ago, operators controlled the processes in their plant manually. This type of process was "valve intensive," or, in other words, required the technician to open and close valves in piping line-ups manually. Basic process instruments improved as the era of automation evolved. One process technician can monitor and control automatically a much larger process from a single control center.

The instruments used in an industrial manufacturing environment include the following (Figure 1-1):
- actuators—mounted on valves
- **alarms**—measure pressure, level, temperature, composition, and flow
- **analyzers**—measure specific components
- **controllers**—pneumatic and electronic, strip chart, and panel mounted
- control valves and other **final control elements**
- computers—control center and console operations, CRTs, **engineer's console,** and supervisory computer
- **control stations**—graphics and **list displays,** programmable logic control (PLC), **distributed control system** (DCS), etc.

- gauges—pressure, level, temperature, and flow
- **transmitters**—pressure, level, temperature, composition, and flow
- **recorders**—pressure, level, temperature, composition, and flow
- **transducers**—converts air signal to electric or vice versa
- primary elements and sensors—displacer and buoyancy float, **thermocouples**, pressure, RTD, and orifice plates
- scales

Symbols and Diagrams

Process technicians use a wide variety of symbols and diagrams to identify the equipment, systems, instruments, and piping found in the CPI. The symbols in this chapter are standardized from information gathered from a wide array of companies. These examples include piping, valves, tanks, vessels, pumps, compressors, motors, turbines, cooling towers, heat exchangers, furnaces, steam generators, reactors, and distillation columns. In addition to these equipment symbols, an assortment of instrumentation, electrical, elevation, foundation, and equipment location symbols are included in this text.

Process instrumentation classes are designed around the concept of the **control loop**, equipment, systems, computers, and the control room. Instruments associated with pressure, temperature, flow, level, and analytical variables are presented in relation to these control systems. When new technicians are walked through the unit for the first time, simple process flow diagrams are used to describe each step. Master flow diagrams or PIDs are used to provide complex and complete information on the operating process. It is difficult to understand how each symbol works in the system without some understanding of the science and technology associated with the various types of instruments and equipment.

Examples of standardized symbols and diagrams can be found in Figure 1-2. Successful students memorize these symbols over a period of weeks and months. Typically, about 200 symbols represent the entire file. It is impossible to progress through an instrumentation class without a good command of the symbols file. This information will be used throughout a technician's career. The language of instrumentation is not difficult to learn, but is grounded in a strange array of symbols and diagrams.

The importance of learning these symbols and diagrams is also linked to careers in the CPI. New technicians are given a comprehensive overview of the plant in which they will be working. These overviews include a crash course in the company's complex systems and processes. A process trainee who has never had this training will find it very difficult, if not impossible, to understand. Learning the language of instrumentation takes a lot longer than one short orientation session.

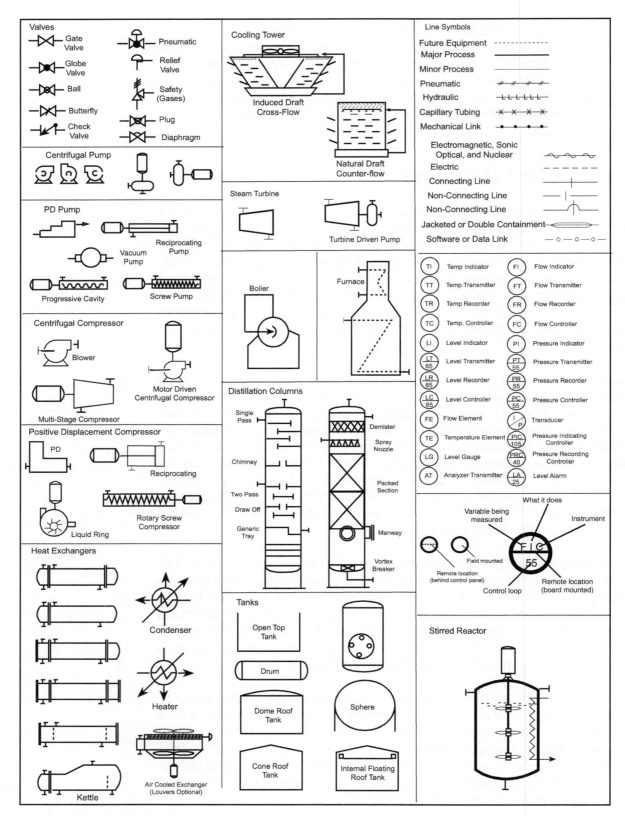

Figure 1-2 *Basic Symbols and Diagrams*

Control Loops

When the curriculum for the process technology program was originally designed, it was decided that an introductory course in process instrumentation should be included in the required core classes. The design of this course would be different from any that was presently being offered at community colleges or universities. During this period, instrumentation courses were taught from the perspective of an instrumentation technician, since several community colleges had well-developed instrumentation technician training programs. The founding members of the process technology program decided that the instrumentation class needed to be taught from a process technician's point of view. This type of course did not exist at that time.

The transition to the development of a process instrumentation course has been difficult since old habits are hard to let go of. Although a number of texts have been developed, unfortunately, each still reads like required course work for an instrumentation program. The needs of a process technician are drastically different from those of an instrument and electrical (I&E) technician, since each approaches the job from a different perspective or point of view. It is important to note, however, that instrumentation is the foundation upon which advanced troubleshooting is based. Basic knowledge of the process equipment and systems combined with modern process control provide the building blocks upon which these skills are developed.

Process technology and modern process instrumentation for process technicians should merge around the concept of the control loop, equipment systems, and the control room. Course work needs to resemble real-time process simulations. It is important to point out that process technicians are uniquely different from other professionals. A process technician operates the equipment and follows the processes using modern instrumentation and electronic equipment. Maintenance technicians are important members of the team who specialize in the maintenance, repair, and installation of equipment, instruments, and electrical and instrument systems. Developing a true process instrumentation course has proven to be very difficult since it would require a paradigm shift from the more traditional approach.

Advances in instrumentation and electronics occur so quickly that it is difficult to keep pace; however, a number of concepts do not change and can be successfully applied to teaching process technicians control instrumentation. A control loop is a collection of instruments that work together to automatically control process variables such as analytical, pressure, temperature, flow, or level. A control loop typically has five parts: primary elements or sensors are coupled to a transmitter that sends a signal to a controller, which compares the process variable to a fixed **set point** and sends a signal to a transducer that converts the signal and sends it to a final control element that is typically a valve. Figure 1-3 illustrates the basic components and physical arrangement of the instruments and equipment.

Figure 1-3
Typical Control Loop

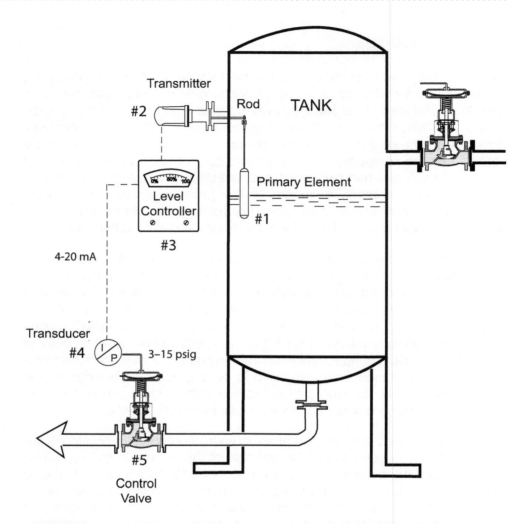

Flow Elements, Instruments, and Control

Industrial manufacturers have carefully evaluated and engineered the complex processes found inside their plants. Flow rates are not just arbitrarily set and allowed to run. In most situations, they are carefully calculated and given a specific range that they can move through. This is accomplished through a relationship between centrifugal pumps and flow control loops. Flow elements are described as devices designed to simulate a situation that can be measured. This mathematical measurement is transmitted to a controller. Examples of flow elements include turbine flow meters, magnetic flow meters, ultrasonic flow meters, vortex flow meters, thermal flow meters, coriolis meters, and orifice plates. Figure 1-4 shows several examples of these types of devices.

Fluid flow control requires the correct application of all five elements in a control loop. A flow control loop includes primary element or sensor, transmitter, controller, transducer, and final control element.

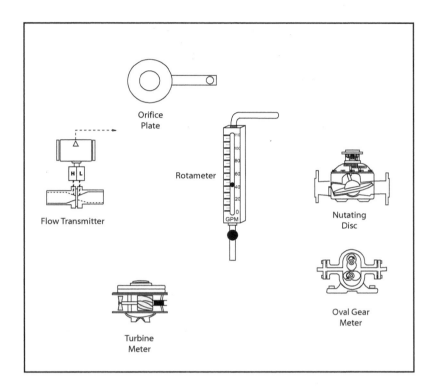

Figure 1-4
Flow Elements

Pressure Elements, Instruments, and Control

The French scientist Blaise Pascal discovered that pressure in fluids is transmitted equally to all distances and in all directions. Pascal formulated laws that describe the effects of pressure within a liquid. This law has many practical applications in hydraulics and presses. The Irish scientist Robert Boyle developed laws that described how the volume of a gas changes when the pressure changes. The higher the gas pressure the closer the gas molecules are and the smaller the volume they occupy. Under ordinary conditions, gas volumes decrease by half when the pressure doubles. Liquids and solids also respond to pressure increases but in much smaller proportions than gases. Liquids and solids are generally considered incompressible.

Pressure is defined as the force per unit area (Pressure = Force ÷ Area) or, in other words, the amount of pressure exerted by fluid on the equipment containing the liquid. Pressure is measured in pounds per square inch (psi) in the English system and kilograms per square centimeter or Pascals in the Metric system.

Two of the most common types of pressure are atmospheric and hydrostatic. Atmospheric pressure is the force exerted on the earth by the weight of the gases that surround it. At sea level, atmospheric pressure is about

14.7 psi (1.3 kPa). This pressure decreases with altitude because of the reduced height (weight) of gas.

Pressure changes the boiling point of chemicals, reaction rates, and the speed at which fluids flow through process pipes and vessels. Pressure changes affect temperature, level, and flow. These changes can impact product quality, so a variety of instruments have been invented and designed to monitor and control pressure. Figure 1-5 shows several examples of these instruments. Examples of process instruments are as follows:

- Pressure elements—sense changes in pressure and convert to mechanical motion
  - bellows **bourdon tube**
  - C type bourdon tube
  - diaphragm capsule pressure elements
  - helical bourdon tube
  - **pressure transmitter**
  - spiral bourdon tube
- Pressure gauges—psia, psig, psiv, psid, bar, pascal, kilopascal, etc.
- Vacuum gauge—expressed in inches of mercury (in. Hg)
- Manometers—can be used to measure pressure or vacuum. A manometer operates under hydrostatic pressure principles; a column of water always exerts a specific force.
- Pressure Transmitter—uses a pressure element to sense pressure and sends a signal to a controller or recorder

Figure 1-5
Pressure Elements

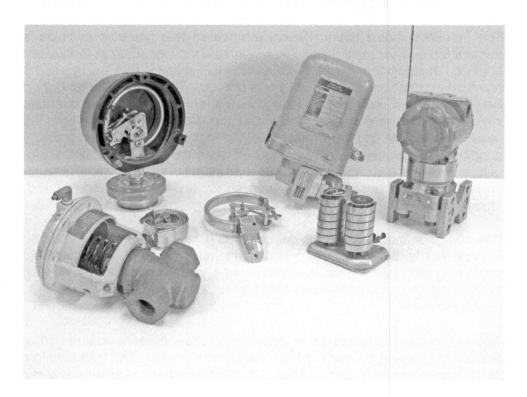

Pressure Gauges

There are four commonly used pressure scales in the manufacturing environment: gauge pressure (psig) zero represents 14.7 psia, absolute pressure (psia) based on zero value for a perfect vacuum, vacuum scale (in. Hg), and differential pressure (psid) based on the difference between two pressure points.

Pounds per Square Inch Gauge

The first and most common pressure gauge is the psig gauge scale. The gauge scale starts with atmospheric pressure as zero and moves up the pressure scale. The zero on this type of scale is actually 14.7 psia. To get psia pressure, 14.7 pounds must be added to the scale for it to represent the total amount of absolute pressure present in the system.

Pounds per Square Inch Absolute

The second type of pressure gauge functions under the psia absolute scale. On this type of scale, a zero reading is used, which takes into account atmospheric pressure. To convert this type of a gauge reading into psig, 14.7 pounds must be subtracted.

Pounds per Square Inch Vacuum

Psig gauges cannot be used with system processes that operate under a vacuum. Negative pressures cause the primary elements to contract beyond design limits. If a psig gauge accidentally encounters a vacuum, the reading scale is compromised (low). Vacuum gauges are designed to operate at less than atmospheric pressure. Vacuum is considered to be anything below atmospheric pressure. Compound gauges can indicate both vacuum and psig readings.

Pounds per Square Inch Differential

Differential pressure is a critical factor in many of the processes found in the CPI. Examples of this include heat exchangers and distillation systems. Differential is calculated by measuring the difference between two pressure points.

Pressure Transmitter

A pressure transmitter uses a pressure element to sense pressure and sends a signal to a controller or recorder. Pressure transmitters use all of the primary pressure elements just discussed. Linkage movement allows the transmitter to transmit a signal that is representative of the pressure to a controller or recorder. Controllers open or close control valves depending on the signal they receive from the transmitter.

Pressure Control

Pressure control uses a primary element or sensor, transmitter, controller, transducer, and final control element. This process is accomplished in a variety of ways depending on the design of the system.

Level Elements, Instruments, and Control

Examples of level elements and instruments are as follows:

- Displacer—buoyancy devices
- Sight glass—transparent tube mounted on the side of a tank
- Float and tape—float rests on the surface of the fluid; tape moves up and down depending on the level
- Conductivity probes—high and low level alarms. Use electricity to complete lower leg circuit. If liquid reaches the higher leg, the circuit is broken. This type of system is designed to keep the level between the high and low conductivity probes. Typically used on nonflammable material
- Capacitance probes—radiation devices, load cells
- D/P cell (transmitter)—converts pressure difference to a level indication. Measures hydrostatic pressure difference between two points on a pressurized vessel
- Continuous level detector gauge—pressure sensitive. Measures hydrostatic pressure in open vessels
- Bubbler system—forces air through a tube that is positioned in the liquid. The liquid's resistance to flow registers on a pressure sensitive level gauge. Measures hydrostatic pressure in open vessels

Figure 1-6 illustrates several examples of these types of devices.

Figure 1-6
Level Elements

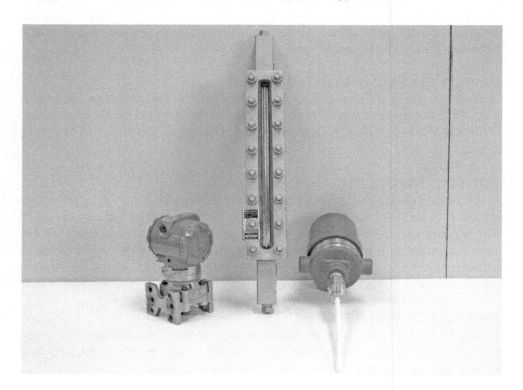

Temperature Elements, Instruments, and Control

Temperature is defined as the degree of hotness or coldness of an object or environment. Process operators use Fahrenheit and Celsius thermometers to measure temperature. Thermometers typically use mercury or alcohol as the expanding or contracting liquid. Fahrenheit scales operate by using 32°F as the freezing point of water and 212°F as the boiling point of water. Celsius uses 0°C as the freezing point of water and 100°C as the boiling point of water.

Bimetallic Thermometer Strip

Local temperature **indicators** usually contain a bimetallic strip that differentially expands with increasing temperature. The deflection is correlated with temperature.

Primary Elements and Sensors

Primary temperature elements and sensors include the following:
- filled thermal bulb and capillary tubing (excluding mercury)
- resistance bulb
- thermocouple

Thermoelectric Temperature Measuring Device

Temperature measuring devices are of two types: RTDs and thermocouples; types J and K are the most common thermocouples.

Resistive Temperature Detector

An RTD is a thermoelectric temperature-measuring device composed of a small platinum or nickel wire encased in a rugged metal tube. The electrical resistance in the wire is influenced by changes in temperature. Temperature changes in an RTD are sensed by an electronic circuit and directed to a temperature indicator.

Thermocouple

Thermocouples are temperature-measuring devices that are composed of two different types of metal. A thermocouple is designed to convert heat into electricity. When heat is applied to the connected ends of a thermocouple, a low-level current is generated. The higher the temperature the greater the voltage generated. Electrical current is detected easily by the associated electronic circuit and converted to a corresponding temperature scale. Thermocouples come in several types: J-type and K-type thermocouples are the most common. Figure 1-7 shows several devices associated with temperature control.

Figure 1-7
Temperature Elements

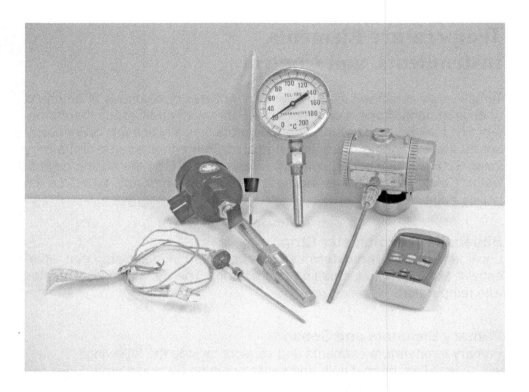

Analytical Elements, Instruments, and Control

Analyzers are devices designed to identify the presence of a substance or the concentration (%) of the substance in a process stream. Analyzers can calculate both quantitative and qualitative variables. Quantitative variables are most closely associated with the amount or percentage of a substance present in a process. Qualitative analyzers simply determine whether a substance is present or not. Examples of this are analyzers that are designed to detect the presence of methane gas.

When an analyzer is properly installed for continuous service, it will first draw in a predetermined quantity or sample that is introduced into the analyzer. After the analyzer has analyzed the sample, it is either safely discarded or returned to the process stream. When analyzers were first introduced into process operations, a number of problems plagued the devices. In most cases technicians ignored the analyzer because it was constantly incorrect. This was the result of a poor design where the analyzer sample line would plug, become contaminated, or rupture. Technicians relied on samples that were sent to the lab or on instruments that could be used to infer whether the system was in control.

Analyzers are devices that are designed to detect the concentration of a specific component or element. For example, an O_2 analyzer is designed to detect how much oxygen is present in the flue gases of a furnace or boiler. Another important component measured in the flue gases is carbon monoxide (CO), a lethal gas. A conductivity meter was used to detect the percentage of suspended solids in the water basin of a cooling tower. A butane analyzer can be installed in a process stream that contains butane, pentane, liquid catalyst, and toluene. In such a mixture, the analyzer will indicate the concentration of butane in relation to the other components. Analyzers are precision instruments that are installed and maintained by the instrumentation department.

Analytical variables are identified as compositional variables such as acid/base (pH) or parts-per-million (ppm). Parts-per-million is related to conductivity, a phenomena where suspended solids conduct an electric current better than a pure liquid with few suspended solids. On the pH scale, a reading of 7 is considered to be neutral. Anything below 7 is considered to be an acid, while any reading above 7 is referred to as a base or caustic. These measurements are critical inside systems such as heat exchangers and cooling towers.

Another analytical device is a gas or liquid chromatograph. Chromatography is a process in which a sample is injected into a column and heated, allowing the various components in the mixture to vaporize at different times. This provides a chart that indicates the physical characteristics associated with the individual components. A carrier gas is admitted at the bottom of the column; it gently tugs the individual components through the column and in front of the detector. A graphical representation of the sample is recorded and attached to the selected sample for analysis.

Introduction to Process Troubleshooting: Methods and Models

A process technician's role can be compared to that of a jet fighter pilot. The pilot not only needs to be familiar with his or her aircraft, but must also be able to use it in a variety of combat situations. The skills needed to make a good fighter pilot are the same skills required to make a good process technician. Troubleshooting skills vary significantly among technicians. Some technicians are content with simply knowing about the equipment and systems they operate and how to start them up and shut them down. Other technicians have the rare ability to move far beyond simply operating the unit. These few technicians are highly valued by their company since they are a critical component to keeping the unit running safely and efficiently. Problems are quickly identified and corrections are applied.

Troubleshooting the operation of process equipment requires a good understanding of basic components and equipment operation. Equipment used in modern manufacturing is run 24 hours a day, 7 days a week, and

52 weeks in a year. Routine maintenance is performed on this equipment during scheduled maintenance. Process technicians should attempt to obtain as much information as possible on the equipment found in their units. Much of this information can be found in technical manuals or operating manuals. Manufacturer information is typically included in engineering specifications, drawings, and equipment descriptions.

Statistical Process Control

Data collection, organization, and analysis can also help to troubleshoot process problems. Check sheets are used to collect large quantities of data. These quantitative data can be organized into graphics or trends to plot process variation or changes. Data analysis utilizes a variety of quality techniques to put all of the parts in place.

Troubleshooting Methods

There are a number of troubleshooting methods that can be used with these models. Methods vary depending on individual educational faculty, consultants, and industry. The basic approach of most methods includes the development of a good educational foundation.

1. Method One: Educational (completed in college program)
   - Know the basics of equipment and technology
   - Understand the math, physics, and chemistry associated with the equipment
   - Study equipment arrangements in systems
   - Study process control instrumentation
   - Operate equipment in complex arrangements
   - Troubleshoot process problems
2. Method Two: Instrumental (completed in college program)
   - Basic understanding of process control instrumentation
   - Basic understanding of the unit process flow plan
   - Advanced training in controller operation—PLC and DCS
   - Troubleshooting process problems
3. Method Three: Experiential (completed on-the-job)
   - Experience in operating specific equipment and systems
   - Familiarity with past problems and solutions
   - Ability to think outside the box
   - Ability to think critically—identify and challenge assumptions
   - Ability to evaluate, monitor, measure, and test alternatives
   - Ability to troubleshoot process problems
4. Method Four: Scientific (requires engineering technology, process technology, experience, and high aptitude)
   - Grounded in principles of mathematics, physics, and chemistry
   - Understands theory-based operations
   - Understands equipment design and operation
   - Views the problem from the outside in

- Utilizes outside information and expertise and reflective thinking
- Generates alternatives, brainstorming, and rank alternatives
- Troubleshoots process problems

Process Troubleshooting Models

One of the highest levels a process technician can achieve is the ability to clearly see the process and to sequentially break down, identify, and resolve process problems. Process troubleshooting has traditionally been considered the area of senior technicians; however, some people believe that successful techniques can be taught to all technicians. Experience has proven over time to be the best teacher on equipment that is manually operated; however, new computer technology provides advanced control instrumentation that can be used to quickly and methodically track down process problems. It is well known that a single problem can have a cascading effect on all surrounding equipment and instrumentation. This phenomenon is commonly associated with primary problems and secondary problems.

Troubleshooting models are attached to equipment and systems presently being studied at every community college and university that teaches process technology. The nine common models used to teach process troubleshooting are as follows:

- distillation model
- reaction model
- separation model
- pump and tank model
- compressor model
- heat exchanger model
- cooling tower model
- boiler model
- furnace model

The nine new models provide the hardware or framework from which the various troubleshooting methodologies are applied. Each model has a complete set of process control instrumentation and equipment arrangements. A complete range of troubleshooting scenarios has been developed and is typically included with these models. Other models used include stripping and adsorption, decanting, and gas and oil recovery.

Summary

Process instrumentation provides a transparent window to the process and enables the technician to detect potential problems earlier and to troubleshoot complex process problems with more detailed information.

Automatic control and modern instrumentation has made it possible for the CPI to operate vast networks of pipes and equipment with a much smaller group of technicians than was required just 10 years ago.

The instruments used in an industrial manufacturing environment include the following:
- actuators—mounted on valves
- alarms—pressure, level, temperature, composition, and flow
- analyzers—measure specific components
- controllers—pneumatic and electronic, strip chart, panel mounted
- control valves and other final control elements
- computers—control center and console operations, CRTs, engineer's console, and supervisory computer
- control stations—graphics and list displays, PLC, DCS, etc.
- gauges—pressure, level, temperature, and flow
- transmitters—pressure, level, temperature, composition, and flow
- recorders—pressure, level, temperature, composition, and flow
- transducers—convert air signal to electric or vice versa
- primary elements and sensors—displacer and buoyancy float, thermocouples, pressure, RTD, and orifice plates
- scales

Process technicians use a wide variety of symbols and diagrams to identify the equipment, systems, instruments, and piping found in the CPI. Examples include piping, valves, tanks, vessels, pumps, compressors, motors, turbines, cooling towers, heat exchangers, furnaces, steam generators, reactors, and distillation columns. In addition to these equipment symbols, an assortment of instrumentation, electrical, elevation, foundation, and equipment location symbols are included in this text.

Process instrumentation classes are designed around the concept of the control loop, equipment, systems, computers, and the control room. Instruments associated with pressure, temperature, flow, level, and analytical variables are presented in relationship to these control systems. Master flow diagrams or PIDs are used to provide complex and complete information on the operating process. Typically, about 200 symbols represent the entire process symbols file. It is impossible to progress through an instrumentation class without good command of the symbols file.

A control loop is a collection of instruments that work together to automatically control process variables such as analytical, pressure, temperature, flow, or level. A control loop typically has five parts: primary elements or sensors are coupled to a transmitter that sends a signal to a controller, which compares the process variable to a fixed set point and sends a signal to a transducer that converts the signal and sends it to a final control element that is typically a valve.

A process technician's role can be compared to that of a jet fighter pilot. The pilot not only needs to be familiar with his or her aircraft, but must also be able to use it in a variety of combat situations. The skills needed to make a good fighter pilot are the same skills required to make a good process technician. Troubleshooting skills vary significantly among technicians. Some technicians are content with simply knowing about the equipment and systems they operate and how to start them up and shut them down. Other technicians have the rare ability to move far beyond simply operating the unit. These few technicians are highly valued by their company since they are a critical component to keeping the unit running safely and efficiently. Problems are quickly identified and corrections are applied.

Pressure is defined as the force divided by the area ($P = F \div A$). There are four commonly used pressure scales in the manufacturing environment: gauge pressure (psig) zero represents 14.7 psia, absolute pressure (psia) based on zero value for a perfect vacuum, vacuum scale (in. Hg), and differential pressure (psid) based on the difference between two pressure points. Changes in pressure will impact the boiling point of a substance.

Examples of level elements and instruments include the following:
- displacer— buoyancy devices
- sight glass—transparent tube mounted on the side of a tank
- float and tape—float rests on the surface of the fluid. Tape moves up and down depending on the level
- conductivity probes—high and low level alarms. Use electricity to complete lower leg circuit. If liquid reaches the higher leg, the circuit is broken. This type of system is designed to keep the level between the high and low conductivity probes. Typically used on nonflammable material
- capacitance probes—radiation devices, load cells
- D/P cell (transmitter)—converts pressure difference to a level indication. Measures hydrostatic pressure difference between two points on a pressurized vessel
- continuous level detector gauge—pressure-sensitive. Measures hydrostatic pressure in open vessels
- bubbler system—forces air through a tube that is positioned in the liquid. The liquid's resistance to flow registers on a pressure-sensitive level gauge. Measures hydrostatic pressure in open vessels

Temperature is defined as the degree of hotness or coldness of an object or environment. Process operators use Fahrenheit and Celsius thermometers to measure temperature. Thermometers typically use mercury or alcohol as the expanding or contracting liquid. Fahrenheit scales operate by using 32°F as the freezing point of water and 212°F as the boiling point of water. Celsius uses 0°C as the freezing point of water and 100°C as the boiling point of water.

Primary temperature elements and sensors include the following:
- filled thermal bulb and capillary tubing (excluding mercury)
- resistance bulb
- thermocouple

Analyzers are devices designed to identify the presence of a substance or the concentration (%) of the substance in a process stream. Analyzers can calculate both quantitative and qualitative variables. Quantitative variables are most closely associated with the amount or percentage of a substance present in a process. Qualitative analyzers simply determine whether a substance is present or not.

There are a number of troubleshooting methods that can be used with these models. Methods vary depending on individual educational faculty, consultants, and industry. The basic approach of most methods includes the development of a good educational foundation.
1. Method One: Educational (completed in college program)
2. Method Two: Instrumental (completed in college program)
3. Method Three: Experiential (completed on-the-job)
4. Method Four: Scientific (requires engineering technology, process technology, experience, and high aptitude)

Process Troubleshooting Models
One of the highest levels a process technician can achieve is the ability to clearly see the process and to sequentially break down, identify, and resolve process problems. Process troubleshooting has traditionally been considered the area of senior technicians; however, some people believe that successful techniques can be taught in local community colleges and universities to technicians.

Troubleshooting models are attached to equipment and systems presently being studied at every community college and university that teaches process technology. The nine common models used to teach process troubleshooting are as follows:
- pump and tank model
- compressor model
- heat exchanger model
- cooling tower model
- boiler model
- furnace model
- distillation model
- reaction model
- separation model

Review Questions

1. Describe the two different types of analyzers and explain how each is designed to work.

2. List the primary elements and sensors associated with flow.

3. List the primary elements and sensors associated with temperature.

4. List the primary elements and sensors associated with pressure.

5. List the primary elements and sensors associated with level.

6. Explain why the skills required to be a jet fighter pilot are similar to those skills required to be a process technician.

7. How do increases and decreases in pressure inside an enclosed vessel impact the boiling point of a substance?

8. Why is it so difficult for educators to develop a pure process instrumentation textbook?

9. What is temperature defined as?

10. Draw a simple control loop. Explain the purpose of each device.

11. Explain how analyzers are used on fired heaters, cooling towers, boilers, and a stirred reactor.

12. Explain the importance of memorizing the process instrumentation symbols and diagram file.

13. List the nine common equipment troubleshooting models presently being studied at most community colleges and universities.

14. List the four troubleshooting methods and explain the special features of each.

15. Identify the troubleshooting method that is most difficult to achieve and explain why.

Process Symbols and Diagrams

LEARNING OBJECTIVES

After studying this chapter, the student will be able to:

- Draw the symbols used to illustrate process equipment.
- Draw the symbols used to illustrate process instruments.
- Draw the symbols used to illustrate piping, tubing, etc.
- Describe a block flow diagram (BFD).
- Draw the symbols used to illustrate electrical symbols.
- Draw a process flow diagram (PFD).
- Explain the primary differences between PFDs and piping and instrumentation drawings (P & IDs).
- Describe an equipment location drawing.
- Describe an electrical drawing.
- Describe an elevation drawing.
- Describe the elements of a foundation drawing.
- Describe the central functions of a controller.
- Describe instrument blocks, bubbles, process variables, process set points, and operating percentages.

Key Terms

Block flow diagram (BFD)—a set of blocks that move from left to right that show the primary flow path of a process.

Process flow diagram (PFD)—a simplified sketch that uses equipment and line symbols to identify the primary flow path through a unit.

Piping and instrumentation drawing (P & ID)—a complex drawing that uses equipment and line symbols, instruments, control loops, and electrical drawings to identify primary and secondary flow paths through the plant. P & IDs may provide operating specifications, temperatures, pressures, flows, levels, analytical variables, and mass relationship data. These documents may also include pipe sizes, equipment specifications, motor sizes, etc.

Field-mounted equipment—instruments or controllers that are mounted near the equipment in the field.

Board-mounted equipment—instruments, gauges, or controllers that are mounted in a control room.

Distributed control system (DCS)—a computer-based system that controls and monitors process variables.

Programmable logic controller (PLC)—a simple stand-alone, programmable computer that could be used to control a specific process or networked with other PLCs to control a larger operation. PLCs are inexpensive, flexible, provide reliable control, and easy to troubleshoot.

Pneumatic actuated valve—a valve that utilizes air to actuate the flow-control element. Internal designs may be piston, vane, or diaphragm.

Electric actuated valve—a valve that utilizes electricity to actuate or move the flow-control device. An example of this type of valve is a solenoid.

Hydraulic actuated valve—a valve that utilizes a hydraulic actuator to position the flow-control element. Internal designs include piston or vane.

Actuator—a device that controls the position of the flow-control element on a control valve by automatically adjusting the position of the valve stem.

Control valve—an automated valve used to regulate and throttle flow; typically provides the final control element of a control loop.

Transmitter—a device used to sense a process variable such as pressure, temperature, composition, or flow and produce a signal that is sent to a controller, recorder, or indicator.

Primary element and sensor—the first element of a control loop. Primary elements and sensors are available in a variety of shapes and designs depending on whether they are to be used with pressure, temperature, level, flow, or analytical control loops. An example of a temperature element is a thermocouple. A flow-control primary element is a turbine meter or an orifice plate. A level element is a displacer. A pressure element is a bourdon tube. Many different types of primary elements or sensors are used in the chemical processing industry.

Transducer—a device used to convert one form of energy into another form, typically electric to pneumatic or vice versa.

Piping—used to convey all kinds of fluids, liquids, or gases.

Controller—an instrument used to compare a process variable with a set point and initiate a change to return the process to the set point if a variance exists.

Recorder—a device that provides a graphical display of a transmitted process.

Electric signal—typically measured in 4–20 mA.

Pneumatic signal—typically measured in 3–15 psig.

Line symbol—is illustrated with major or minor lines, pneumatic, electrical, hydraulic, capillary tubing, mechanical links, electromagnetic, and jacketed. A variety of other line symbols may be used.

Equipment symbol—there are equipment symbols for every piece of equipment found in industry. Examples include valves, pumps, compressors, heat exchangers, steam turbines, gas turbines, cooling towers, furnaces, boilers, reactors, distillation columns, drums and tanks, and separators.

Electrical symbol—includes motors, voltmeters, relays, ammeters, transformers, breakers, fuses, motor-control centers, switches, and contacts. As with other symbols, a much larger symbol file exists for electrical symbols and diagrams.

Elevation drawing—shows the location of process equipment in relation to existing structures and relative elevations. Elevation drawings provide valuable information to the technician in finding exactly where a piece of equipment would be located on the second or fifth floor of a building.

Equipment location drawing—shows the exact floor plan for location of equipment in relation to the plan's physical boundaries.

Foundation drawing—concrete, wire mesh, and steel specifications that identify width, depth, and thickness of footings, support beams, and foundation.

Legend—a block of information used by a company to define its symbol file, abbreviations, prefixes, and specialized equipment. Although a standard exists, many companies have slight variations that are peculiar to their company. These slight modifications are easy to adapt to if a technician is familiar with the standard symbol file.

Introduction to Process Schematics

Process technicians, engineering technicians, chemists, and engineers use process schematics to describe the primary flows and equipment used in their facilities. The maintenance department and construction division also use these drawings and refer to them each time work is performed in a specific section. Process schematics can vary from very simple to complex. These drawings simplify complex processes and provide a window through which a technician can study **piping,** instrumentation, equipment, and locations. Process schematics include symbols, **block flow diagrams (BFDs), process flow diagrams (PFDs),** and **piping and instrumentation drawings (P & IDs).** These drawings are used by every level of the plant and are a critical document.

Block Flow Diagrams

Simple BFDs can be used to show technical and nontechnical people how material moves from one location to another. Figure 2-1 illustrates the basic components of a BFD. BFDs provide a simplistic set of sequences that

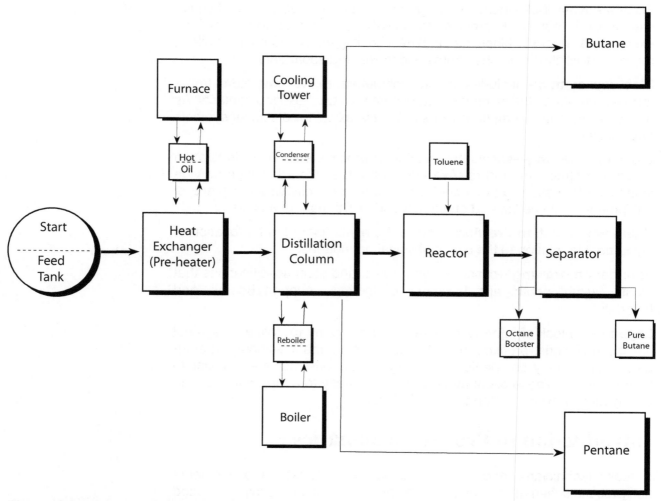

Figure 2-1 *Block Flow Diagram*

move from left to right and show the primary flow path of a process. In reality, these drawings are helpful to individuals who are not expected to operate the unit. The best set of teaching tools for a new trainee is a good trainer, a good PFD, and training materials that accurately reflect the standard operating procedures of the plants.

Process Flow Diagrams

PFDs and process instrumentation drawings (P & IDs) are used to outline or explain the complex flows, equipment, instrumentation, and equipment layouts in a process unit. New technicians are required to study a simple flow diagram of their assigned operating system. PFDs typically include the major equipment and piping path the process takes through the unit. As operators learn more about symbols and diagrams, they graduate to the much more complex P & ID.

Some symbols are common among plants while others change depending upon the company. In other words, there may be two different symbols used to identify a centrifugal pump or a valve. Some standardization of process symbols and diagrams is taking place. The symbols used in this chapter reflect a wide variety of petrochemical and refinery operations.

Simple BFDs and PFDs are typically used to describe the primary flow path through a unit. The PFD contains design process data, including flow rates, pressures, temperatures, equipment duties, catalyst data, etc. Like a well-illustrated road map, a PFD can show intricate details of an operating unit that cannot be easily noticed during a walk through. Process technicians are expected to read simple flow diagrams within hours of starting their initial training. Technicians will graduate to complex P & IDs over the course of their training.

Figure 2-2 shows the basic relationships and flow paths found in a process unit. It is easier to understand a simple flow diagram if it is broken down into sections: feed, preheating, the process, and the final products. This simple left to right approach allows a technician to first identify where the process starts and where it will eventually end. The feed section includes the feed tanks, mixers, piping, and valves. In the second step, the process flow is gradually heated up for processing. This section includes heat exchangers and furnaces. In the third section, the process is included. Typical examples found in the process section could include distillation or reaction. The process area is a complex collection of equipment that works together in a system. The process is designed to produce products that will be sent to the final section.

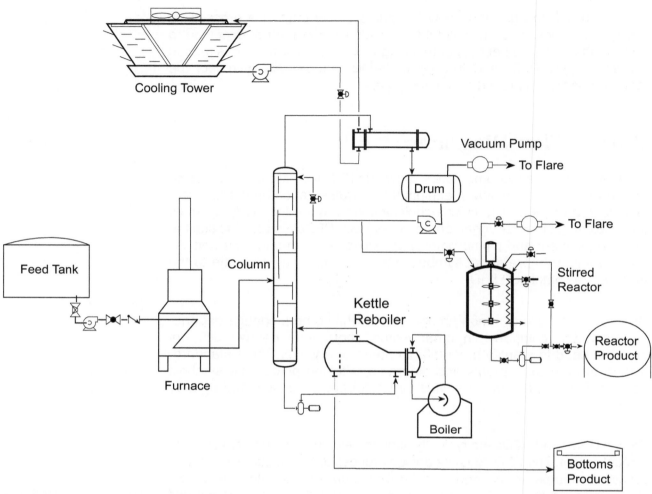

Figure 2-2 *Process Flow Diagram*

Process Flow Diagrams

A process flow diagram is a simple illustration that uses process symbols to describe the primary flow path through a unit. A PFD provides a quick snapshot of the operating unit. Flow diagrams include all primary equipment and flows. A process technician, engineer, chemist, or safety or quality manager can use this document to trace the primary flow of chemicals through the unit. Secondary or minor flows are not included. Complex control loops and instrumentation are not included. The flow diagram is used for customer information, visitor information, and new employee training.

Simple Flow Diagrams

A simple flow diagram provides a quick snapshot of the operating unit. Flow diagrams include all primary equipment, flows, and numbers. A technician can use this document to trace the primary flow of chemicals through the unit. Secondary or minor flows are not included. Complex control loops and instrumentation are not included. The flow diagram is used for visitor information and new employee training.

A typical plant will be composed of hundreds of small processes or systems. PFDs are used to illustrate these systems. New employees typically learn a series of simple flow processes as they move into more complex equipment arrangements. If adequate time and correct instruction are given, it is possible to learn the most complicated process easily. Although the concepts of life-long learning are clearly found in the chemical processing industry, it is possible to qualify on a typical job post in 3 to 4 months. Many engineers and technicians will continue to learn and improve as they move from one assignment to the next.

Instrument Symbols

Circles or bubbles are used to indicate an instrument. The information inside the circle identifies the instrument, what it does, and the variable being measured. A line drawn horizontally through the circle indicates that it is remotely located. This typically indicates that the device is located in a control room. A dashed line indicates that the instrument is located in a control room behind the control panel. Figure 2-3 illustrates what the information in a bubble means. Occasionally numbers are located in the circle that may indicate the number of the control loops. This could indicate which system the device is in or any variety of special applications used by the company.

There are a large variety of instruments and instrument symbols used in the chemical processing industry. The variables that instruments measure are pressure, temperature, level, flow, and analytical variables. If the

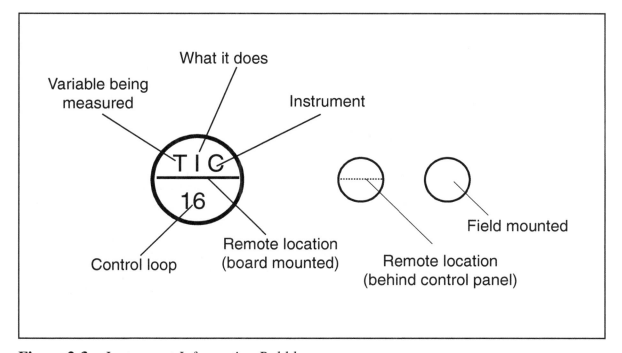

Figure 2-3 *Instrument Information Bubble*

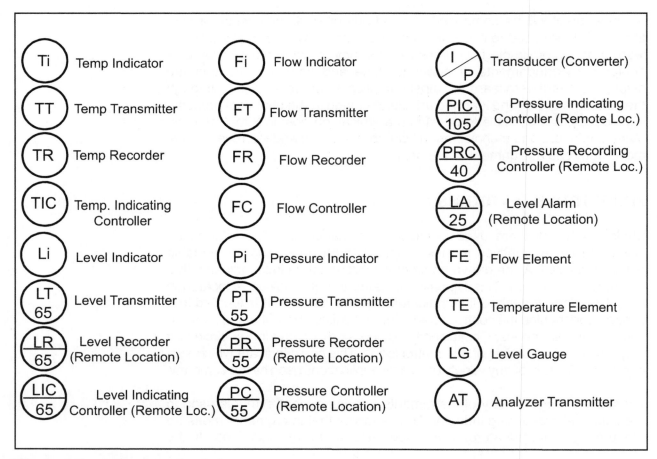

Figure 2-4 *Instrument Symbols*

information in a circle starts with the letter "L," it is typically some type of level instrument. If the first letter in the circle is "A," it is an analytical instrument. Figure 2-4 shows a list of simple instruments. These instruments will be covered in more detail later in this chapter.

Piping or Line Symbols

Each plant has a standardized file for its piping symbols. Process technicians should carefully review the piping symbols for major and minor flows: electric, pneumatic, capillary, and hydraulic tubing, and future equipment. The major flow path through a unit illustrates the critical areas a new technician should concentrate on. A variety of other symbols are also included on the piping. These include valves, strainers, flexible spool, filters, flanges, removable spool pieces, blinds, insulation, piping size, expansion joint, vent cover, in-line mixer, eductor, pulsation dampener, exhaust head, pressure rating, material codes, and steam traps. Figure 2-5 illustrates the basic **line symbols** used by process technicians. It should be noted that not every symbol will be covered here.

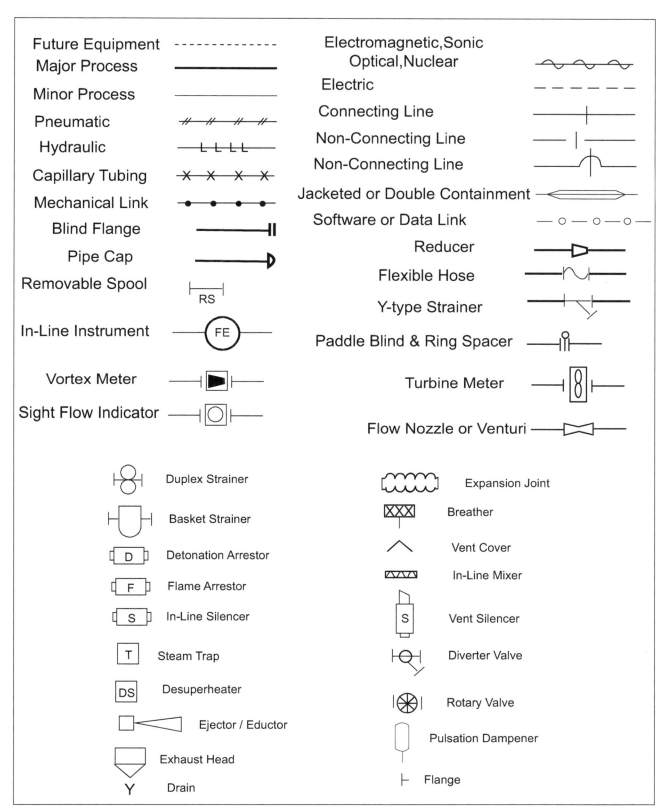

Figure 2-5 *Piping and Auxiliary Symbols*

Valve Symbols

In a refinery or chemical plant, there are thousands of different valves. These valves come in a variety of designs and sizes. Modern process control is linked to the operation of a device called an automated valve. An automated valve has an **actuator** mounted on the valve, which is designed to open, close, or throttle flow through the pipe. A process system is a network of piping and valves. Valves are designed to control the flow of liquids and gases through the system. There is a corresponding symbol for each valve you will study. The most common valves are:

- gate valves
- ball valves
- plug valves
- diaphragm valves
- needle valves
- safety valves
- pneumatically operated valves
- electrically operated valves
- pinch valves
- globe valves
- butterfly valves
- check valves
- angle valves
- knife valves
- relief valves
- hydraulically operated valves
- solenoid valves, and
- motor operated valves.

Figure 2-6 has a list of valve symbols that are frequently used by technicians. Each valve has a unique purpose and design. It is important to understand how these devices operate. The ability to recognize a symbol allows a technician to understand how it is being used in the unit in which the technician is assigned to operate. Using a valve incorrectly can result in damage to the valve or injury to those assigned to operate the unit. Some valves are designed for on–off service, some are designed for throttling, and some are referred to as quick opening or quarter turn. Although the basic functions of valves are similar, it is still important to know how to safely and correctly operate each of the different designs found in a given unit.

Flow Symbols

The basic flow elements most commonly used include orifice plates, turbine meter, rotameter, venturi tube, mass flow meter, magmeter, pitot tube, flow nozzle, annubar tube, nutating disk meter, weir and flume, ultrasonic flow meter, vortex flow meter, coriolis meter, and oval gear meter. The letters that represent these elements are "FE," which are written inside the bubble. The term "Fi," for flow indicator, is also used. A flow indicator may appear to be a simple gauge that provides relatively accurate measurements. For example,

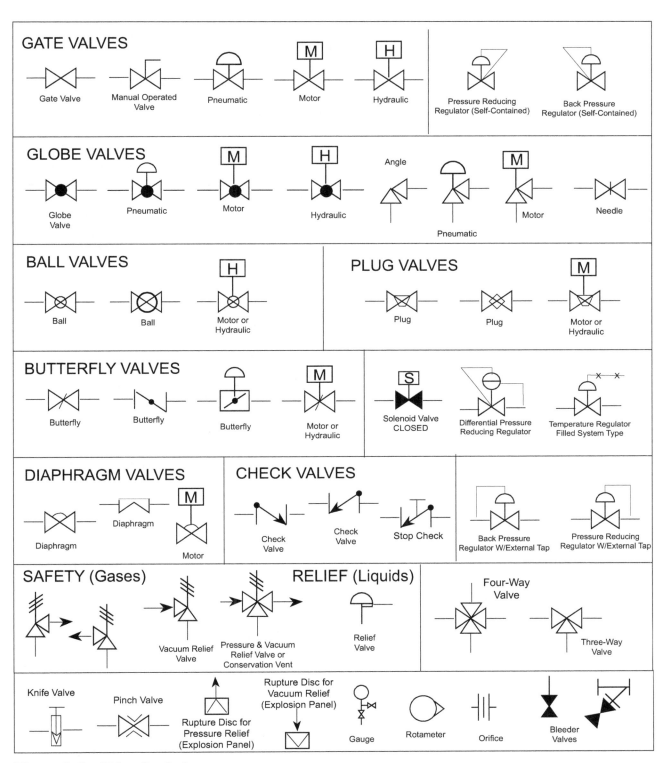

Figure 2-6 *Valve Symbols*

some flow indicators measure flow in gallons per minute (gpm). Figure 2-7 shows the instrument symbol for flow indicator and flow element and shows how each symbol is used in an instrument system. A common mistake is to use the symbol "Fi" as the primary element in a flow-control loop.

Fluid flow can be classified as turbulent, laminar, or stream-lined flow. Centrifugal pumps require laminar flow on the suction side of the system. Turbulent flow is useful in places on the discharge side of the pump. Fluid velocity inside a pipe changes depending upon the diameter of the pipe. These changes also have a direct relationship with pressure. In pipes where the diameter suddenly gets smaller, pressure will increase at the inlet point. Fluid velocity will increase through the narrow passage, resulting in a pressure drop. Bends in the piping will cause the liquid to become turbulent. Turbulent flow tends to transfer heat energy better.

Other factors that will affect the fluid velocity are density, viscosity, and temperature increases or decreases. Many flow elements work directly with **transmitters.** These flow transmitters are designed to send a signal to a **controller.** An example of this is a differential pressure transmitter, which is working with an orifice plate. The transmitter is typically mounted below the line to keep the legs full of liquid. A high-pressure and a low-pressure side are located on each side of a pressure capsule. The orifice plate creates an artificial low-pressure situation inside the pipe, which can be measured and transmitted to a controller.

Fluid flow is typically controlled with a flow-control loop (FIC). In this system, each of the five elements of the control loop will be flow-related and labeled appropriately. It is also possible to have other flow symbols such as flow recorders (FR), high-flow alarms (HFA), and low-flow alarms (LFA).

Pressure Symbols

The most common pressure devices are pressure gauges, manometers, and pressure transmitters. Pressure gauges are represented on a P & ID as pressure indicators (Pi). A circle with "PE" is the symbol for a pressure element. A diaphragm in a DP cell is an example of a PE. A simple equation can be used to determine pressure: $P = F$ (force) $\div A$ (area). Pressure gauges have a variety of internal designs. Some of these designs use a bourdon tube, diaphragm capsule, bellows, diaphragm, spiral, or helical. Basic laws of physics cause the device to respond to pressure changes. These responses or movements are transferred by a mechanical linkage to an indicator. For example, a bourdon tube is a hollow tube shaped like a hook. As pressure is admitted into the hollow curved tube, it attempts to straighten out. A mechanical linkage transfers this travel to an indicator that shows the pressure in pounds per square inch (psi), pounds per square inch absolute (psia), pounds per square inch vacuum (psiv), or pounds per square inch differential (psid). Figure 2-8 shows the basic instrument symbols used to show pressure.

LOCATION (ACCESSIBILITY)	DISCRETE INSTRUMENTS	DCS	PLC
FIELD MOUNTED INSTRUMENTS 1. Located near device	◯	▢◯	◇
REMOTE LOCATION INSTRUMENTS 1. Central Control Room 2. Video Display	⊖	▢⊖	◇
REMOTE LOCATION INSTRUMENTS 1. Not available to Process Technician	◌	▢◌	◇
REMOTE & FIELD INSTRUMENTS 1. Secondary Control Room 2. Video Display 3. Field or Local Control Panel	⊜	▢⊜	◇
REMOTE LOCATION INSTRUMENTS 1. Not available to Process Technician 2. Secondary Control Room	◌	▢◌	◇

FLOW INSTRUMENTS

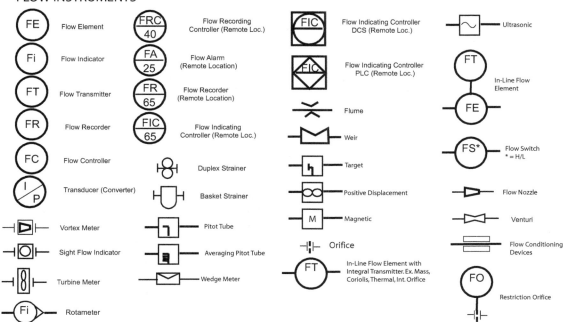

Figure 2-7 *Flow Instrument Symbols*

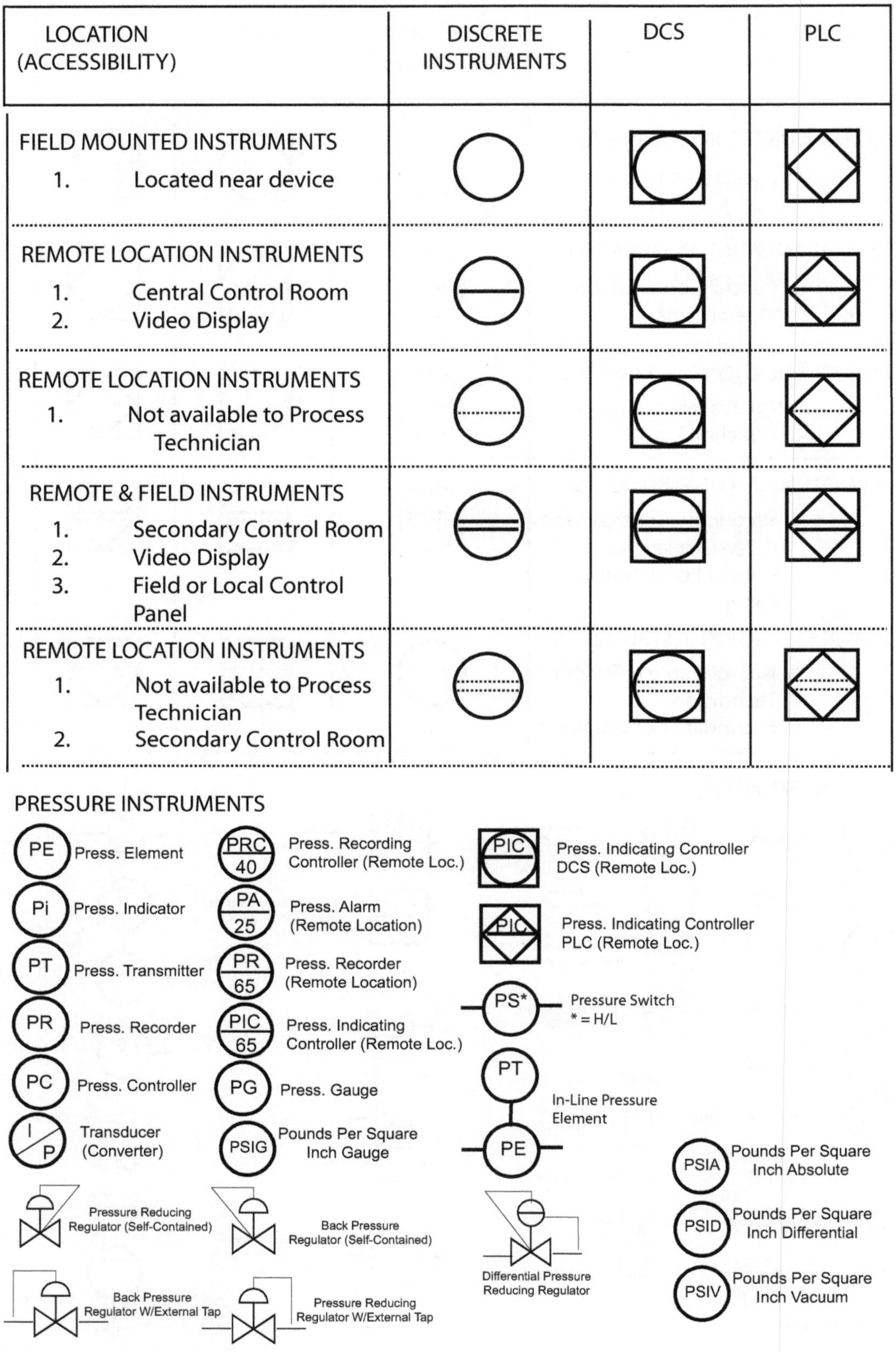

LOCATION (ACCESSIBILITY)	DISCRETE INSTRUMENTS	DCS	PLC
FIELD MOUNTED INSTRUMENTS 1. Located near device			
REMOTE LOCATION INSTRUMENTS 1. Central Control Room 2. Video Display			
REMOTE LOCATION INSTRUMENTS 1. Not available to Process Technician			
REMOTE & FIELD INSTRUMENTS 1. Secondary Control Room 2. Video Display 3. Field or Local Control Panel			
REMOTE LOCATION INSTRUMENTS 1. Not available to Process Technician 2. Secondary Control Room			

PRESSURE INSTRUMENTS

Figure 2-8 *Pressure Instrument Symbols*

Pressure is typically controlled with a pressure-control loop (PIC). In this system, each of the five elements of the control loop will be pressure-related and labeled appropriately. It is also possible to have other pressure symbols such as pressure recorders (PR), high-pressure alarms (HPA), and low-pressure alarms (LPA).

Level Symbols

Level measurement is directly linked to pressure measurements. Controlling the level in a tank, vessel, reactor, or distillation column is an important concept. The common instruments used to measure level are level gauges (LG), displacer bulbs coupled to a level transmitter, capacitance probes, a bubbler system, load cells, and differential pressure transmitter (ΔP cell). As the level in a tank increases, the pressure at the bottom of the tank increases. A simple equation, H (height of liquid above point being calculated) $\times$ 0.433 $\times$ specific gravity = pounds per square inch. If the liquid is in an enclosed tank, things like vapor pressure or gas pressure must be added to the total pressure. Figure 2-9 illustrates the basic symbols used to illustrate level. Level indicators are very common on process drawings and are represented with an instrument symbol that looks like a circle with "Li" written inside. The symbol "LE" is used to describe a level element. An example of an "LE" is a displacer bulb.

Level is typically controlled with a level-control loop (LIC). In this system, each of the five elements of the control loop will be level-related and labeled appropriately. It is also possible to have other level symbols such as level recorders (LR), high-level alarms (HLA), and low-level alarms (LLA).

Temperature Symbols

Temperature symbols are typically represented as "Ti" for temperature indicator or "TE" for temperature element. A temperature indicator is typically a gauge or a thermometer. A temperature gauge has a bimetallic strip inside, which differentially expands with increasing temperature, creating a deflection that is correlated with temperature. These instruments are located in the field beside the equipment. Temperature elements are typically thermoelectric temperature measuring devices: thermocouples, thermal bulbs, resistance bulbs, or resistive temperature detectors (RTDs). Temperature is defined as the degree of hotness or coldness of an object or environment. The four most common temperature scales used are Fahrenheit, Rankin, Celsius, and Kelvin. Each of these temperature scales has its own system for measuring temperature. Industrial temperature indicators and temperature elements can be operated in any of these systems. Figure 2-10 provides a graphical illustration of the symbols used for temperature.

Temperature is typically controlled with a temperature-control loop (TIC). In this system, each of the five elements of the control loop will be temperature-related and labeled appropriately. It is also possible to have other temperature

LOCATION (ACCESSIBILITY)	DISCRETE INSTRUMENTS	DCS	PLC
FIELD MOUNTED INSTRUMENTS 1. Located near device			
REMOTE LOCATION INSTRUMENTS 1. Central Control Room 2. Video Display			
REMOTE LOCATION INSTRUMENTS 1. Not available to Process Technician			
REMOTE & FIELD INSTRUMENTS 1. Secondary Control Room 2. Video Display 3. Field or Local Control Panel			
REMOTE LOCATION INSTRUMENTS 1. Not available to Process Technician 2. Secondary Control Room			

LEVEL INSTRUMENTS

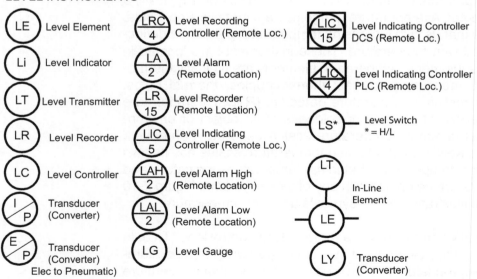

Figure 2-9 *Level Instrument Symbols*

LOCATION (ACCESSIBILITY)	DISCRETE INSTRUMENTS	DCS	PLC
FIELD MOUNTED INSTRUMENTS 1. Located near device	○	◉	◈
REMOTE LOCATION INSTRUMENTS 1. Central Control Room 2. Video Display	⊖	⊖	⊟
REMOTE LOCATION INSTRUMENTS 1. Not available to Process Technician	⊖	⊖	⊟
REMOTE & FIELD INSTRUMENTS 1. Secondary Control Room 2. Video Display 3. Field or Local Control Panel	⊜	⊜	⊟
REMOTE LOCATION INSTRUMENTS 1. Not available to Process Technician 2. Secondary Control Room	⊜	⊜	⊟

TEMPERATURE INSTRUMENTS

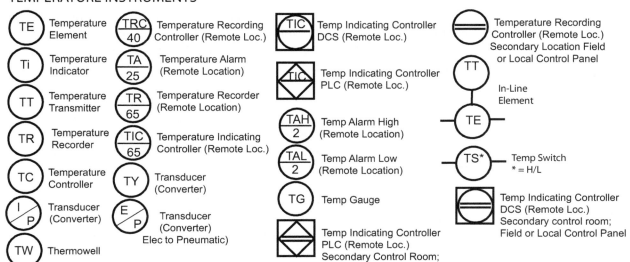

Figure 2-10 *Temperature Instrument Symbols*

symbols such as temperature recorders (TR), high-temperature alarms (HTA), and low-temperature alarms (LTA).

Analytical Symbols

Analytical symbols are used to identify the use of quantitative or qualitative analyzers. Some analyzers check for the presence of a specific substance while others check for the composition or percentage of the chemical in a process stream. Analyzers are used in the following systems:

- pH (acid/base)—cooling tower basin
- oxidation reduction potential (ORP)—free electron concentration
- conductivity (parts per million)—cooling tower blow down
- O_2 analyzer—flue gases of boiler or furnace or incinerator
- CO analyzer—flue gases of boiler or furnace or incinerator
- color analyzer—ensures color specifications on products
- chromatography—measures concentration of components
- opacity analyzer—identifies concentration of particulates in a matter stream
- butane analyzer—measures concentration of butane
- other analyzers—measure concentration or test for the presence of a specific chemical
- other analyzers—mass spectrometers and total carbon analyzers.

Analyzers provide an open window on the molecular level to the chemical composition of a substance. Like other control loops, an analytical control loop can be designed to automate a manual process. For example, the blow down on a cooling water basin can use a conductivity meter as the primary element, a transmitter, a controller, a **transducer,** and a **control valve** located on the blow-down line. When the parts per million exceed operational specifications, the control valve opens and transfers a specific volume of water to the holding lagoons. This lost water is replaced by fresh water from the make-up system. The symbol for an analyzer used in this way is "AE," or analyzer element. Figure 2-11 illustrates the basic symbols used to describe analyzers.

Since a variety of analytical variables are controlled with an analytical control loop (AIC), each of the five elements of the control loop will be analytical-related and labeled appropriately. It is also possible to have other analytical symbols such as analytical recorders (AR), high-analyzer alarms (HAA), and low-analyzer alarms (LAA).

Equipment Symbols

Pumps and tanks are available in a variety of designs and shapes. Process symbols are designed to graphically display the process unit. Common pump and tank symbols can be found in Figure 2-12. Tanks may be designed as bins, drums, dome roof, open top, internal floating roof, double wall, cone

LOCATION (ACCESSIBILITY)	DISCRETE INSTRUMENTS	DCS	PLC
FIELD MOUNTED INSTRUMENTS 1. Located near device			
REMOTE LOCATION INSTRUMENTS 1. Central Control Room 2. Video Display			
REMOTE LOCATION INSTRUMENTS 1. Not available to Process Technician			
REMOTE & FIELD INSTRUMENTS 1. Secondary Control Room 2. Video Display 3. Field or Local Control Panel			
REMOTE LOCATION INSTRUMENTS 1. Not available to Process Technician 2. Secondary Control Room			

ANALYTICAL INSTRUMENTS

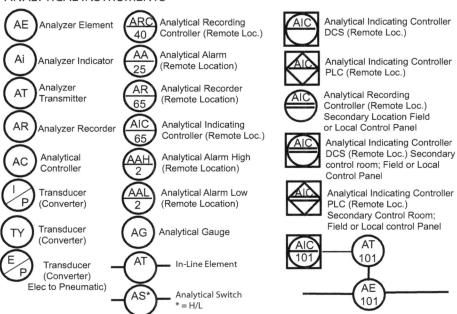

Figure 2-11 *Analytical Instrument Symbols*

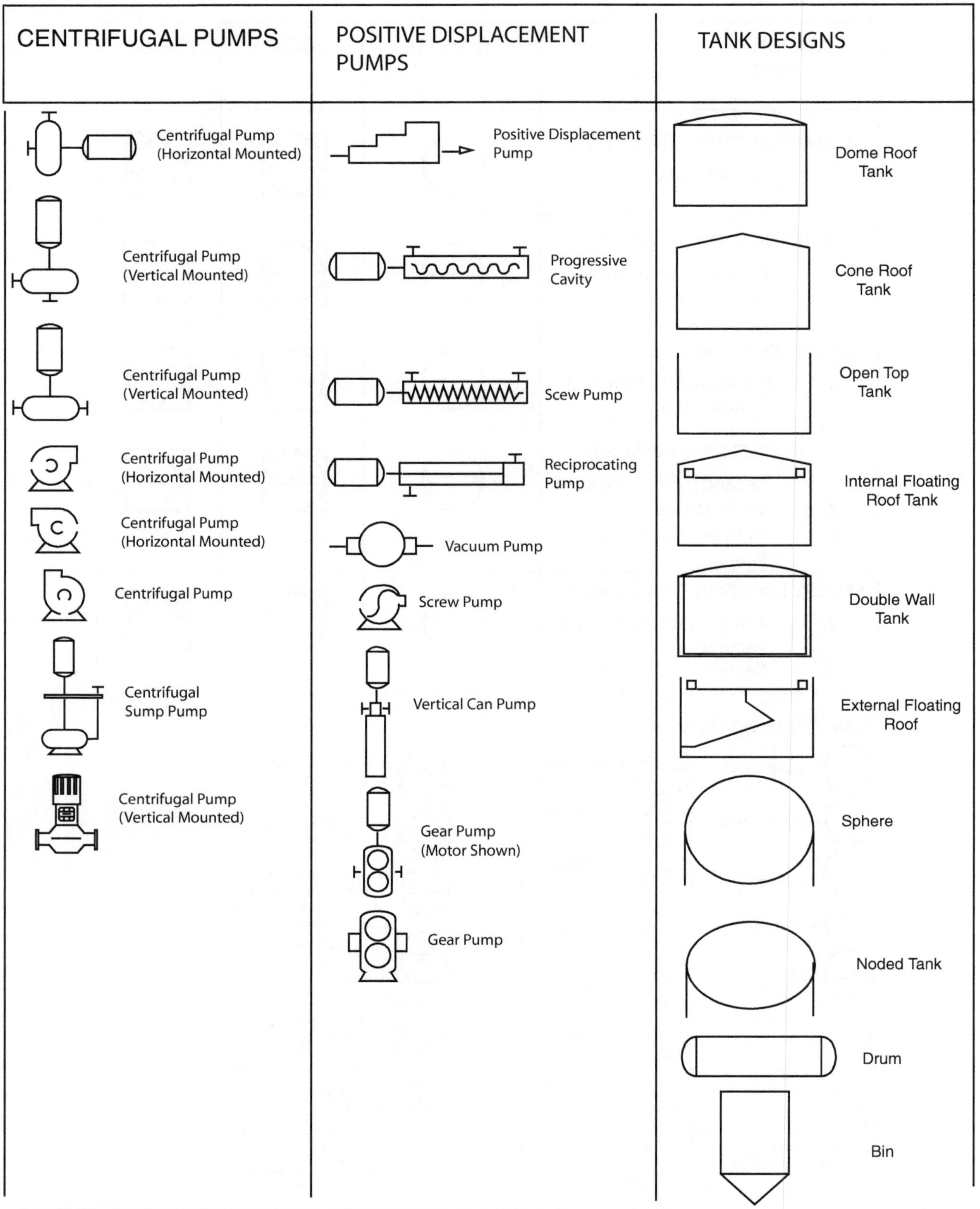

CENTRIFUGAL PUMPS	POSITIVE DISPLACEMENT PUMPS	TANK DESIGNS
Centrifugal Pump (Horizontal Mounted)	Positive Displacement Pump	Dome Roof Tank
Centrifugal Pump (Vertical Mounted)	Progressive Cavity	Cone Roof Tank
Centrifugal Pump (Vertical Mounted)	Scew Pump	Open Top Tank
Centrifugal Pump (Horizontal Mounted)	Reciprocating Pump	Internal Floating Roof Tank
Centrifugal Pump (Horizontal Mounted)	Vacuum Pump	Double Wall Tank
Centrifugal Pump	Screw Pump	External Floating Roof
Centrifugal Sump Pump	Vertical Can Pump	Sphere
Centrifugal Pump (Vertical Mounted)	Gear Pump (Motor Shown)	Noded Tank
	Gear Pump	Drum
		Bin

Figure 2-12   *Pump and Tank Symbols*

roof, external floating roofs, and a variety of spherical-shaped tanks. The materials used in these tanks can be plastic, carbon steel, stainless steel, specialty alloys, or any variety of materials. Tanks have special designs that are shown on a P & ID and on some symbol charts. Each tank is different and has a variety of pressure and temperature requirements.

Pumps come in two basic designs: dynamic and positive displacement. Dynamic pumps include the very popular centrifugal and axial pumps. Positive displacement (PD) pumps have two categories: rotary and reciprocating. Reciprocating pumps typically include piston, plunger, and diaphragm. These pumps operate with a back and forth motion, displacing a predetermined amount of liquid. Rotary pumps use a rotary motion that displaces a specific amount of fluid on each rotation. Rotary pumps include the following designs: lobe, gear, vane, and screw. A pump and tank system includes piping, valves, instruments, and control loops as well as the pump and tank. These systems are designed to be circulated. Compressors and pumps share a common set of operating principles. The dynamic and PD families share common categories. The symbols for compressors may closely resemble a pump. In most cases, the symbol is slightly larger in the compressor symbol file. In the multistage centrifugal compressor, the symbol clearly describes how the gas is compressed prior to being released. This is in sharp contrast to the steam turbine symbol, which illustrates the opposite effect as the steam expands while passing over the rotor. Modern P & IDs show the motor symbol connected to the driven equipment. This equipment may be a pump, compressor, mixer, or generator. Figure 2-13 illustrates the standard symbols for compressors, steam turbines, and motors.

Heat exchangers and cooling towers are two types of industrial equipment that share a unique relationship. A heat exchanger is a device used to transfer heat energy between two process flows. The cooling tower performs a similar function; however, cooling towers and heat exchangers use different scientific principles to operate. Heat exchangers transfer heat energy through conductive and convective heat transfer while cooling towers transfer heat energy to the outside air through the principle of evaporation. Figure 2-14 and Figure 2-15 illustrate the standard symbols used for heat exchangers and cooling towers.

The symbol for a heat exchanger clearly illustrates the flows through the device. It is important for a process technician to be able to recognize the flow paths of the shell inlet and outlet and the tube inlet and outlet. A heat exchanger with an arrow drawn through the body illustrates whether the tube-side flow is being used to heat or cool the shell-side fluid. The downward direction indicates heating while the upward direction indicates cooling. Heat exchangers come in a variety of designs from simple to complex. Standard symbols have been developed for shell and tube heat exchangers, air-cooled heat exchangers, spiral heat exchangers, and plate and frame heat exchangers.

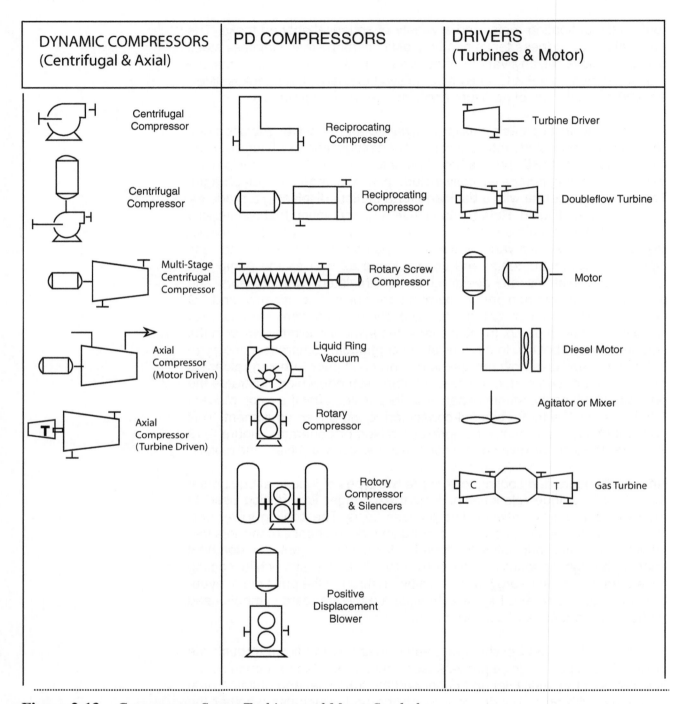

Figure 2-13 *Compressor, Steam Turbine, and Motor Symbols*

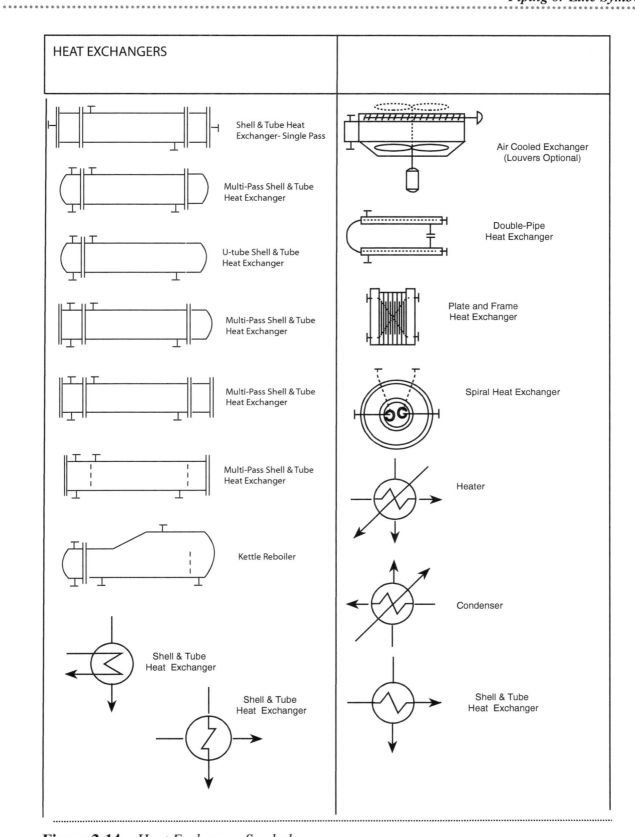

Figure 2-14 *Heat Exchanger Symbols*

Figure 2-15
*Cooling Tower
Symbols*

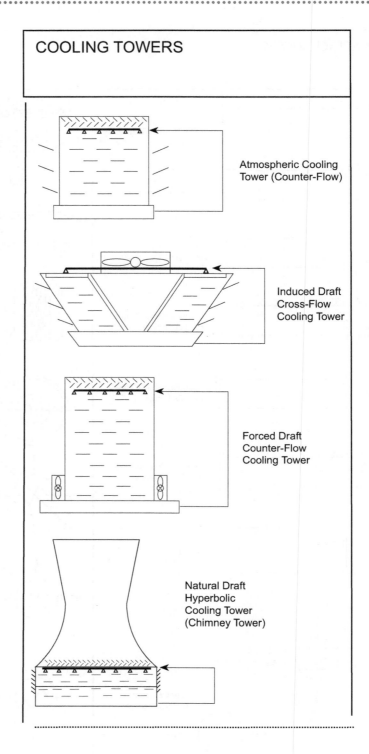

COOLING TOWERS

Atmospheric Cooling
Tower (Counter-Flow)

Induced Draft
Cross-Flow
Cooling Tower

Forced Draft
Counter-Flow
Cooling Tower

Natural Draft
Hyperbolic
Cooling Tower
(Chimney Tower)

The symbol for a cooling tower is designed to resemble the actual device in the process unit. Cooled product flows out of the bottom of the tower and to the processing units, while hot water returns to a point located above the fill. Cooling towers can be found in the following designs: forced draft, natural draft, induced draft, and atmospheric draft. All of these designs have similarities and differences. The symbol will not show all of the various components of the cooling tower system; however, it will provide a technician with a good foundation in cooling tower operation with enough information to clearly see the process.

On a typical P & ID, distillation columns, reactors, boilers, and furnaces will be drawn as they visually appear in the plant. The standard symbols file for these devices can be found in Figure 2-4. If a proprietary process includes several types of equipment not typically found on a standard symbol file, the designer will draw the device as it visually appears in the unit. Figure 2-16 illustrates standard symbols used for boilers and furnaces. The basic components and operation of boilers and furnaces are covered in equipment textbooks. A furnace is designed to heat up large volumes of feedstock for chemical processing in a reactor or distillation column. Feed enters a furnace in the cooler convection section. Heat transfer is primarily through radiant and convective processes. The hottest spot in the furnace is the firebox. Tubes in the firebox are referred to as radiant tubes. A boiler is similar to a furnace; however, it is designed to boil water for steam generation. This process changes the internal arrangement of the boiler. Water can absorb a tremendous amount of heat. When it changes state, it expands to many times its original volume. A typical water tube boiler has a large upper steam-generating drum, a lower mud drum, and a series of tubes designed to provide natural circulation of the steam back to the generating drum.

Distillation columns come in two basic designs: plate and packed. Flow arrangements vary from process to process. The symbols allow the technician to identify primary and secondary flow paths. The two standard symbols for distillation columns can be found in Figure 2-17. Distillation is a process designed to separate various components in a mixture by boiling point. A distillation column is the central component of a much larger system. This system typically includes all of the **equipment symbols** found in this chapter. Plate distillation columns include sieve trays, valve trays, and bubble-cap trays. Packed columns are filled with packing material, rings, saddles, sulzer, intalox, teller rosette, or panapak.

Reactors are stationary vessels and can be classified as batch, semibatch, and continuous batch. Some reactors use mixers to blend the individual components. Reactor design is dependent on the type of service it will be used in. Some of these processes include alkylation, cat cracking, hydrodesulfurization, hydrocracking, fluid coking, reforming, polyethylene, mixed xylenes, and many other processes. A reactor is designed to allow

Figure 2-16
*Boilers and Furnace
Symbols*

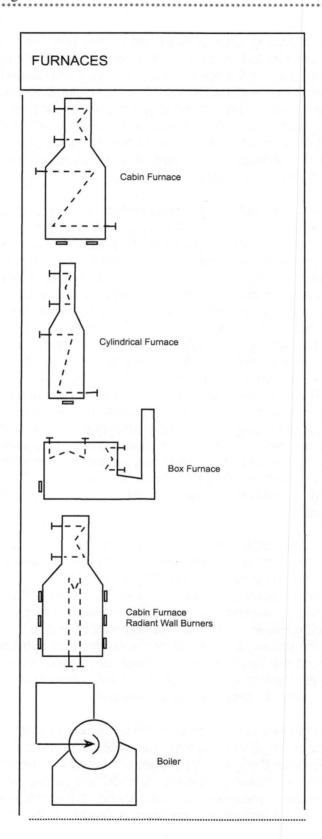

FURNACES

Cabin Furnace

Cylindrical Furnace

Box Furnace

Cabin Furnace
Radiant Wall Burners

Boiler

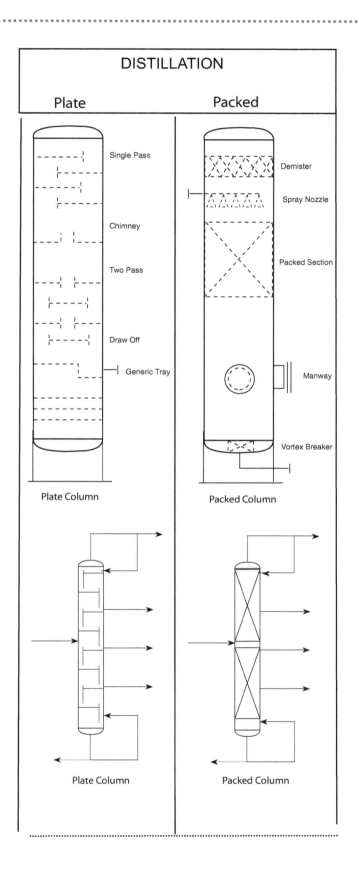

Figure 2-17
Distillation Symbols

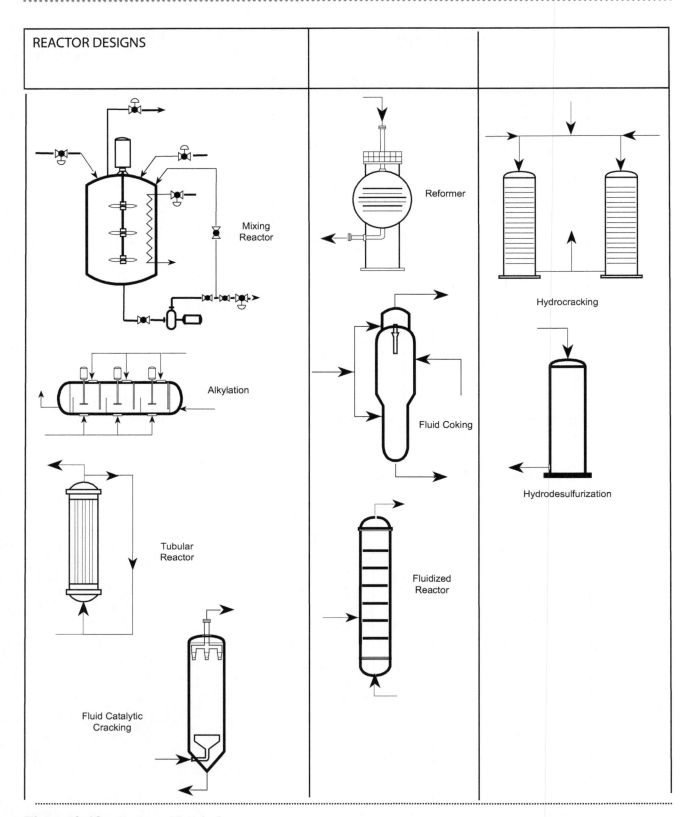

Figure 2-18 *Reactor Symbols*

chemicals to mix together under specific conditions to make chemical bonds, break chemical bonds, or make and break chemical bonds to form new products. The process it is being used for will determine the shape and design of a reactor. Figure 2-18 shows several examples of reactor symbols.

Piping and Instrumentation Drawings

A P & ID is a complete representation of the process and facilities. The P & ID includes a graphical illustration of the piping, equipment, and instrumentation. Modern process control includes a complex array of electronics and instrumentation that is carefully illustrated for the engineer or technician. This master flow document provides a complete picture of each primary element or sensor, transmitter, controller, converter, and final control element. These documents also show how all of these devices are connected to and communicate with each other.

P & ID Components
The basic components of a P & ID are (1) the process legends, (2) **foundation drawings,** (3) **elevation drawings,** (4) electrical drawings, (5) **equipment location drawings,** (6) complex flow diagrams, and (7) the P & IDs that include all of the major flows, minor flows, process instrumentation, and operating information.

Process technicians use P & IDs to identify all of the equipment and piping found on their units. New technicians use these drawings during their initial training period. Knowing and recognizing these symbols is important for a new technician. Technicians who have not applied themselves to the study and application of process symbols will eventually regret it. This lack of knowledge has already cost a number of process technicians their jobs. Apparently, memorizing the prices on the menus at fast food facilities and wearing a Burger King or McDonald's uniform is easier than memorizing the technical language of symbols and diagrams. Figure 2-19 shows a reasonable cross-section of some of these symbols and diagrams. Unfortunately, a large number of prospective college students end their short venture into process technology at this point, refusing to memorize the required symbols.

The chemical processing industry has assigned a corresponding symbol to each type of valve, pump, compressor, steam turbine, heat exchanger, cooling tower, basic instrumentation, reactor, distillation column, furnace, or boiler. Virtually all of the process equipment has a corresponding symbol. There are symbols used to represent major and minor process lines; pneumatic, hydraulic, or electric lines; and a wide variety of **electrical symbols.** These symbols and lines can be arranged to show how process flow enters and moves through the unit. This type of simple drawing is

51

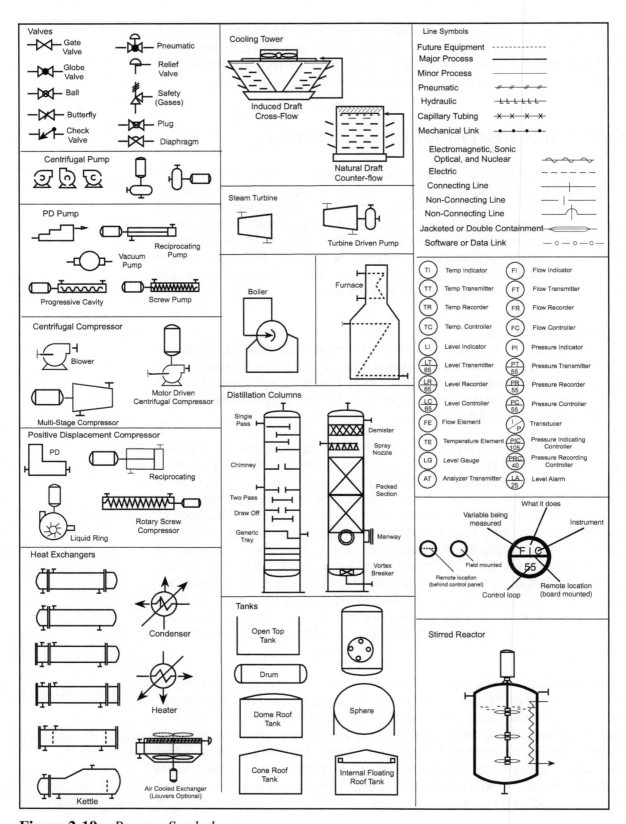

Figure 2-19 *Process Symbols*

called a PFD. This document is very popular with new employees for on-the-job training.

A process instrumentation and piping drawing (P & ID) is a complex representation of the various units found in a plant. To read a P & ID, an understanding of the equipment, instrumentation, and technology is needed. Some of this equipment includes piping, valves, pumps, tanks, compressors, steam turbines, process instrumentation, heat exchangers, cooling towers, furnaces, boilers, reactors, and distillation columns. The next step in using a P & ID is to memorize the process symbol list of the plant. This information can be found on the process legend. P & IDs have a variety of elements. Some of these elements include flow diagrams, equipment layouts, elevation plans, electrical layouts, title blocks and **legends,** footings, and foundation drawings.

Process diagrams can be broken down into two major categories: flow diagrams and P & IDs. As discussed earlier, a flow diagram is a simplified illustration that uses process symbols to describe the primary flow path through a unit. A P & ID is a complex diagram that uses process symbols to describe a process unit. Instrumentation symbols are shown on a P & ID as circles. Inside the circle, information is included that tells the process technicians what type of instrument is represented. Figure 2-20 includes examples of typical instrument symbols. The P & ID includes a graphic representation of the equipment, piping, and instrumentation. Modern process control can be clearly inserted into the drawing to provide a process technician with a complete picture of electronics and instrumentation. Process operators can look at their process and see how the engineering department has automated the unit. Pressure, temperature, flow, level, and analytical control loops are all included in the unit P & ID or master flow drawing. The following drawing is a simplified overview of the systems and processes found in Chapters 8–16. If all of these drawings could be put together, the combined system would represent this chapter's P & ID, the multivariable unit. These systems include the pump, compressor, heat exchanger, cooling tower, boiler, furnace, distillation, reactor, and separation systems.

Loop Diagrams

A loop diagram traces all instrument connections between the field instrument and the control room panel. This includes instrument airlines, wiring connections at field junction boxes and control room panels, and front connections. During repair projects, loop diagram drawings are used by maintenance technicians; however, the assigned process technician is held responsible for any and all of the information on the document. It is important to point out that the P & ID is actually a set of drawings used by a variety of individuals or groups; however, since the equipment owner is typically the process technician, greater demands are made of her. The old saying

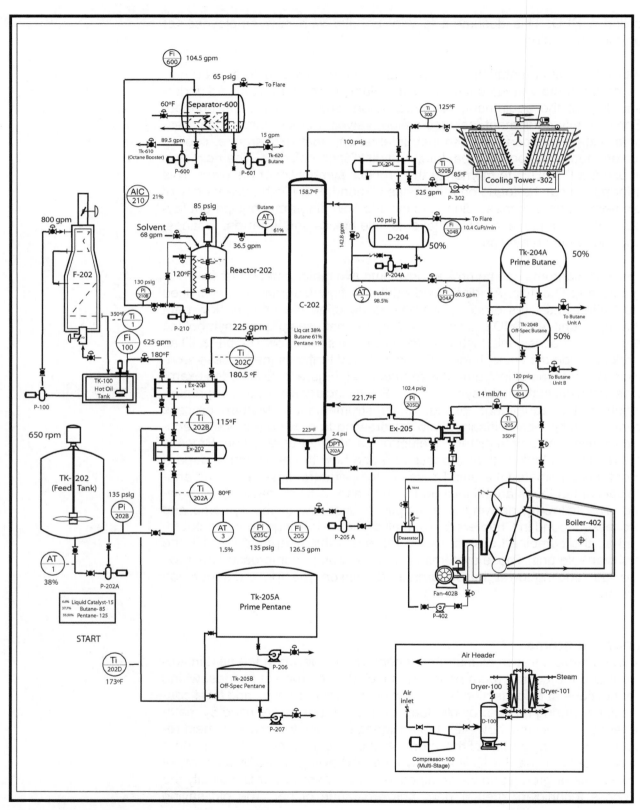

Figure 2-20 *Piping and Instrumentation Drawing*

"Where much is given, much is expected" is very true of the responsibilities of the next generation of process technicians.

Process Legends

The process legend provides the information needed to interpret and read the P & ID. Process legends are found at the front of the P & ID. The legend includes information about piping, instrument and equipment symbols, abbreviations, title block, drawing number, revision number, approvals, and company prefixes. Figure 2-21 shows what a simple process legend looks like. At the present time, symbol and diagram standardization is not complete. Many companies use their own symbols file to display unit drawings. Unique and unusual equipment will also require a modified symbols file.

Foundation Drawings

The construction crew pouring the footers, beams, and foundation uses foundation drawings. Concrete and steel specifications are designed to support equipment, integrate underground piping, and provide support for exterior and interior walls. Process technicians do not typically use foundation drawings; however, the drawings are useful when questions arise about piping that disappears under the ground and when new equipment is being added. Figure 2-22 illustrates what a typical foundation drawing looks like.

Elevation Drawings

Elevation drawings convey a graphical representation that shows the location of process equipment in relation to existing structures and the ground level. In a multistory structure, the elevation drawing provides the technician with information about equipment operation and location. This information is important for making rounds, checking equipment, developing checklists, catching samples, and performing start-ups and shut downs. The elevation plan in Figure 2-23 illustrates equipment and structure locations. An elevation drawing resembles a process where the outer sheathing on the building is removed, exposing the equipment as it is hung on the structural steel and fixed in place on the elevated floor. This view provides information about where tanks or vessels connect, how long the lines are, at what angle the equipment is suspended, etc. Elevation drawings also help new technicians determine the best order to fill out or complete a checklist or equipment locations. Trainees will typically go to the eighth floor to inspect a piece of equipment, then get on the elevator or staircase if an elevator is not provided, and go to the ground floor to check on another piece of equipment, only to discover that the next reading or sample is on the seventh floor.

Electrical Drawings

Electrical drawings include symbols and diagrams that depict an electrical process system. Electrical drawings show unit electricians where power transmission lines run and places where it is stepped down or up for operational purposes. A complex P & ID is designed to be used by a variety of

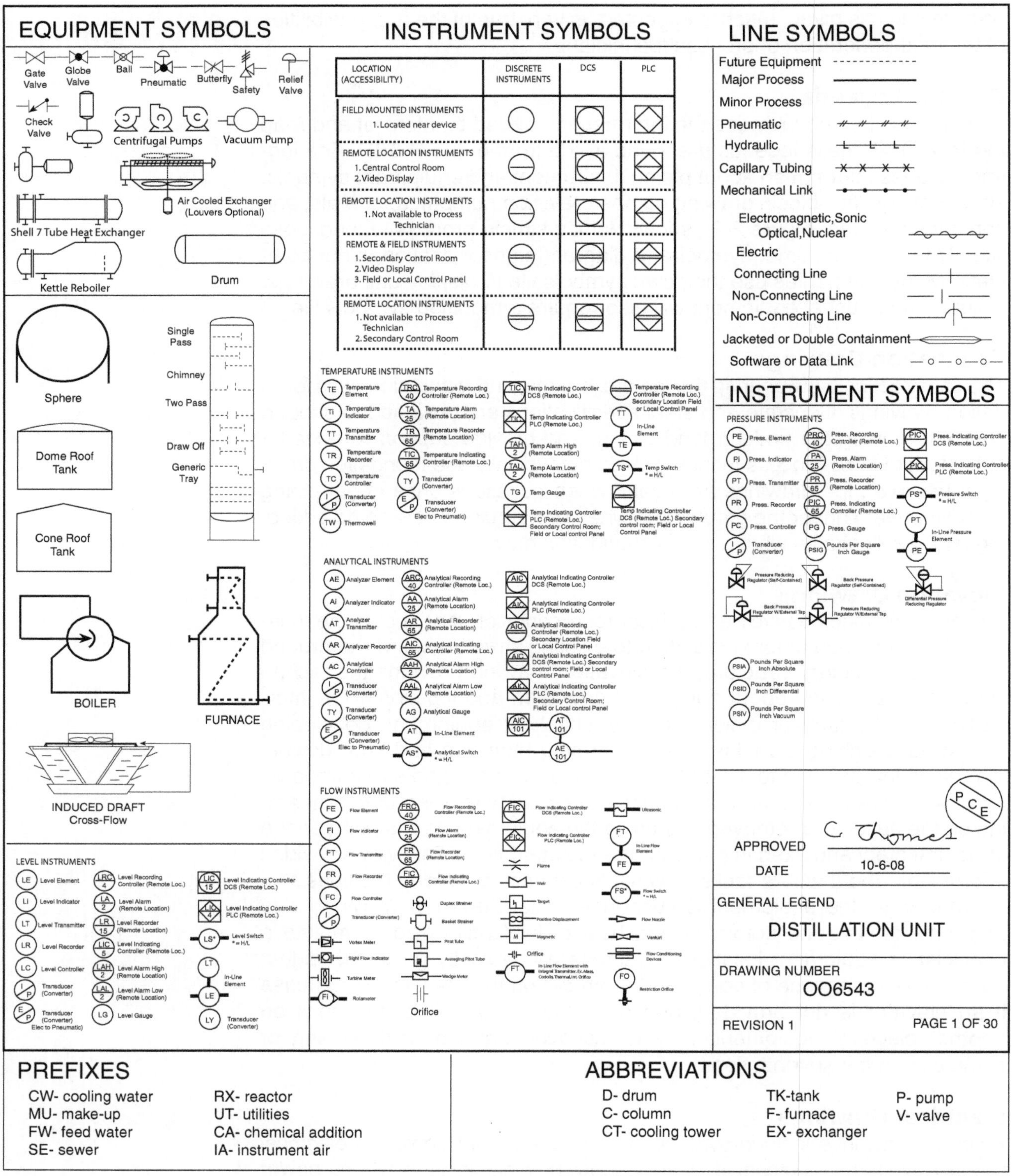

EQUIPMENT SYMBOLS

Gate Valve · Globe Valve · Ball · Pneumatic · Butterfly · Safety · Relief Valve

Check Valve · Centrifugal Pumps · Vacuum Pump

Shell 7 Tube Heat Exchanger · Air Cooled Exchanger (Louvers Optional)

Kettle Reboiler · Drum

Sphere

Single Pass · Chimney · Two Pass · Draw Off · Generic Tray

Dome Roof Tank

Cone Roof Tank

BOILER · FURNACE

INDUCED DRAFT Cross-Flow

LEVEL INSTRUMENTS

LE Level Element	LRC 4 Level Recording Controller (Remote Loc.)	LIC 15 Level Indicating Controller DCS (Remote Loc.)
LI Level Indicator	LA 4 Level Alarm (Remote Location)	LIC 4 Level Indicating Controller PLC (Remote Loc.)
LT Level Transmitter	LR 15 Level Recorder (Remote Location)	
LR Level Recorder	LIC 6 Level Indicating Controller (Remote Loc.)	LS* Level Switch * = H/L
LC Level Controller	LAH 2 Level Alarm High (Remote Location)	LT In-Line Element
I/P Transducer (Converter)	LAL 2 Level Alarm Low (Remote Location)	LE
E/P Transducer (Converter) Elec to Pneumatic)	LG Level Gauge	LY Transducer (Converter)

INSTRUMENT SYMBOLS

LOCATION (ACCESSIBILITY)	DISCRETE INSTRUMENTS	DCS	PLC
FIELD MOUNTED INSTRUMENTS 1. Located near device			
REMOTE LOCATION INSTRUMENTS 1. Central Control Room 2. Video Display			
REMOTE LOCATION INSTRUMENTS 1. Not available to Process Technician			
REMOTE & FIELD INSTRUMENTS 1. Secondary Control Room 2. Video Display 3. Field or Local Control Panel			
REMOTE LOCATION INSTRUMENTS 1. Not available to Process Technician 2. Secondary Control Room			

TEMPERATURE INSTRUMENTS

TE Temperature Element	TRC 40 Temperature Recording Controller (Remote Loc.)	TIC Temp Indicating Controller DCS (Remote Loc.)
TI Temperature Indicator	TA 25 Temperature Alarm (Remote Location)	TIC Temp Indicating Controller PLC (Remote Loc.)
TT Temperature Transmitter	TR 65 Temperature Recorder (Remote Location)	TT In-Line Element
TR Temperature Recorder	TIC 65 Temperature Indicating Controller (Remote Loc.)	TE
TC Temperature Controller	TY Transducer (Converter)	TS* Temp Switch * = H/L
I/P Transducer (Converter)	E/P Transducer (Converter) Elec to Pneumatic)	
TW Thermowell	TG Temp Gauge	

ANALYTICAL INSTRUMENTS

AE Analyzer Element	ARC 40 Analytical Recording Controller (Remote Loc.)	AIC Analytical Indicating Controller DCS (Remote Loc.)
Ai Analyzer Indicator	AA 25 Analytical Alarm (Remote Location)	AIC Analytical Indicating Controller PLC (Remote Loc.)
AT Analyzer Transmitter	AR 65 Analytical Recorder (Remote Location)	AIC Analytical Indicating Controller (Remote Loc.) Secondary Location Field or Local Control Panel
AR Analyzer Recorder	AIC 65 Analytical Indicating Controller (Remote Loc.)	AIC Analytical Indicating Controller DCS (Remote Loc.) Secondary control room; Field or Local Control Panel
AC Analytical Controller	AAH 2 Analytical Alarm High (Remote Location)	AIC Analytical Indicating Controller PLC (Remote Loc.) Secondary Control Room; Field or Local control panel
I/P Transducer (Converter)	AAL 2 Analytical Alarm Low (Remote Location)	
TY Transducer (Converter)	AG Analytical Gauge	
E/P Transducer (Converter) Elec to Pneumatic)	AT In-Line Element	AT 101
	AS* Analytical Switch * = H/L	AE 101

FLOW INSTRUMENTS

FE Flow Element	FRC 40 Flow Recording Controller (Remote Loc.)	FIC Flow Indicating Controller DCS (Remote Loc.)
FI Flow Indicator	FA 25 Flow Alarm (Remote Location)	FIC Flow Indicating Controller PLC (Remote Loc.)
FT Flow Transmitter	FR 65 Flow Recorder (Remote Location)	FT In-Line Flow Element
FR Flow Recorder	FIC 65 Flow Indicating (Remote Loc.)	FE
FC Flow Controller		FS* Flow Switch * = H/L
I/P Transducer (Converter)		

Orifice

LINE SYMBOLS

Future Equipment	------------
Major Process	
Minor Process	
Pneumatic	
Hydraulic	
Capillary Tubing	
Mechanical Link	
Electromagnetic, Sonic Optical, Nuclear	
Electric	
Connecting Line	
Non-Connecting Line	
Non-Connecting Line	
Jacketed or Double Containment	
Software or Data Link	—o—o—o—

INSTRUMENT SYMBOLS

PRESSURE INSTRUMENTS

PE Press. Element	PRC 40 Press. Recording Controller (Remote Loc.)	PIC Press. Indicating Controller DCS (Remote Loc.)
Pi Press. Indicator	PA 25 Press. Alarm (Remote Location)	PIC Press. Indicating Controller PLC (Remote Loc.)
PT Press. Transmitter	PR 65 Press. Recorder (Remote Location)	PS* Pressure Switch * = H/L
PR Press. Recorder	PIC 65 Press. Indicating Controller (Remote Loc.)	PT In-Line Pressure Element
PC Press. Controller	PG Press. Gauge	PE
I/P Transducer (Converter)	PSIG Pounds Per Square Inch Gauge	

Pressure Reducing Regulator (Self-Contained) · Back Pressure Regulator (Self-Contained) · Differential Pressure Reducing Regulator

Back Pressure Regulator W/External Tap · Pressure Reducing Regulator W/External Tap

PSIA Pounds Per Square Inch Absolute

PSID Pounds Per Square Inch Differential

PSIV Pounds Per Square Inch Vacuum

APPROVED ___C. Thomas___

DATE ___10-6-08___

GENERAL LEGEND

DISTILLATION UNIT

DRAWING NUMBER
OO6543

REVISION 1 · PAGE 1 OF 30

PCE

PREFIXES

CW- cooling water	RX- reactor
MU- make-up	UT- utilities
FW- feed water	CA- chemical addition
SE- sewer	IA- instrument air

ABBREVIATIONS

D- drum	TK-tank	P- pump
C- column	F- furnace	V- valve
CT- cooling tower	EX- exchanger	

Figure 2-21 *Process Legend*

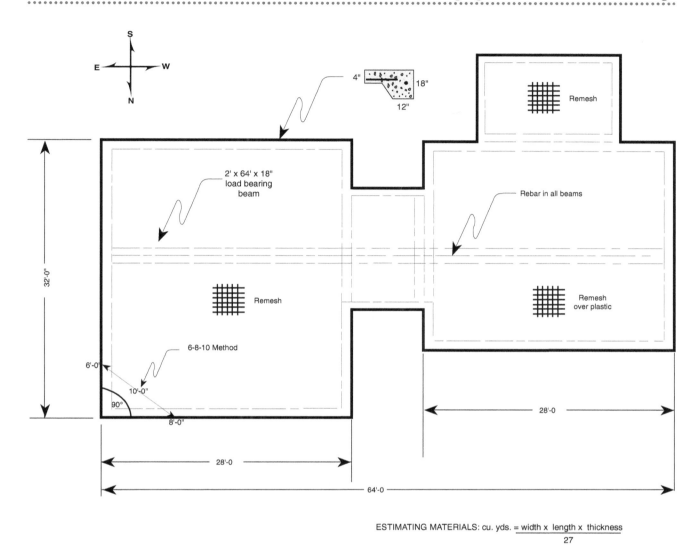

Figure 2-22 *Foundation Plans*

crafts. The primary users of the document after plant start-up include process technicians and instrument, electrical, mechanical, and safety engineers.

A process technician typically traces power to the unit from a motor control center (MCC). The primary components of an electrical system include the MCC, motors, transformers, breakers, fuses, switchgears, starters, and switches. Specific safety rules are attached to the operation of electrical systems. The primary safety system is the isolation of hazardous energy "lock-out, tag-out." Process technicians are required to have training in this area. Figure 2-24 shows the basic symbols and flow path associated with an electrical drawing. Electrical lines are typically run in cable trays to switches, motors, ammeters, substations, and control rooms.

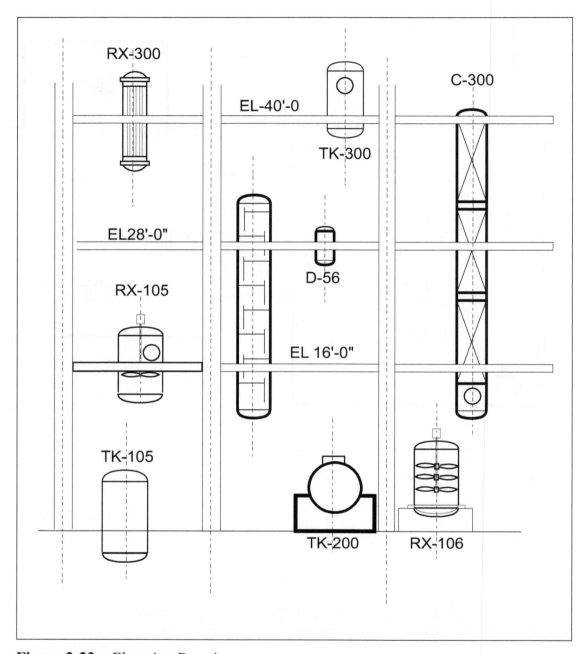

Figure 2-23 *Elevation Drawing*

A transformer is a device used by industry to convert high voltage to low voltage. The electrical department always handles problems with transformers. Electrical breakers are designed to interrupt current flow if design conditions are exceeded. Breakers are not switches and should not be turned on or off. If a tripping problem occurs, the technician should call for an electrician. Fuses are devices designed to protect equipment from excess current. A thin strip of metal will melt if design specifications are

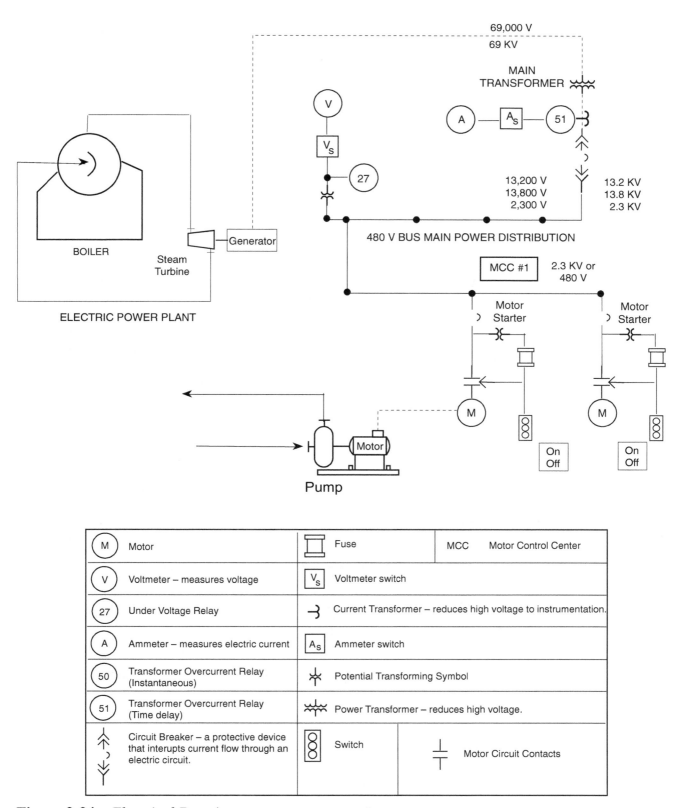

Figure 2-24 *Electrical Drawings*

exceeded. During operational rounds, technicians check the ammeters inside the MCC for current flow to their electrical systems. Voltmeters are electrical devices used to monitor voltage in an electrical system and are also checked during routine rounds.

Electrical One-Line Drawings

Like the piping in process systems, the wiring in electrical diagrams shows a flow path for distributing power throughout the unit and to all electrical equipment. These diagrams show the different voltage levels in the unit, electrical equipment such as transformers, circuit breakers, fuses, and motors, and the required horsepower. They also include start-stop switches, emergency circuits, and MCCs. Process technicians can use these diagrams to trace a system from the power source to the load.

Equipment Location Drawings (Plot Plan)

Equipment location drawings show the exact floor plan location of equipment in relation to the physical boundaries of the plant. Figure 2-25 illustrates this layout. Location drawings provide benefits similar to those of the elevation drawings. One of the most difficult concepts to explain to a new process technician is the scope and size of modern chemical processing. Most chemical plants and refineries closely resemble small cities with well-defined blocks and areas connected by a highway of piping and equipment. Equipment

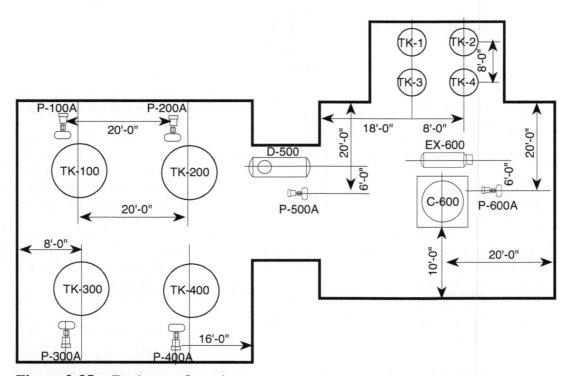

Figure 2-25   *Equipment Location*

location drawings provide information about the neighborhood. The entire P & ID provides a three-dimensional look at the various operating units in a plant.

Summary

Process technicians use process schematics to describe the primary flows and equipment used in their facilities. These drawings simplify complex processes and provide a window through which a technician can study piping, instrumentation, equipment, and locations. Process schematics include symbols, BFDs, PFDs, and P & IDs. These drawings are used by every level of the plant and are a critical document.

Block Flow Diagrams
Simple BFDs can be used to show technical and nontechnical people how material moves from one location to another. BFDs provide a simplistic set of sequences that move from left to right and show the primary flow path of a process.

Instrument Symbols
Circles or bubbles are used to indicate an instrument. The information inside the circle identifies the instrument, what it does, and the variable being measured. A line drawn horizontally through the circle indicates that it is remote located. This typically indicates that the device is located in a control room. A dashed line indicates that the instrument is located in a control room behind the control panel.

There are a large variety of instruments and instrument symbols used in the chemical processing industry. The variables that instruments measure are pressure, temperature, level, flow, and analytical variables. If the information in a circle starts with the letter "L," it is typically some type of level instrument. If the first letter in the circle is "A," it is an analytical instrument.

Piping and Auxiliary Symbols
Each plant has a standardized file for its piping symbols. Process technicians should carefully review the piping symbols for major and minor flows; electric, pneumatic, capillary, and hydraulic tubing; and future equipment. The major flow path through a unit illustrates the critical areas a new technician should concentrate on. A variety of other symbols are also included on the piping. These include valves, strainers, filters, flanges, spool pieces, blinds, insulation, piping size, pressure rating, material codes, and steam traps.

Valve Symbols

In a refinery or chemical plant, there are thousands of different valves. These valves are available in a variety of designs and sizes. The most common valves are:

- gate valves
- ball valves
- plug valves
- diaphragm valves
- needle valves
- safety valves
- pneumatically operated valves
- electrically operated valves
- pinch valves
- globe valves
- butterfly valves
- check valves
- angle valves
- knife valves
- relief valves
- hydraulically operated valves
- solenoid valves, and
- motor operated valves.

Flow Instruments

The basic flow elements most commonly used include orifice plates, turbine meter, rotameter, venturi tube, mass flow meter, magmeter, pitot tube, flow nozzle, annubar tube, nutating disk meter, weir and flume, ultrasonic flow meter, vortex flow meter, coriolis meter, and oval gear meter. The letters that represent these elements are "FE" written inside the bubble. The term "Fi" for flow indicator is also used. A flow indicator may appear to be a simple gauge that provides relatively accurate measurements.

Pressure Instruments

The most common pressure devices are pressure gauges, manometers, and pressure transmitters. Pressure gauges are represented on a P & ID as pressure indicator (Pi). A circle with "PE" is the symbol for a pressure element. A diaphragm in a DP cell is an example of a PE.

Level Instruments

Level measurement is directly linked to pressure measurements. The common instruments used to measure level are LG, displacer bulbs coupled to a level transmitter, capacitance probes, a bubbler system, load cells, and differential pressure transmitters (ΔP cell).

Temperature Instruments

Temperature symbols are typically represented as "Ti" for temperature indicator or "TE" for temperature element. A temperature indicator is typically a gauge or a thermometer. A temperature gauge has a bimetallic strip inside, which differentially expands with increasing temperature, creating a deflection that is correlated with temperature. These instruments are located in the field beside the equipment. Temperature elements are typically thermo-electric temperature measuring devices: thermocouples, thermal bulbs, resistance bulbs, or RTDs.

Analytical Instruments

Analytical symbols are used to identify the use of quantitative or qualitative analyzers. Some analyzers check for the presence of a specific substance while others check for the composition or percentage of the chemical in a process stream. Analyzers provide an open window on the molecular level to the chemical composition of a substance. Analyzers are used in the following systems:

- pH (acid/base)—cooling tower basin
- oxidation reduction potential (ORP)—free electron concentration
- conductivity (parts per million)—cooling tower blow down
- O_2 analyzer—flue gases of boiler or furnace or incinerator
- CO analyzer—flue gases of boiler or furnace or incinerator
- color analyzer—ensures color specifications on products
- chromatography—measures concentration of components
- opacity analyzer—identifies concentration of particulates in a matter stream
- butane analyzer—measures concentration of butane
- other analyzers—measure concentration or test for presence of a specific chemical
- other analyzers—mass spectrometers, total carbon analyzers.

Process Equipment

There are symbols for every piece of equipment found in the chemical processing industry. Even exotic equipment has a symbol. It may not be well-known, but it exists. Basic examples of equipment symbols are:

- tanks and vessels
- valves
- compressors
- motors
- cooling towers
- furnaces
- reactors
- steam traps, strainers
- pulsation dampeners
- piping
- pumps
- steam and gas turbines
- heat exchangers
- boilers

- distillation columns
- separators
- removable spool-y-strainers, and
- eductors.

Process Flow Diagram—A Training Tool

PFDs and P & IDs are used to outline or explain the complex flows, equipment, instrumentation, electronics, elevations, footings, and foundations in a process unit. A process instrumentation and piping drawing (P & ID) is a complex representation of the various units found in a plant. A simple PFD is typically used to describe the primary flow path through a unit. Process diagrams can be broken down into two major categories: flow diagrams and P & IDs. A flow diagram is a simplified illustration that uses process symbols to describe the primary flow path through a unit. A P & ID is a complex diagram that uses process symbols to describe a process unit. Symbols and diagrams have been developed for most pieces of industrial equipment, process flows, and instrumentation.

Elements of a Piping and Instrumentation Drawing

The basic components of a P & ID are (1) the process legends, (2) foundation drawings, (3) elevation drawings, (4) electrical drawings, (5) equipment location drawings, (6) complex flow diagrams, and (7) the P & IDs that include all of the major flows, minor flows, process instrumentation, and operating information.

Review Questions

1. Describe the purpose of a process schematic.

2. Explain how BFDs can help teach a new technician her job.

3. Describe the basic elements of a PFD.

4. Draw the symbols for a gate valve, globe valve, and automatic valve.

5. Draw the symbols for a centrifugal pump and a PD pump.

6. Draw the symbols for a blower and a reciprocating compressor.

7. Draw the symbols for a steam turbine and a centrifugal compressor.

8. Draw the symbols for a heat exchanger and a cooling tower.

9. Draw the symbols for a packed column and a plate column.

10. Draw the symbols for a furnace and a boiler.

11. Draw a pressure and flow-control loop.

12. Draw a level and temperature control loop.

13. Draw a simple PFD using the symbols from the previous questions.

14. What information is obtained from a loop diagram?

15. What information is available on electrical one-line diagrams?

16. What information is available on plot-plan drawing?

17. Describe the basic elements of a P & ID.

18. Explain how process instruments are illustrated on a drawing.

19. List the different types of piping and auxiliary equipment.

20. List the different types of tanks.

chapter 3

Understanding Process Equipment

LEARNING OBJECTIVES

After studying this chapter, the student will be able to:

- Identify and describe the valves used in industry.
- Describe the various types of storage and piping used in the chemical processing industry.
- Identify the operation and primary components of a centrifugal pump.
- Explain the operation and types of positive displacement pumps.
- Describe dynamic and positive displacement compressors.
- Describe how a steam turbine works.
- Describe the purpose of seals, bearings, and lubrication systems.
- Describe the major components and operation of the various types of heat exchangers.
- Describe the key components and operation of a cooling tower.
- Describe the basic components of a steam generation system.
- Describe the important aspects of a fired heater system.
- Describe the various components of a stirred reactor.
- List the primary components of a distillation column.

Key Terms

Compressors—come in two basic designs: (1) positive displacement—rotary and reciprocating and (2) dynamic—axial and centrifugal. A compressor is designed to accelerate or compress gases.

Process instruments—control processes and provide information about pressure, temperature, levels, flow, and analytical variables.

Pumps—are used to move liquids from one place to another. Pumps come in two basic designs: (1) positive displacement—rotary and reciprocating and (2) dynamic—axial and centrifugal.

Steam turbines—are used as drivers to turn pumps, compressors, and electric generators. A steam turbine is an energy conversion device that converts steam energy (kinetic energy) to useful mechanical energy.

Tanks and pipes—store and hold fluids. Tank designs include spherical, open top, floating roofs, drums, and closed tanks.

Boilers—are primarily designed to boil water and generate steam for industrial applications. Boilers are classified into two groups: (1) water tube or (2) fire tube. Steam generation systems produce high-, medium-, and low-pressure steam for industrial use.

Heat exchangers—are energy transfer devices designed to convey heat from one substance to another. Basic designs include pipe coil, shell and tube, air-cooled, plate and frame, and spiral.

Cooling towers—are devices used by industry to remove heat from water. A box-shaped collection of wooden slats and louvers direct airflow and break up water as it cascades from the top of the water distribution system. Cooling towers are classified by the way they produce airflow or by the way the air moves in relation to the downward flow of water. Basic designs include atmospheric, natural, forced, and induced drafts.

Fired heaters—consist of a battery of tubes that pass through a firebox. Fired heaters or furnaces are commercially used to heat up large volumes of crude oil or hydrocarbons. Basic designs include cylinder, cabin, and box.

Distillation columns—separate chemical mixtures by boiling points. A distillation column is a collection of stills stacked one on top of the other. Distillation columns fall into two distinct classes: (1) plate and (2) packed.

Reactors—are devices used to convert raw materials into useful products through chemical reactions. There are five reactor designs commonly used in the chemical processing industry: stirred reactors, fixed-bed reactors, fluidized-bed reactors, tubular reactors, and furnace reactors.

Valves

Gate Valve

A gate valve places a movable metal gate in the path of a process flow in a pipeline. Gate valves are available in two designs: (1) rising stem and (2) nonrising stem. Located at the top of a closed gate valve is the handwheel. The handwheel is attached to a threaded stem. As the handwheel is

Figure 3-1
Gate Valve

turned counterclockwise, the stem in the center of the handwheel begins to rise. This lifts the gate out of the valve body and allows product to flow. The basic components of a gate valve are illustrated in Figure 3-1.

Globe Valve

A globe valve places a movable metal disk in the path of a process flow. This type of valve is most commonly used for throttling service. The disk is designed to fit snugly into the seat and stop flow. Globe and gate valves have very similar component lists, as illustrated in Figure 3-2. Process fluid enters the globe valve and is directed through a 90° turn to the bottom of the seat and disk. As the fluid passes by the disk, it is evenly dispersed. Globe valves are designed to be installed in high-use areas. Globe valves can be found in the following designs: typical globe valve with ball, plug, or composition element; needle valve; and angle valve.

Ball Valve

Ball valves take their name from the ball-shaped, movable element in the center of the valve. Unlike the gate and the globe valves, a ball valve does not lift the flow control device out of the process stream; instead, the hollow ball rotates into the open or closed position. Ball valves provide very little restriction to flow and can be opened 100% with a quarter turn on the valve handle. A typical ball valve is illustrated in Figure 3-3. In the closed position, the port is turned away from the process flow. In the open position, the port lines up perfectly with the inner diameter of the pipe. Larger valves require handwheels and gearboxes to be opened, but most only require a quarter turn on a handle.

Figure 3-2
Globe Valve

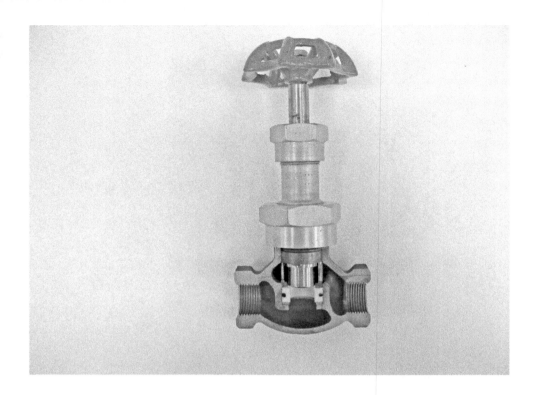

Figure 3-3
Ball Valve

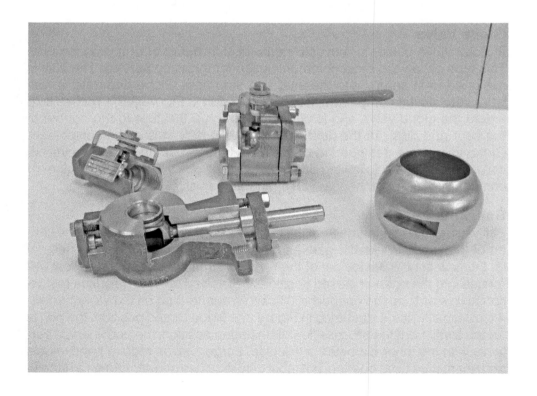

Automatic Valves

The chemical processing industry uses a complex network of automated systems to control its processes. The smallest unit in this network is called a control loop. Automatic valves can be controlled from remote locations, making them invaluable in modern processing. Figure 3-4A and B is an

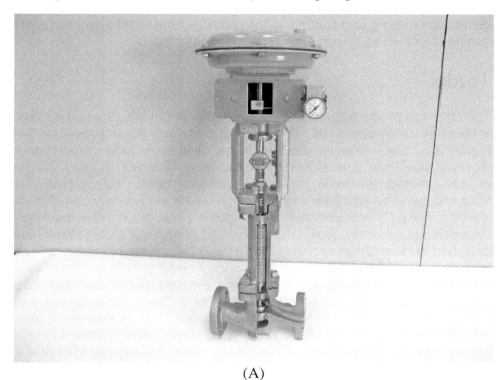

(A)

Figure 3-4A & B
Pneumatic Operated Valve

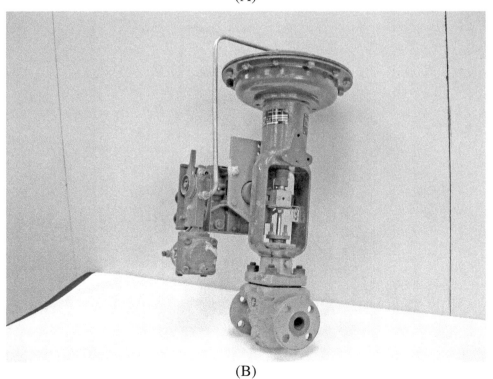

(B)

illustration of a pneumatically operated valve. Any of the valves studied in this chapter can be automated. To automate a valve, a device known as an actuator is installed. The actuator controls the position of the flow control element by moving and controlling the position of the valve stem. Actuators can be classified as pneumatic, hydraulic, or electric.

Tanks

The chemical processing industry uses a variety of tanks, drums, bins, and spheres to store chemicals. The most popular designs are illustrated in Figure 3-5. The materials used in these designs include carbon steel, stainless steel, iron, specialty metals, and plastic. Common storage designs include spheres, spheroids, horizontal cylindrical tanks (drums), bins, and fixed and floating roofs. Tanks, drums, and vessels are typically classified as low pressure (LP), high pressure (HP), liquid service, gas service, insulated, steam traced, or water cooled. Spherical and spheroidal storage tanks are designed to store gases or pressures above 5 psig. Spheroid tanks are flatter than spherical tanks. Horizontal cylindrical tanks or drums can be used for pressures between 15 and 1000+ psig. Floating roof storage tanks are used for materials near atmospheric pressure. In the basic design, a void forms between the floating roof and the product, forming a constant seal. The primary purpose of a floating roof is to reduce vapor losses and hold stored fluids. In areas of heavy snowfall, an internal floating roof is used with an external roof since the weight of the snow would affect the seal.

Figure 3-5
Tank Designs

Pumps

Dynamic **pumps** can be classified as centrifugal or axial. Centrifugal pumps move liquids by centrifugal force. The primary principle involves spinning the liquid in a circular rotation that propels it outward and into a discharge chute known as a volute. Centrifugal force and the design of the volute add energy or velocity to the liquid. As the liquid leaves the volute, it begins to slow down, creating pressure. Fluid pressure moves the process through the pipes. The basic components of a centrifugal pump include the casing, motor or driver, coupling, volute, suction eye or inlet, impellers, wear rings, seals, bearings, discharge port, and suction and discharge gauge (Figure 3-6).

Positive displacement (PD) pumps displace a specific volume of fluid on each stroke or rotation. PD pumps can be classified as rotary or reciprocating. Rotary pump designs include screw pumps (progressive cavity and screw), gear pumps (internal and external), vane pumps (sliding and flexible), and lobe pumps. Rotary pumps displace fluids with gears and rotating screw elements. Reciprocating pumps move fluids by drawing them into a chamber on the intake stroke and positively displacing them with a piston or diaphragm on the discharge stroke (Figure 3-7).

Figure 3-6
Centrifugal Pump

Figure 3-7
PD Pump

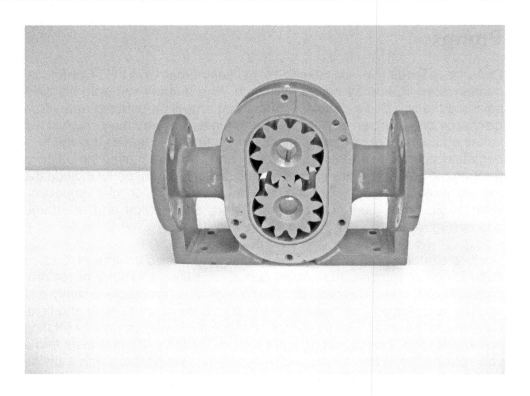

Compressors

The operation and design of a **compressor** can usually be classified in two groups: PD or dynamic. Dynamic compressors operate by accelerating the gas and converting the energy to pressure. This type of compressor has two basic designs: centrifugal and axial. Centrifugal compressors operate by adding centrifugal force to the product stream. The design and application of centrifugal compressors accelerates the velocity of the gases. This velocity or kinetic energy is converted to pressure as the gas flow leaves the volute and enters the discharge pipe. Centrifugal compressors can deliver much higher flow rates than PD compressors. The basic components of a centrifugal compressor include the casing, motor or driver, coupling, volute, suction eye or inlet, impellers, wear rings, seals, bearings, discharge port, and suction and discharge gauge (Figure 3-8).

Positive Displacement Compressors

PD compressors operate by trapping a specific amount of gas and forcing it into a smaller volume. They are classified as rotary or reciprocating. PD compressors and PD pumps operate under similar conditions. The primary difference is that compressors are designed to transfer gases while pumps move liquids. The rotary compressor design includes a rotary screw, sliding vane, lobe, and liquid ring. Reciprocating compressors include a piston and diaphragm.

Figure 3-8
Centrifugal Compressor

Steam Turbine

A steam turbine is a device (driver) that converts kinetic energy (steam energy of movement) to mechanical energy. **Steam turbines** have a specially designed rotor that rotates as steam strikes it. This rotation is used to operate a variety of shaft-driven equipment. Steam turbines are used primarily as drivers for pumps, compressors, and electric power generation. Figure 3-9 illustrates the internal components of an impulse steam turbine.

Electricity and Motors

The chemical processing industry uses three-phase motors to operate pumps, compressors, fans, blowers, and other electrically driven equipment. Three-phase motors come in three basic designs: squirrel-cage induction motors, wound rotor induction motors, and synchronous motors. The primary difference is in the rotor. The direction of rotation in a motor is determined by strong magnetic fields. A typical motor is composed of stator windings, rotor and shaft, bearings and seals, conduit box, frame, fan, lubrication system, and shroud. Figure 3-10 illustrates the location of these components.

Figure 3-9A & B
Steam Turbine

(A)

(B)

Figure 3-10
Typical Motor

Heat Exchangers

The most widely used and inexpensive solution to heat transfer is a heat exchanger system. A heat exchanger allows a hot fluid to transfer energy in the form of heat to a cooler fluid without their physically coming into contact with each other. This device can provide heat or cooling to a process. Common names used to describe a heat exchanger include heater, preheater, cooler, and condenser.

A typical shell and tube heat exchanger is composed of a series of tubes surrounded by a shell. The tubes are typically connected to a fixed tube sheet and are supported by a series of internal baffles. The shell and tube cylinder has a water box or head securely attached on the inlet and outlet side. A tube inlet and outlet admits fluid through the tubes. A shell inlet and outlet admits flow through the shell. Figure 3-11 illustrates the typical layout for a shell and tube heat exchanger.

Another type of heat exchanger is a kettle reboiler. Reboilers are energy balance devices attached to **distillation columns** to help control the temperature. Kettle reboilers have a specially designed vapor-disengaging cavity that removes the lighter components of the bottom stream. These lighter fractions are returned to the bottom of the column. A kettle reboiler has five connections, two on the tube side and three on the shell side. Steam or hot oil flows through the tube side and provides the heat source. Flow rate is carefully controlled and frequently linked to the bottom temperature control

Figure 3-11
Shell and Tube Heat Exchanger

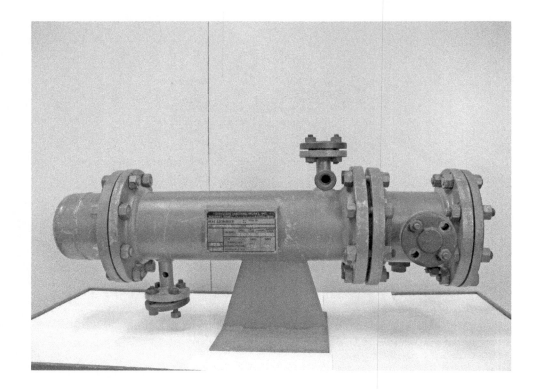

system. The shell side has three nozzles: a liquid product feed line, a vapor return line to the column, and a heavy liquid out product line. A kettle reboiler can be used to (1) control the liquid level at the bottom of the column, (2) control the temperature of the column, and (3) help control product purity in the bottom of the column. Figure 3-12 shows what a kettle reboiler looks like.

Cooling Tower

A cooling tower is a simple device used by industry to remove heat from water and is the central feature in a cooling water circuit. Water is stored in the bottom of the cooling tower in a basin. A pump takes suction off the bottom of the basin and pumps it into the plant, where it absorbs heat before being returned to the cooling tower. A cooling tower is a large rectangular or box-shaped device filled with wooden slats and louvers that direct airflow and break up water as it falls from the top of the water distribution header. The internal design of the tower ensures good air and water contact. Hot water enters the top of the cooling tower and is discharged through a water distribution system over a matrix of splash boards or fill. As the hot water hits the flat boards, it spreads out and cool air flows over it, transferring heat energy away. The fill or splash boards enhance liquid–air contact while the drift eliminators reduce the amount of water lost from the tower because of excess airflow. When water changes to vapor, it takes

Figure 3-12
Kettle Reboiler

heat energy with it, leaving behind the cooler liquid. Evaporation is the primary means of heat transfer. Cool water collected in the basin is pumped back into the plant.

The basic components of a cooling tower include a water basin, pump, internal framework used to support the fill, fan, water distribution system, drift eliminators, louvers, and water make-up system at the base. Figure 3-13 shows a typical cooling tower.

Figure 3-13
Induced Draft Cooling Tower

Boilers

The chemical processing industry uses **boilers** to produce high-, medium-, and low-pressure steam. Steam is one of the most important utilities required for process operations. A water tube boiler consists of an upper steam generating drum and a lower mud drum connected by three types of tubes: downcomer, riser, and steam-generating tubes. These drums and

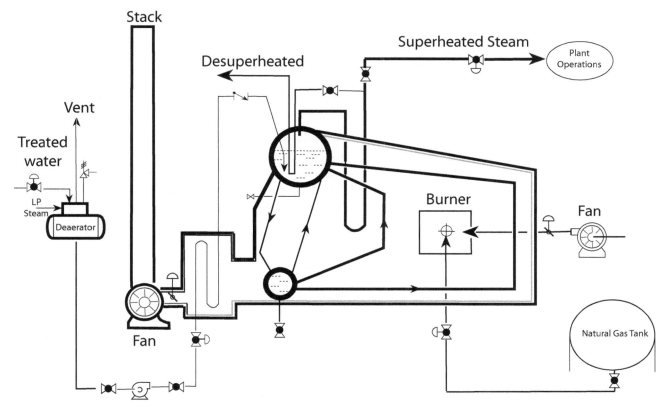

Figure 3-14 *Water Tube Boiler*

tubes are surrounded by a furnace and a series of specially designed burners. The lower mud drum and water tubes are completely filled with water while the upper steam generating drum is only partially filled. This vapor cavity allows steam pressure to build and to collect and pass out of the header. Water is carried through tubes that flow near and around the burners. As heat is applied to the water-generating tubes and drums, water circulates around the boiler, down the downcomer tube, into the lower drum, and rises up the riser tube and steam-generating tubes of the furnace. During normal operations, HP steam is superheated and sent to the main steam header. Figure 3-14 shows all of the basic components of a boiler.

Furnaces

A fired heater or furnace is a device used primarily to heat up large quantities of hydrocarbons. Furnaces heat up raw materials so they can produce products such as gasoline, oil, kerosene, plastic, and rubber. **Fired heaters** consist essentially of a battery of pipes or tubes that pass through a firebox. These tubes run along the inside walls and roof of a furnace. The heat released by the burners is transferred through the tubes and into the

Figure 3-15
Furnace

process fluid. The fluid remains in the furnace long enough to reach operating conditions before exiting and being shipped to the processing unit.

The basic components of a cabin-fired heater include a tough metal shell surrounding a firebox, convection section, and stack. The inside of the furnace is lined with a special refractory material (brick, blocks, peep stones, and gunite) designed to reflect heat. A battery of tubes pass through the convection and radiant sections and into a common insulated header that passes out of the furnace. A series of burners are located at the bottom of the furnace or on the sides. Fluid flow is carefully balanced through the tubes to prevent the damage to the equipment or to the product. Airflow and oxygen content are controlled through primary, secondary, and damper adjustments. Figure 3-15 illustrates the basic layout of a cabin furnace.

Reactors

A reactor is a device used to convert raw materials into useful products through chemical reactions. **Reactors** are designed to operate under a variety of conditions. These devices combine raw materials with catalyst, gases, pressure, or heat. The shape and design of a reactor are dictated by the application it will be used in. Figure 3-16 is an illustration of a simple mixing reactor.

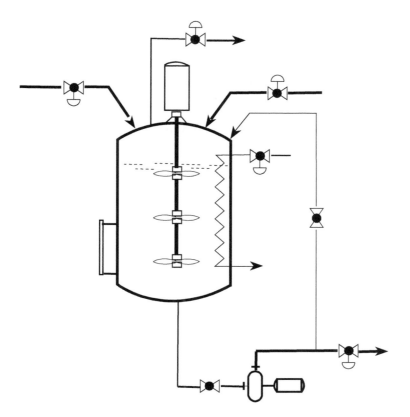

Figure 3-16 *Stirred Reactor*

There are a number of critical process variables associated with reactor operation. Some of these include temperature, pressure, concentration of reactants, catalysts, and time. As the temperature increases, molecular activity increases. Since a chemical reactor is designed to make chemical bonds, break chemical bonds, or make and break chemical bonds, temperature is carefully controlled. When the pressure is increased, molecules are moved closer together. When this process is combined with heat, a higher number of collisions can be achieved. The more collisions, the more chemical reactions occur within a specific amount of time. The speed at which two or more chemicals will react doubles for each 10°C increase in temperature.

The concentration of reactants in the reactor has a significant impact on how fast a reaction will occur. Stirred reactors are designed to enhance molecular contact. Reaction time can also provide the contact that reactants need in order to produce the desired products. In some cases, a catalyst may be used to speed up the reaction.

Distillation

A distillation column is a series of stills placed one on top of the other. As vaporization occurs, the lighter components of the mixture move up the tower and are distributed on the various trays. The lightest component goes out of the top of the column in a vapor state and is passed over the cooling coils of a shell and tube condenser. As the hot vapor comes in contact with the coils, it condenses and is collected in the overhead accumulator. Part of this product is sent to storage while the other part is returned to the tower as reflux.

Distillation is a process by which a substance is separated from a mixture at its boiling point. During the distillation process, a mixture is heated until

Figure 3-17
Distillation Column

it vaporizes, then condensed on the trays or at various stages of the column, where it is drawn off and collected in a variety of overhead, side-stream, and bottom receivers. The condensed liquid is referred to as the distillate while the liquid that does not vaporize in a column is called the residue.

During column operation, raw materials are pumped to a feed tank and mixed thoroughly. Mixing is usually accomplished with a pump-around loop or a mixer. This mixture is pumped to a feed preheater or furnace where the temperature of the fluid mixture is brought up to operating conditions. Preheaters are usually shell and tube **heat exchangers** or fired furnaces. This fluid is admitted onto the feed tray of the distillation column. Part of the mixture vaporizes as it enters the column while the rest begins to drop into the lower sections of the tower.

Heat balance on the tower is maintained by a device known as a reboiler. Reboilers take suction off the bottom of the tower. The heaviest components of the tower are pulled into the reboiler and stripped of smaller molecules. The stripped vapors are returned to the column and allowed to separate in the tower.

The basic components of a plate distillation column include a feed line, feed tray, rectifying or enriching section, stripping section, downcomer, reflux line, energy balance system, overhead cooling system, condenser, preheater, reboiler, accumulator, feed tank, product tanks, bottom line, top line, side-stream, and an advanced instrument control system. Figure 3-17 illustrates the basic components of a distillation system.

Summary

Tanks and pipes store and hold fluids. Tank design will vary depending on the service it will be used in. Pipe size and design will determine flow rates, pump and valve sizes, turbulent or laminar flow, and instrument type and automation. Valves are devices used to control the flow of fluids. Valves come in a variety of shapes, sizes, and designs that throttle, stop, or start flow. The more common designs include gate, globe, ball, and automated valves.

Pumps are primarily used to move liquids from one place to another. They come in two basic designs: PD and dynamic. PD pumps can be classified as rotary or reciprocating. Reciprocating pumps are characterized by a back and forth motion, while rotary pumps move in a circular rotation. Dynamic pumps can be classified as centrifugal or axial. The centrifugal pump uses centrifugal force to move liquids, while the axial pump pushes liquids along a straight line. Compressors come in two basic designs: PD—rotary and reciprocating and dynamic—axial or centrifugal. A compressor is designed to accelerate or compress gases. Compressors are closely related to pumps.

Steam turbines are used as drivers to operate pumps, compressors, and electric generators. HP steam is directed into buckets designed to operate a rotor and provide rotational energy. Steam turbines provide the same feature that electric motors are designed to do. A typical motor is composed of stator windings, rotor and shaft, bearings and seals, conduit box, frame, fan, lubrication system, and shroud. Steam turbines and motors are two of the most popular devices used by industry as drivers.

Heat exchangers transfer energy in the form of heat between two fluids without their physically coming into contact with each other. A typical shell and tube heat exchanger has a tube side flow and a shell side flow. Heat energy is transferred to the cooler stream as they pass each other in the exchanger. A standard exchanger has a shell, tubes, tube sheet, shell inlet and outlet, tube inlet and outlet, and baffles. Another type of heat exchanger is a kettle reboiler. Reboilers are energy balance devices attached to distillation columns to help control the temperature.

A cooling tower is a simple device used by industry to remove heat from water. Heat exchangers and **cooling towers** typically work together to remove heat from a variety of industrial applications. A cooling tower is a box-shaped collection of multilayered wooden slats and louvers that direct airflow and break up water as it cascades from the top of the tower or water distribution system. The internal design of the tower ensures good air and water contact. Hot water transfers heat to cooler air as it passes through the tower. Sensible heat accounts for 10% to 20% of the heat transfer in a cooling tower. Evaporation accounts for 80% to 90% of the heat transfer in a cooling tower. The principle of evaporation is the most critical factor in cooling tower efficiency.

Cooling towers are classified by (1) how they produce airflow and (2) the direction the airflow takes in relation to the downward flow of water. Airflow is produced naturally or mechanically. Mechanical drafts are created by fans located on the side or top of the cooling tower. Airflow into and through a tower is either cross-flow or counterflow. Cross-flow goes horizontally across the downward flow of water before exiting the system. When the air is forced to move vertically upward against the downward flow of water, it is referred to as counterflow. The basic components of a cooling tower include a water basin, pump, and water make-up system at the base of the cooling tower. The internal frame is made of pressure-treated wood or plastic and is designed to support the internal components of the tower. Some of these components include the fill or splash boards and drift eliminators. The fill or splash boards enhance liquid–air contact while the drift eliminators reduce the amount of water lost from the tower because of excess airflow. Louvers on the side of the cooling water tower let air into the device.

A hot water distribution system is typically located on the top of the cooling tower fill. A fan may be used to enhance airflow through the cooling tower.

Boilers are used by industrial manufacturers to produce steam. Steam is used to drive turbines and provide heat to process equipment. A typical steam generation system includes super high pressure (SHP) steam generation and distribution, HP steam (400–800 psig), MP steam (200 psig), and LP steam (50 psig). SHP steam can be as high as 1200 psig. Water tube boilers are typically designed for large industrial applications. A water tube boiler consists of an upper steam generating drum and a lower mud drum connected by three sets of tubes: downcomers, risers, and steam generating tubes. A furnace surrounds and provides heat to the drums and tubes. The lower drum and water tubes are completely filled with water while the upper drum is only partially filled. This vapor cavity allows steam to collect and pass out of the header. As heat is applied to the water-generating tubes and drums, water circulates around the boiler, down the downcomer tube, into the mud drum, and rises up the riser tube and steam-generating tubes of the furnace. During normal operations, steam rises into the upper drum and moves to the steam header. Lost water in the boiler is replaced by the make-up water line.

The chemical processing industry uses fired heaters to heat large quantities of crude oil or other hydrocarbon feed stocks up to operating temperatures for processing. Fired heaters consist of a battery of tubes that pass through a firebox. Fired heaters or furnaces are commercially used to heat up large volumes of crude oil or hydrocarbons. The basic components of a cabin-fired heater include a tough metal shell surrounding a firebox, convection section, and stack. The inside of the furnace is lined with a special refractory material (brick, blocks, peep stones, and gunite) designed to reflect heat. A battery of tubes pass through the convection and radiant sections and into a common insulated header that passes out of the furnace. A series of burners are located on the bottom of the furnace or on the sides.

Reactors are used to combine raw materials, heat, pressure, and catalysts in the right proportions. There are a number of critical process variables associated with reactor operation. Some of these include temperature, pressure, concentration of reactants, catalysts, and time. The basic components of a reactor include a shell, a heating or a cooling device, two or more product inlet ports, and one outlet port. A mixer may be used to blend the materials together.

A distillation process is a complex arrangement of systems that includes a cooling tower system, a pump and feed system, a preheat system, product storage system, compressed air system, steam generation system, and a complex instrument control system. Distillation towers separate chemical mixtures by boiling points. A distillation tower is a collection of stills stacked one on top of the other. As chemical mixtures enter the tower, they are separated by boiling point and distributed on different trays or sections of the tower.

Review Questions

1. Describe the purpose of valves in industrial applications.

2. Describe the basic operation and components of a centrifugal pump.

3. Describe how a steam turbine works.

4. Describe the basic operation and components of an electric motor.

5. List the basic tank and storage designs.

6. List the components of a simple shell and tube heat exchanger and explain its primary operation.

7. List the components of a cooling tower and explain its primary operation.

8. List the basic components of a boiler and explain its primary purpose.

9. List the basic components of a furnace and explain its primary purpose.

10. Describe how a typical distillation column works.

11. What are the basic components of a stirred reactor? Explain how it works.

Introduction to Control Loops

LEARNING OBJECTIVES

After studying this chapter, the student will be able to:

- Identify the five elements of a control loop.
- Draw a level control loop.
- Draw a pressure control loop.
- Draw a flow control loop.
- Draw a temperature control loop.
- Draw an analytical control loop.
- Describe the various controller modes: rate reset, proportional, and how each compliments the other.
- Describe manual, automatic, and cascade control features.

Key Terms

Control loop—a collection of instruments that work together to automatically control a process. This includes a primary element or sensor, a transmitter, a controller, a transducer, and a final control element.

Cascade control—a term used to describe how one control loop controls or overrides the instructions of another control loop in order to achieve a desired set point.

Controller modes—include proportional (P), proportional plus integral (PI), proportional plus derivative (PR), and proportional integral derivative (PID). Proportional control is primarily used to provide gain where little or no load change typically occurs in the process. PI is used to eliminate offset between set point and process variables. PI works best where large changes occur slowly. PR is designed to correct fast changing errors and reduce overshooting the set point. PD works best when frequent small changes are required. PID is applied where massive rapid load changes occur. PID reduces swinging between the process variable and set point.

The rate or derivative mode—enhances controller output by increasing the output in relation to the changing process variable. As the process variable approaches the set point, the rate or derivative mode relaxes, providing a braking action that prevents overshooting the set point. The rate responds aggressively to rapid changes and passively to smaller changes in the process variable.

The reset or integral mode—is designed to reduce the difference between the set point and process variable by adjusting the controller's output continuously until the offset is eliminated. The reset mode responds proportionally to the size of the error, the length of time that it lasts, and the integral gain setting.

The proportional band—on a controller describes the scaling factor used to take a controller from 0% to 100% output. If the proportional band is set at 50% and the amount of lift off the seat for the final control element (globe valve) is four inches, the control valve will open 2 inches if the maximum valve opening is 4 inches.

Range—is defined as the portion of the process controlled by the controller. For example, the temperature range for a controller may be limited to 80–140°F.

Span—is the difference (Δ) between the upper and lower range limits.

The primary purpose of a controller—is to receive a signal from a transmitter, compare this signal to a set point, and adjust the (final control element) process to stay within the range of the set point. Controllers come in three basic designs: pneumatic, electronic, and electric.

Auto/man control—controllers can be operated in manual or automatic. During plant start-up, the controller is typically placed in the manual position. Manual control affects the position of the control valve. It does not respond to process load changes. After the process is stable, the operator places the controller in automatic mode and allows the controller to supervise the control loop function. At this point, the controller will attempt to open and close the control valve to maintain the set point.

Basic Elements of a Control Loop

Process technicians use instrumentation to control a variety of automated processes. The key component of automatic control is the **control loop.** A control loop is defined as a group of instruments that work together to control a process. These instruments typically include a transmitter coupled with a sensing device or primary element, a controller, a transducer, and a control valve. Process plants contain many control loops that are used to maintain pressure, temperature, flow, and level and composition (Figure 4-1).

The basic elements of a control loop are:
- Measurement device—primary elements and sensors
  - flow—orifice plate and flow nozzle
  - level—float and displacer
  - pressure—helix, spiral, and bellows
  - temperature—thermocouple, thermal, and resistance bulb
- Transmitter—a device designed to convert a measurement into a signal. This signal will be transmitted to another instrument.
  - pressure transmitter—tubing to process
  - temperature transmitter—tubing to process

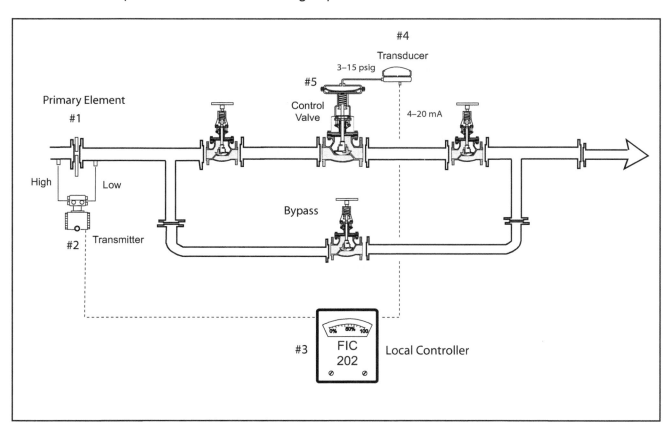

Figure 4-1 *Typical Control Loop*

- ○ flow transmitter—differential pressure (DP) cell, high- and low-pressure taps
- ○ level transmitter—hooked to float or displacer
- Controller—a device designed to compare a signal to a set point and transmit a signal to a final control element.
  - ○ recording
  - ○ indicating
  - ○ blind
  - ○ strip chart
  - ○ scale, pointers, and switches
- Transducer—a device designed to convert an air signal to an electric signal or an electric signal to a pneumatic signal. Sometimes referred to as an I to P or as a converter. For electric current (4–20 mA), for instrument air (3–15 psig), or "e" to "p"—voltage (0–5 V) to air (3–15 psig).
  - ○ air signal to an electric signal
  - ○ electric signal to a pneumatic signal
- Final control element—the part of a control loop that actually makes the change to the process.
  - ○ control valve
  - ○ damper
  - ○ governor for speed control.

Process Variables

The process variables that a technician is responsible for controlling include temperature, flow, level, pressure, and analytical variables. It only stands to reason that a study of the scientific principles associated with these variables is required for any new trainee. This study should also include the instrumentation specific to the process and the equipment associated with it inside the system. Technicians can spend a lifetime studying the unique features associated with the many processes found in the chemical processing industry.

Flow Control Loop and Simple Instruments

Flow control loops are typically designed, so a measurement of the flow rate is taken first and then the flow is interrupted or controlled down stream. Flow control loops start at the primary element and work their way back to the final control element. Flow control primary elements could include orifice plates, flow nozzles, pitot tubes, annubar pitots, magmeters, turbine meters, mass flow meters, nutating discs, oval gears, venturi tubes, vortex meters, target flow meters, and integral orifice flow meters. The most common primary element is the orifice plate. Orifice plates artificially create a high-pressure/low-pressure situation that can be measured by the transmitter. Primary

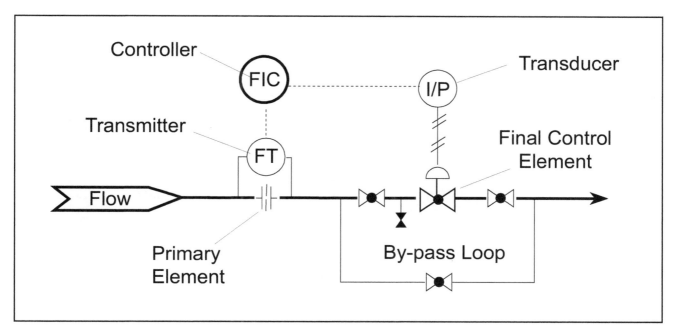

Figure 4-2 *Flow Control Loop*

elements are typically used in conjunction with a transmitter. Figure 4-2 shows an example of a flow control loop. Although it appears that the primary element is interrupting the flow, this is not the case. Increased velocity across the orifice plate compensates for the restriction. The transmitted signal is sent to a controller that compares the incoming signal with the desired set point. If a change is required, the controller will send a signal to a final control element.

Flow Instruments
Flow transmitters use differential pressure (DP) to measure flow rate. Fluid flow through a pipe can be related to pressure differences inside a pipe when flow restrictive devices such as orifice plates, venturi tubes, or flow nozzles are installed. When fluid flow encounters a restriction in a pipe, the pressure increases in front of the restriction. Fluid velocity through the restriction increases. The pressure on the other side of the restriction drops.

A DP cell is used to measure the difference between the pressure on the inlet and outlet side of the restrictive device. The DP cell usually is connected to a transmitter that sends a signal to a controller. Controllers send signals to control valves to open or close depending on the comparison of the signal from the field with the flow rate set point (Figure 4-3).

Nutating Disc Meter
The nutating disc meter is composed of a counter nutating disc (resembling a spinning top), flow inlet, and flow outlet. Nutating disc meters measure fluid flow directly by counting the rotations of the disc as fluid passes through it (Figure 4-4).

Figure 4-3
*(A) Flow Transmitter
and Primary Element,
(B) Flow Transmitter,
and (C) Orifice
Primary Element*

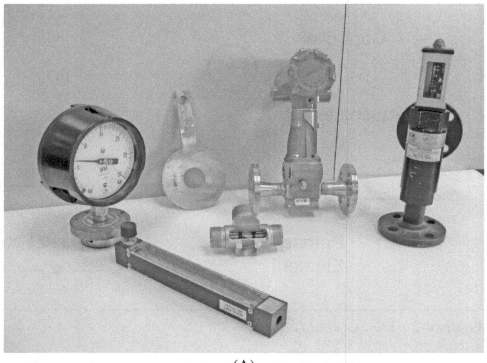

(A)

(B)

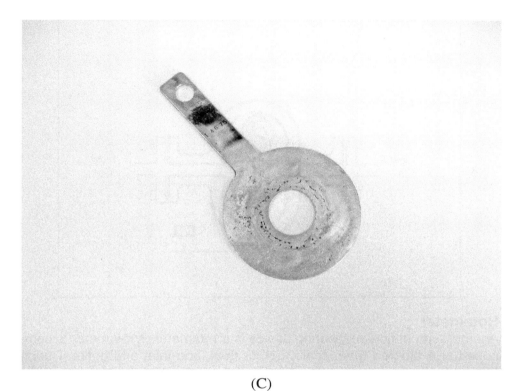

Figure 4-3
(continued)

(C)

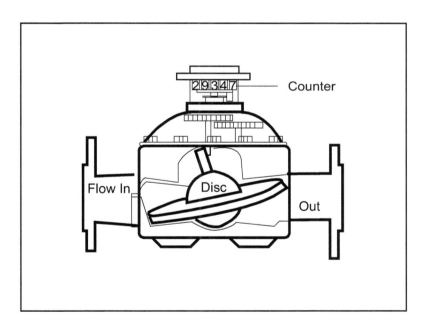

Figure 4-4
Nutating Disc Meter

Oval Gear Meter

An oval gear meter has an internal structure that resembles a lobe-pump. The lobe-shaped elements rotate as fluid passes through the internal chamber. The rotation of the gears is used to calculate the total flow rate (Figure 4-5).

Figure 4-5
Oval Gear Meter

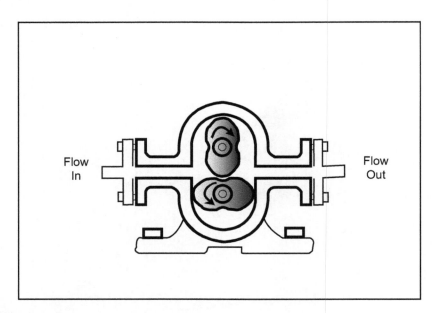

Rotameter

Another type of flow-measuring device is a rotameter. A rotameter is composed of a tapered tube, scale, ball or float, and inlet and outlet. During operation, flow enters a tapered tube at the bottom of the rotameter and lifts the ball off its seat. The ball provides a constant restriction to the flow and corresponds to the flow rate on the scale that runs the length of the tube. The higher the flow rate, the higher the ball rises in the tube. Fluid flows around the ball and out of the top of the rotameter (Figure 4-6).

Figure 4-6
Rotameter

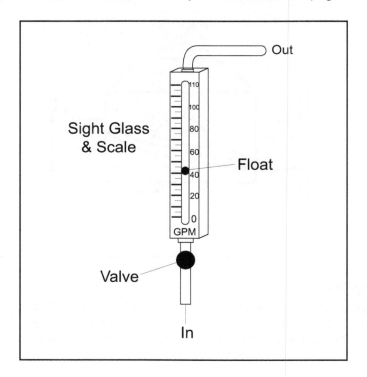

Turbine Flow Meter

Turbine flow meters usually consist of a section of pipe with a rotor mounted in the pipe and a sensor on the outside of the pipe. As fluid enters the turbine flow meter, the turbine blades begin to rotate. The speed of the rotation is proportional to the velocity of the fluid. It is important to understand that flow velocity and flow rate are not the same. Flow velocity is defined as the actual speed of the fluid, while flow rate is defined as the total quantity of liquid that passes a specific point. Turbine flow meters are accurate over a wide range of flows (Figure 4-7).

Figure 4-7
(A and B) Turbine Flow Meter

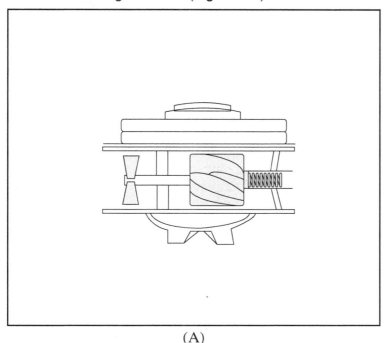

(A)

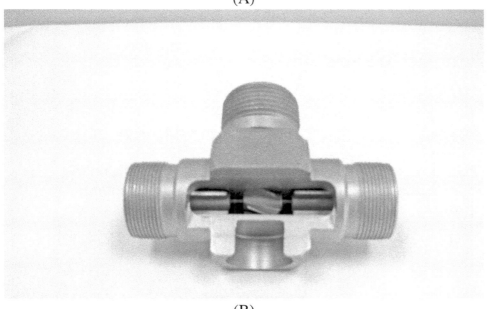

(B)

Pressure Control Loop and Simple Instruments

A pressure control loop design uses the five elements of the control loop. The area to be controlled is a special vapor disengaging cavity that allows vapors or gases that are compressible to be controlled at a set point. The one area that changes consistently is the first or primary elements and sensors. Pressure control loops use devices to detect pressure changes. These primary elements are typically expansion-type devices. Primary pressure elements include C-bourdon tubes, helical, spiral, bellows, pressure capsule, or diaphragm. Figure 4-8 includes a pressure element, pressure transmitter, controller, transducer, and control valve. An electric signal (4–20 mA) connects the PE, PT, and PIC to the transducer, where the signal is changed to an air signal (3–15 psig) that corresponds to changes required by the controller's set point.

Pressure changes can affect temperature, level, and flow. Pressure changes the boiling point of chemicals, reaction rates, and the speed at which fluids flow through piping. These changes can impact product quality, so a variety of instruments have been invented and designed to monitor and control pressure. The instruments discussed in this chapter include:

- Pressure gauges—psia and psig
  - vacuum gauge expressed in inches of mercury (in. Hg)
- manometer
- Pressure elements—sense changes in pressure and convert to mechanical motion
  - bellows bourdon tube
  - C-type bourdon tube

Figure 4-8
Pressure Control Loop

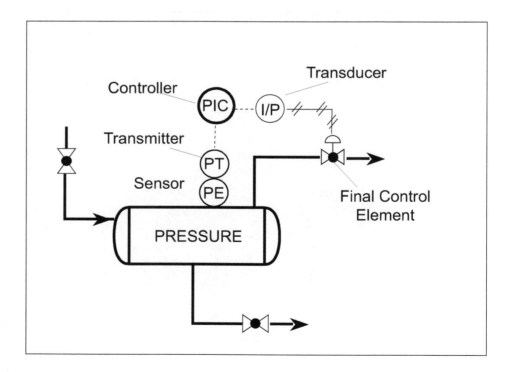

- ○ diaphragm capsule pressure element
- ○ helical bourdon tube
- ○ pressure transmitter
- ○ spiral bourdon tube.

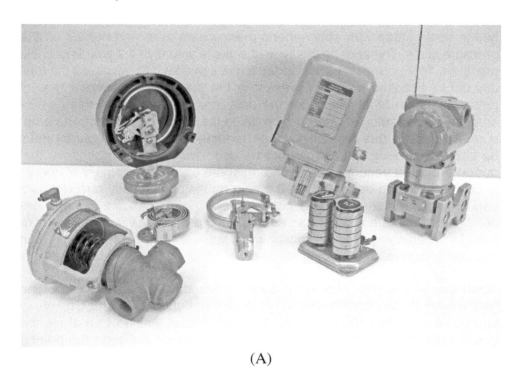

(A)

Figure 4-9
*(A) Pressure Elements
(photo) and
(B) Pressure Indicator*

(B)

Pressure Gauges

There are three commonly used pressure scales in the manufacturing environment: gauge pressure (psig) zero represents 14.7 psia, absolute pressure (psia) based on zero value for a perfect vacuum, and vacuum scale (in. Hg).

The first and most common pressure gauge is the psig gauge scale. The gauge scale starts with atmospheric pressure as zero and moves up the pressure scale. The zero on this type of scale is actually 14.7 psi. To get psia pressure, 14.7 pounds must be added to the scale for it to represent the total amount of absolute pressure present in the system. The second type of pressure gauge functions under the psia absolute scale. On this type of scale, a zero reading is used, which takes into account atmospheric pressure. To convert this type of a gauge reading to psig, 14.7 pounds must be subtracted. Psig gauges cannot be used with system processes that operate under a vacuum. Negative pressures cause the primary elements to contract beyond design limits. If a psig gauge accidentally encounters a vacuum, the reading scale is compromised (low). Vacuum gauges are designed to operate at less than atmospheric pressure. Vacuum is considered to be anything below atmospheric pressure. Compound gauges can indicate both vacuum and psig readings.

Note: Gauge readings should be taken by standing directly in front of the gauge face. If you position yourself to the left or right of the center, an effect known as "parallax" occurs. Parallax is an optical illusion that shifts the gauge face reading left or right of actual. The space between the pointer and the face of the gauge causes the parallax problem. Do not stand directly in front of the gauge when opening the valve that admits pressure to the gauge.

Manometer

A manometer is a device that can be used to measure pressure or vacuum. A manometer operates under hydrostatic pressure principles; a column of water always exerts a specific force. The liquid level of the water indicates the pressure.

There are three basic types of manometers:
- U-tube manometer—measures pressure in units of inches of water. Add the inches displaced on the inlet leg plus the inches above the zero on the outlet leg.
- Well manometer—reads scale directly.
- Inclined manometer—reads scale directly, expands the scale to make it easier to read small changes.

Primary Pressure Elements

Primary pressure elements are the specific part of a pressure instrument designed to sense changes in pressure and convert it to mechanical motion. Pressure elements are connected to mechanical linkages and

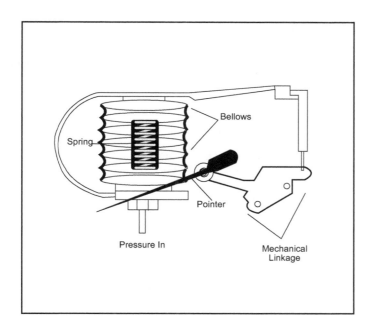

Figure 4-10
*Pressure Elements—
Bellows*

scale indicators. The higher the pressure, the greater is the movement by the mechanical linkage.

Bellows Pressure Element

A bellows pressure element consists of an accordion-type bellows, a spring that resists expansion of the bellows, a pressure inlet, a mechanical linkage, and a pointer. This type of device can be used to measure a variety of pressures. During operation, a bellows pressure element admits flow into the bellows. As the bellows expands, tension on the spring increases and the mechanical linkage moves. This movement operates the pointer on the gauge. If the pressure is reduced, the spring forces the bellows back to its original position (Figure 4-10).

Bourdon Tubes

The most common type of pressure element is a bourdon tube. Bourdons are available in a variety of shapes and designs. The most common are C-type bourdon tube, helical-type bourdon tube, and spiral-type bourdon tube.

C-Type Bourdon Tubes

C-type bourdon tubes take their name from their C-shaped, hollow pressure element. C-type bourdons are composed of a C-shaped hollow tube, a pressure inlet attached to one end of the tube, a mechanical linkage attached to the tip of the bourdon tube, and a pointer. During operation, the tube expands and contracts in response to pressure changes. This process is sometimes referred to as "elastic deformation." This expansion moves the mechanical linkage and pointer. Bourdons measure a wide variety of pressures, including vacuum.

Helical- and spiral-type bourdons operate the same way the C-type bourdon does. The main difference is in the actual shape of the pressure element.

In an automatic control system, a spiral-type bourdon can be connected to a transmitter. As the spiral element responds to pressure changes, the transmitter sends a signal to the controller, which sends a signal to the control valve (Figure 4-11).

Figure 4-11
(A and B) Pressure Elements—Bourdons

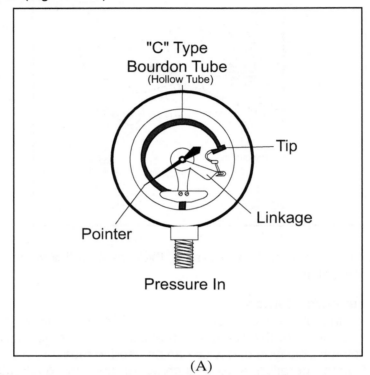

(A)

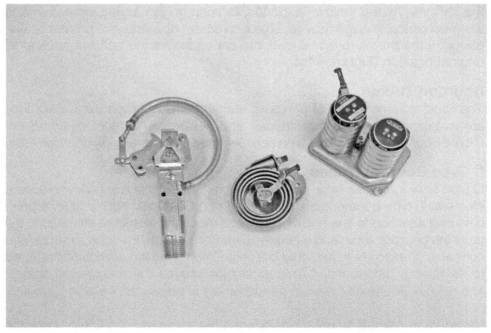

(B)

102

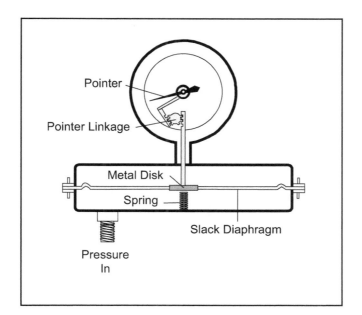

Figure 4-12
Pressure Elements—Diaphragm

Diaphragm Capsule Pressure Element

The third type of pressure element is a diaphragm. Diaphragms come in two basic types: diaphragm capsule pressure element and slack diaphragm pressure element. The metallic diaphragm capsule pressure element is comprised of a metal cup covered by a flexible metal plate, a pressure inlet line to the cup, a mechanical linkage, and a pointer. Diaphragm capsule pressure elements are designed to measure small pressure changes. During operation, the dome of the cup flexes up or down. This movement is transferred to the pointer proportionally. This means that a little movement on the dome can equal a lot of travel on the pointer. The key elements of slack diaphragm pressure elements are:

- a flexible diaphragm attached to a spring
- a pressure inlet
- a mechanical linkage
- a pointer.

Slack diaphragm pressure elements are designed to operate under very low pressures, 0–0.5 psi.

Pressure Transmitter

A pressure transmitter uses a pressure element to sense pressure and sends a signal to a controller or recorder. Pressure transmitters use all of the primary pressure elements just discussed. Linkage movement allows the transmitter to transmit a signal that is representative of the pressure to a controller or recorder. The controller opens or closes control valves depending on the signal it receives from the transmitter (Figure 4-13).

Figure 4-13
(A, B, and C)
Pressure Transmitter

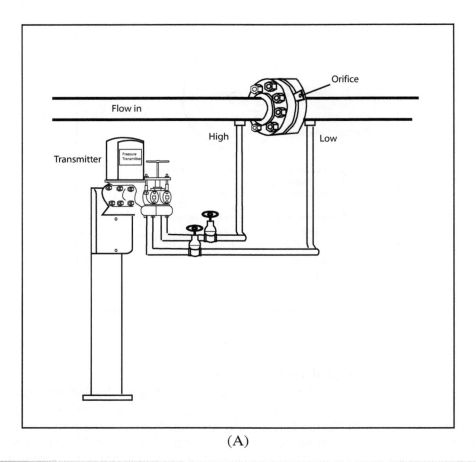

(A)

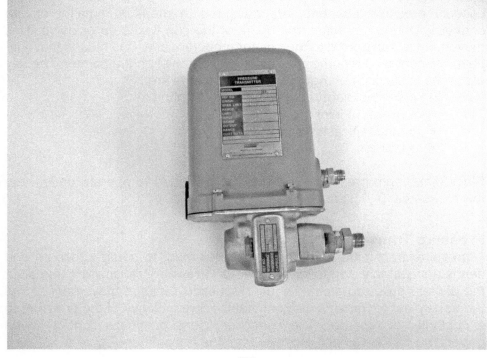

(B)

Figure 4-13
(continued)

(C)

Temperature Control Loop and Simple Instruments

Temperature control can control the amount of heating or cooling received by a substance. Examples include steam to a kettle reboiler, hot oil to a heat exchanger (preheater), and cooling water to a condenser (heat exchanger). It can also control the amount of natural gas flowing to a burner. The burner provides heat to the boiler or furnace. A hundred other applications could be described for controlling the temperature of a substance; however, the most common is by controlling the flow. Figure 4-14 is a simple layout of a temperature control loop. Notice the location of the primary element and the transmitter and how the electric signal is sent to the controller. The controller compares the signal to an incoming electric or pneumatic set point. If a change is required, it is sent to the final control element. The primary sensors used to detect temperature are resistive temperature detectors (RTDs), thermocouples, and thermistors, often called temperature elements. Temperature elements are linked to transmitters. A 4–20 mA signal is sent to a controller that compares it to a set point. Controllers may be located in the field near the equipment or in a control room. The controller sends an electric signal to a transducer that is located on or near the final control element or valve. This will eliminate a process lag or delay. The transducer converts the signal to a 3–15 psig pneumatic signal. The pneumatically actuated

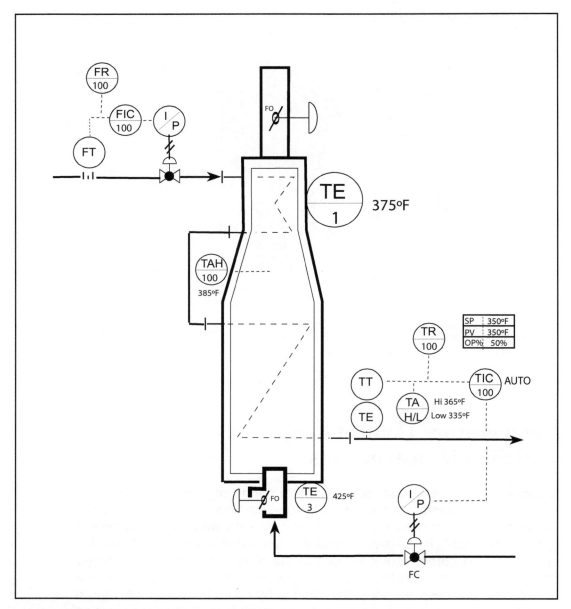

Figure 4-14 *Temperature Control Loop*

control valve opens and closes depending upon the signal. Temperature is controlled by reducing or increasing the opening on the valve.

Temperature Measurement

Process operators are required to closely monitor the temperatures of process streams. When heat energy is applied to an area, molecular activity increases. This increased activity transfers energy from molecule to molecule in an effort to return to its original temperature. As this sharing process occurs, the following changes take place:

- Pressure increases in an enclosed environment.
- Temperature increases.
- Density changes.
- Expansion occurs.

Temperature is defined as the degree of hotness or coldness of an object or environment. Two commonly used scales are Fahrenheit and Celsius. Fahrenheit scales operate by using 32°F as the freezing point of water and 212°F as the boiling point of water. Celsius uses 0°C as the freezing point of water and 100°C as the boiling point of water. Process operators use Fahrenheit and Celsius thermometers to measure temperature. Local temperature indicators usually contain a bimetallic strip that differentially expands with increasing temperature. The deflection is correlated with temperature. Bimetallic thermometers can **range** from 300 to 800°F.

Primary Elements and Sensors
Primary temperature elements and sensors include:

- filled thermal bulb and capillary tubing (excluding mercury)
- resistance bulb
- thermocouple.

Thermoelectric Temperature Measuring Device
Temperature measuring devices come in two types: RTDs and thermocouples. Types J and K are the most common.

Resistive Temperature Detector (RTD)
An RTD is a thermoelectric temperature measuring device. RTDs are composed of a small platinum or nickel wire encased in a rugged metal tube. The electrical resistance in the wire is influenced by changes in temperature. Temperature changes in an RTD are sensed by an electronic circuit and directed to a temperature indicator.

Thermocouple
Thermocouples are temperature-measuring devices that are composed of two different types of metal. A thermocouple is designed to convert heat into electricity. When heat is applied to the connected ends of a thermocouple, a low-level current is generated. The higher the temperature, the greater the voltage generated. Electrical current is detected easily by the associated electronic circuit and converted to a corresponding temperature scale. Thermocouples come in several types, J-type and K-type thermocouples being most common. Type K is preferred for higher temperature measurements (Figure 4-15 and Figure 4-16).

Figure 4-15
(A) Resistive Temperature Detector (RTD), (B) Thermocouple— Temperature Element, (C) Thermocouple Wire and Temperature Element, (D) Thermocouple Wire and RTD, and (E) Temperature Indicator

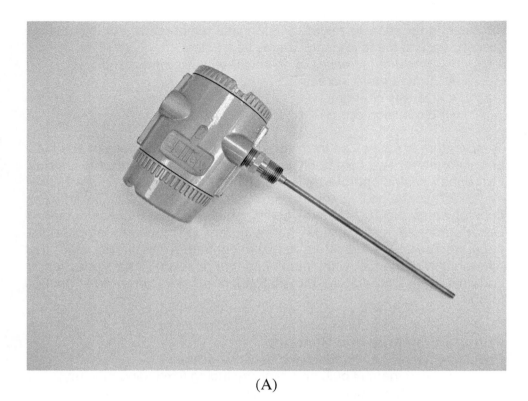

(A)

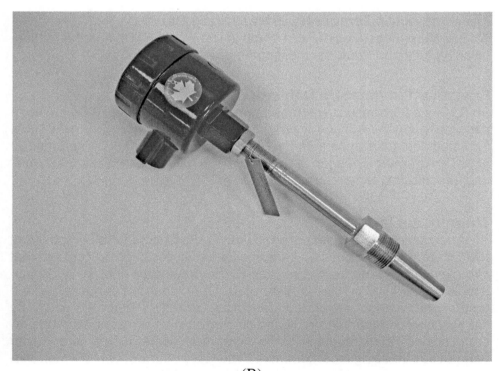

(B)

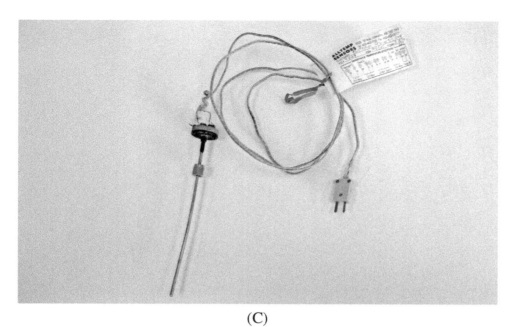

Figure 4-15
(continued)

(C)

THERMOCOUPLE

RTD

200

0 400

C

200

0 400

C

Electronic Circuit

Thermal Well

Different Metals

Metal Wire Platinum or Nickel

(D)

Figure 4-15
(continued)

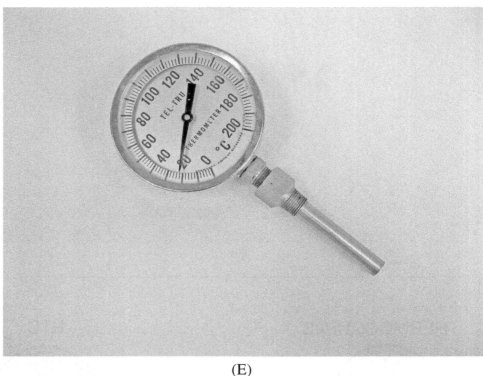

(E)

Level Control Loop and Simple Instruments

The following example of a level control loop can easily be applied to level control on any vessel or tank. Level control uses floats and float gauges, displacers, tapes and tape gauges, DP transmitters, bubblers, load cells, capacitance probes, electromagnetic measuring devices, and nuclear measuring instruments. Figure 4-17 illustrates what a standard control loop looks like.

Principles of Level Measurement
Process technicians use fixed reference points, typically vessel taps, on which to base level measurements. The lower tap represents zero level and the upper tap is 100%. Correct level readings and control help make modern processing possible and profitable.

Level measurements can be listed as:
- Continuous level measurement—monitors level continuously.
- Single-point level detection—readings are taken from single or multiple points at a vessel. Used to turn equipment on or off (valves, pumps, compressors, motors, and alarms), and to detect high and low process levels.

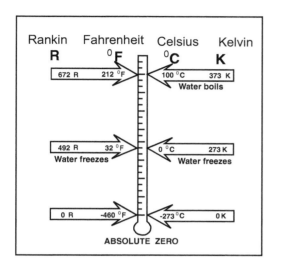

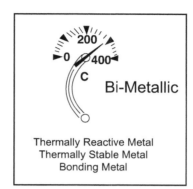

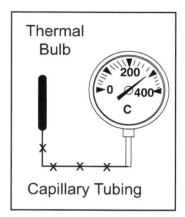

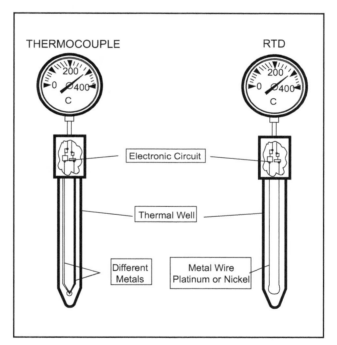

Figure 4-16 *Temperature Measurement*

Level measurement devices can be classified as direct or indirect. Direct instrumentation is in physical contact with the surface of the fluid. Direct level measurement equipment may calculate the product surface level from a specific point of reference. Indirect instrumentation incorporates pressure changes that respond proportionally to level changes. Industrial processes that require continuous level measurements are referred to as single-point level detection and continuous level measurement.

Figure 4-17
Level Control Loop

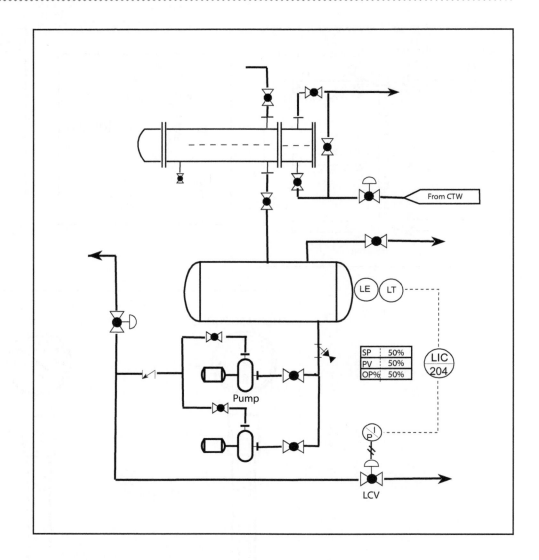

Direct Level Measurement Instruments
- Sight glass—transparent tube mounted on the side of a tank
- Float and tape—float rests on the surface of the fluid. Tape moves up and down depending on the level
- Displacer—buoyancy devices
- Conductivity probes—high- and low-level alarms. Use electricity to complete lower leg circuit. If liquid reaches the higher leg, the circuit is broken. This type of system is designed to maintain the level between the high and low conductivity probes. Typically used on nonflammable material
- Capacitance probes—radiation devices, load cells.

Indirect Level Measurement

- DP cell—converts pressure difference to a level indication. Measures hydrostatic pressure difference between two points on a pressurized vessel.
- Continuous level detector gauge—pressure-sensitive. Measures hydrostatic pressure in open vessels.
- Bubbler system—forces air through a tube that is positioned in the liquid. The liquid's resistance to flow registers on a pressure-sensitive level gauge. Measures hydrostatic pressure in open vessels.

Figures 4-18A and B shows a sight glass that is classified as a direct level measurement instrument.

Figure 4-19 is another example of a direct level measurement instrument. A capacitance probe can be used as part of a high-level alarm system inside a tank or a vessel. When the liquid touches the probe, the system is activated.

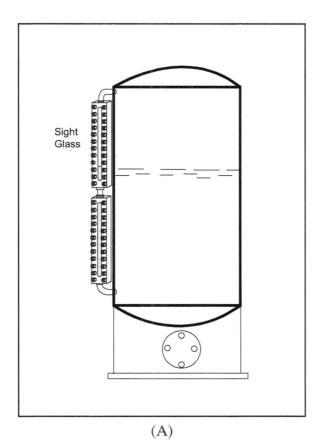

(A)

Figure 4-18
(A and B) Sight Glass

Figure 4-18
(continued)

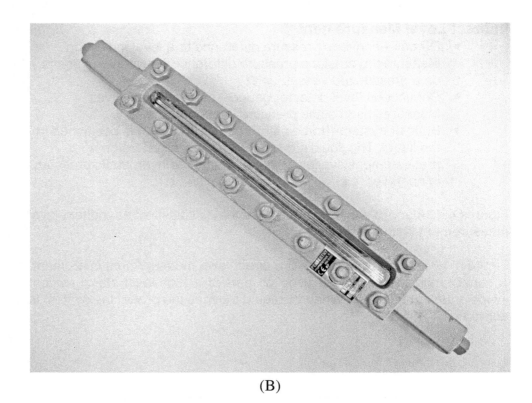

(B)

Figure 4-19
Capacitance Probe

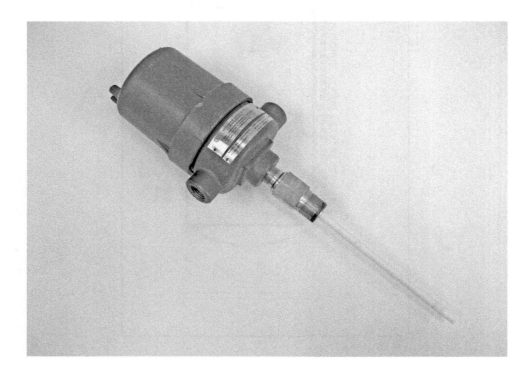

Analytical Control Loop and Simple Instruments

Process analyzers are available in a variety of designs and applications. An analytical control loop performs the same way as any of the other four mentioned control loops. As with the other control loops, the primary element of an analytical control loop is some type of analyzer. Figure 4-20 shows how an analytical control loop can be used to control a process variable on a cooling tower system. Analytical control loops can be found on furnaces, boilers, cooling towers, reactors, distillation systems, and in many more industrial applications.

Cascaded Control Loop Example

A cascaded control loop utilizes a series of unique features in modern process control. In a cascaded control loop, two different control variables

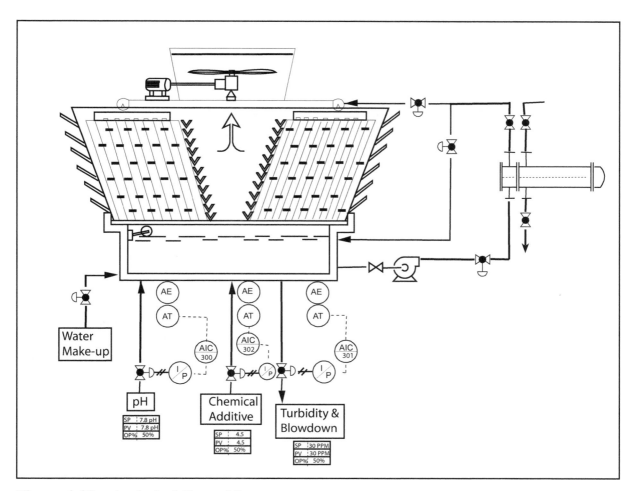

Figure 4-20   *Analytical Control Loop*

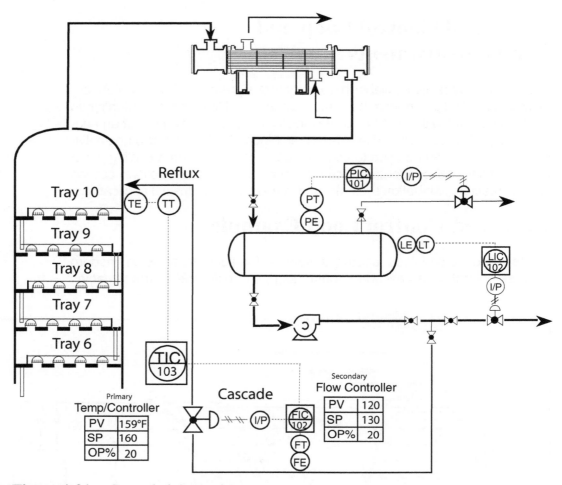

Figure 4-21 *Cascaded Control Loop*

work together to control a critical variable. For example, in some cases temperature control is more important than flow. In this case, a cascaded control loop can be used. A cascaded control loop has one primary or "master" controller, and one secondary or "slave" controller. The master controller is wired to the secondary controller and has the ability to override the initial or original set point. During the operation, the secondary controller operates with all five elements of the control system. The master controller has a (1) primary element, (2) transmitter, and (3) master controller electrically connected to the other control loop. Figure 4-21 shows a typical cascaded control loop.

Primary Elements and Sensors

The primary elements and sensors associated with flow include:
- orifice plate
  - concentric
  - eccentric bore off center orifice plate

- ◦ segmental plate half moon opening orifice plate
- ◦ quadrant-edged orifice plate
- ◦ conical orifice plate
- flow nozzle
- pitot tube
- annubar pitot
- oval gear
- nutating disc
- magmeter
- turbine meter
- mass flow meter
- DP transmitter
- integral orifice flow meter
- venturi tube
- vortex flow meter
- target flow meter
- rotameter.

The primary elements and sensors associated with level include:
- float and float gauge
- tape and tape gauge
- sight glass or level gauge
- DP transmitter
- bubbler
- displacer
- load cell
- capacitance probe
- electromagnetic measuring device
- nuclear level instrument.

The primary elements and sensors associated with temperature are:
- thermocouple
- thermistor
- bimetallic temperature gauge
- RTD
- thermometer
- thermal filled bulb.

The primary elements and sensors associated with pressure are:
- manometer
- C-shaped bourdon tube
- spiral bourdon tube
- helical bourdon tube
- bellows tube
- pressure capsule
- DP cell.

The primary elements and sensors associated with analytical variables are:

- pH meter (acid/base)
- oxidation reduction potential (ORP)—electron-free concentrations
- conductivity—parts per million, turbidity
- chromatography
- oxygen meter
- quantitative analyzer—compositional
- qualitative analyzer—existence of specific chemical
- mass spectrometer
- total carbon analyzer
- monitor and detector
- color
- opacity.

Transmitters

Transmitters and Control Loops

ΔP cell transmitters can be found in two basic designs: pneumatic and electronic. Controllers are typically mounted between 400 (closed loop) and 1000 (open loop) ft. from the transmitter. The signal from an electronic transmitter is proportional to the difference in the high- and low-pressure legs. Standard output signals are 4–20 mA, 10–50 mA, and 1–5 V. The 10–50 mA transmitter is used because it has a higher tolerance to outside interference than the 4–20 mA. Pneumatic transmitters require a 20–30 psig air supply in order to run the standard 3–15 psig output. Table 4-1 shows the relationship between air and electric current. This relationship is important to understand the operation of an automatic control valve.

DP (ΔP) cells function by running a high- and low-pressure tap to each side of an internal twin diaphragm capsule. Pressure changes cause the diaphragms to move. This process increases or decreases the signal to the controller. Figure 4-22 illustrates how a DP transmitter operates.

Smart transmitters are another type of transmitter frequently found in the chemical processing industry. This type of transmitter is very reliable and

Table 4-1 *Pneumatic Electric Comparative Chart—Air-to-Open*

PSI	4–20 mA	Valve position	10–50 mA	1–5 V
3	4	Closed	10	1
6	8	25	20	2
9	12	50	30	3
12	16	75	40	4
15	20	100	50	5

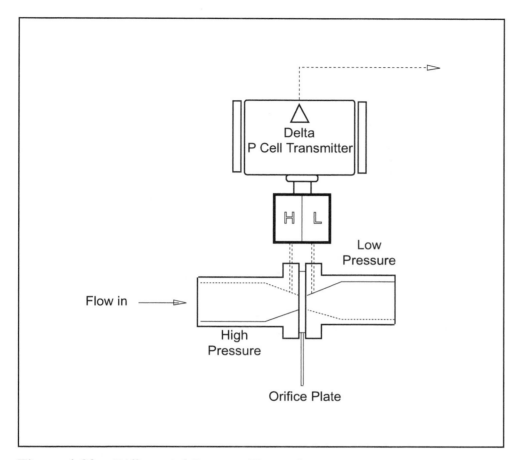

Figure 4-22 *Differential Pressure Transmitter*

does not need constant attention. Smart transmitters have an internal diagnostic system that warns the operator if a problem is about to occur. This type of transmitter can be used with liquid or gas service, pressure, viscosity, temperature, flow, and level. Several features of the smart transmitter include speed, reliability, internal diagnostics, strong digital signal, and remote calibration capabilities.

Controllers

Field-Mounted Controllers
When automatic control was first added to process equipment, the controllers were mounted next to the equipment. This was due to lag time issues associated with pneumatics. There are still small systems in service today that utilize field-mounted controllers. Over time, a series of smaller control rooms were developed and located near major systems that had been automated. These were often referred to as sound islands or "dog houses." Today, modern process has virtually no lag time and can operate facilities

from a single control room over great distances. Recently, a group of instrument consultants in Hong Kong were able to open and close an automated valve located in a small pilot plant at a community college in Baytown, Texas.

Programmable Logic Controllers (Control Room)

A programmable logic controller is a modern control system that combines microprocessor features with software configurable controllers. The basic components of this system include a processor CPU module, a mounting rack, power supply, user-defined plug in input/output modules, and a communication interface module. This type of system requires minimal space, is extremely reliable, is reprogrammable, and has high computational ability. Another attractive feature allows laptop computers to interface and program the system.

Distributive Control Systems (Control Room)

Distributive control systems (DCS) combine some of the most innovative technologies into an interactive network of intelligent microprocessors, application software, and communication networks. The hardware for a DCS includes a host CPU or PLC, intelligent field devices (transmitters, controllers, and control valves), remote CPU, and keyboard. This type of system offers the highest level of operator interaction. Modern process control utilizes remote operation and monitoring of equipment and systems, as illustrated in Figure 4-23.

Figure 4-23
(A) Computer Console Operator and (B) Field-Mounted Controller

(A)

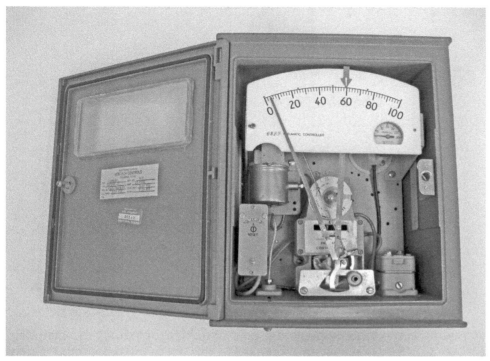

Figure 4-23
(continued)

(B)

Transducers

A transducer is a device designed to change an electric signal into an air signal or vice versa. Air is required to operate, open, close, or throttle pneumatic valves. Pneumatic valves are commonly used in modern process control and as the final control element in a control loop. An electric signal (4–20 mA) can be converted by the transducer to a (3–15 psig) signal.

Final Control Elements and Control Loops

Automatic Valves

Final control elements are typically classified as automated valves; however, motors or other electrical devices can be used. The final control element is the last link in the modern control loop and is the device that actually makes the change in the process. Automatic valves will open or close to regulate the process. Control loops usually have (1) a sensing device, (2) a transmitter, (3) a controller, (4) a transducer, and (5) an automatic valve. Automatic valves can be controlled from remote locations, making them invaluable in modern processing. To automate a valve, a device known as an actuator is installed. The actuator controls the position of the flow control element by moving and controlling the position of the valve stem. Figure 4-24 illustrates some of the values process technicians look at when operating control systems.

Figure 4-24
*Instrument
Troubleshooting
Blocks*

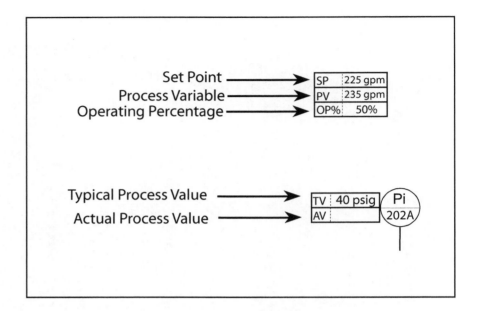

Actuators come in three basic designs: pneumatic, electric, and hydraulic.
1. Pneumatically (air) operated—this is the most common type of actuator. Pneumatic actuators convert air pressure to mechanical energy. They can be found in three designs: (1) diaphragm, (2) piston, and (3) vane.

Three Types of Pneumatic Actuators   Diaphragm—the diaphragm actuator is a dome-shaped device that has a flexible diaphragm running through the center. It is typically mounted on the top of the valve. The center of the diaphragm in the dome is attached to the stem. The valve position (on or off) is held in place by a powerful spring. When air enters the dome on one side of the flexible diaphragm, it opens, closes, or throttles the valve depending on the design.
*   Piston—the piston actuator uses an airtight cylinder and piston to move or position the stem. Commonly found in use with automated gate valves or slide valves. Used where a lot of stem travel is needed.
*   Vane—vane actuators direct air against paddles or vanes.
2. Electrically operated—converts electricity to mechanical energy. Examples: (1) solenoid valve and (2) motor-driven actuator.

Two Types of Electric Actuators   Solenoid—solenoid valves are designed for on–off service. The internal structure of a solenoid resembles a globe valve. The disc rests in the seat, stopping the flow. The stem is attached to a metal core or armature that is held in place by a spring. A wire coil surrounds the upper spring and stem. When the wire coil is energized, a magnetic field is set up causing the armature to lift, compressing the spring. The armature is held in place until the current stops.

- Motor—a motor-driven actuator is attached to the stem of a valve by a set of gears. Gear movement controls the position of the stem.

3. Hydraulically operated—converts liquid pressure to mechanical energy. The hydraulic actuator uses a liquid-tight cylinder and piston to move or position the stem. Commonly found in use with automated gate valves or slide valves. Used where a lot of stem travel is needed.

Common terminology for actuators:

- Air to open—spring to close. Fails in the closed position if air system goes down. Air line is typically located on the bottom of the dome.
- Air to close—spring to open. Fails in the open position if air system goes down. Air line is typically located on the top of the dome.
- Double-acting—no spring—air lines located on both sides of the dome.
- Fails in last position—air pressure to diaphragm is locked on instrument air failure.

The most common type of automated valve is a globe valve because of its versatile, on–off, or throttling feature. Control loops use on–off or throttling type valves to regulate the flow of fluid in and out of a system. Automatic valves can be used to control (1) pressure, (2) temperature, (3) flow, and (4) level and composition.

Automatic valves fall into the following categories:

- Control valve—air operated, electrically operated, and hydraulically operated.
- Spring- or weight-operated—spring-operated valves hold the flow control element in place until pressure from under the disc grows strong enough to lift the element from the seat. Example: check valve.

When a process technician studies troubleshooting for the first time, the concept of fail open and fail closed must be discussed. Figure 4-25 illustrates how the operating percentage on the control valve changes depending on whether it is a fail open or fail closed valve. Figure 4-25B illustrates a control valve and Figure 4-25C shows the basic internal components.

Control Modes

Controllers and Control Modes

The primary purpose of a controller is to receive a signal from a transmitter, compare this signal to a set point, and adjust the (final control element) process to stay within the range of the set point. Controllers come in three basic designs: pneumatic, electronic, and computer. Electronic controllers were first introduced in the early 1960s. Before this time, only pneumatic controllers were available. Pneumatic controllers require a clean air supply pressure of 20–30 psig. Several of the more attractive features in electronic controllers are the reduction of lag time in process changes, low installation expense, and ease of installation.

With the widespread use of the PC, a number of applications were found for controller use. DCS software in computers replaced pneumatic and

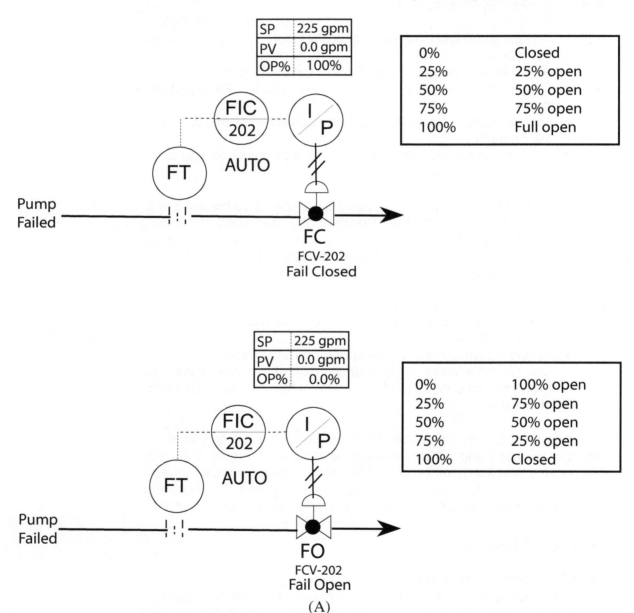

Figure 4-25   *(A) Fail Open and Fail Closed, (B) Control Valve, and (C) Control Valve Element*

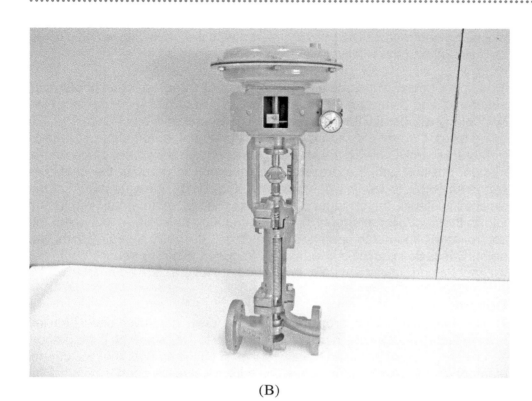

Figure 4-25
(continued)

(B)

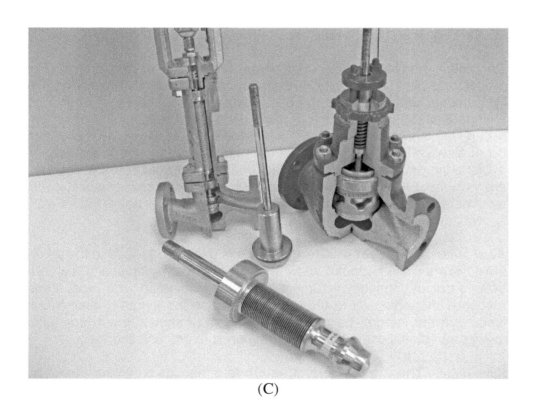

(C)

electronic controllers. The primary reason was the ease with which it could be installed and the relatively few wires required to do it.

Most modern plants include a combination of all three systems: pneumatic, electronic, and computer. When looking at a diagram of a control loop, the symbology will identify the type of controller (pneumatic, electronic, or DCS) being used. Controllers can be operated in manual, automatic, or cascade (remote) control. During plant start-up, the controller is typically placed in the manual position until the process has lined out. In manual, the controller sends an output to the final control element, as set by the operator. After the process is stable, the operator places the controller in automatic mode and allows the controller to supervise the control loop function. At this point, the controller will attempt to open and close the control valve to maintain the set point. **Cascade control** is a term used to describe how a second controller will reset a controller's set point in order to achieve a desired outcome.

Controller Modes

Proportional Band **The proportional band** on a controller describes the scaling factor used to take a controller from 0% to 100% output. If the proportional band is set at 50% and the amount of lift off the seat for the final control element (globe valve) is four inches, the control valve will open 2 inches if the maximum valve opening is 4 inches. Range is defined as the portion of the process controlled by the controller. For example, the temperature range for a controller may be limited to 80–140°F. **Span** is the difference (Δ) between the upper and lower range limits. This value is always recorded as a single number. For example, the difference between 80 and 140 is 60.

Controller modes include proportional (P), proportional plus integral (PI), proportional plus derivative (PR), and proportional integral derivative (PID). Proportional control is primarily used to provide gain where little or no load change typically occurs in the process. PI is used to eliminate offset between set point and process variables. PI works best where large changes occur slowly. PR is designed to correct fast changing errors and reduce overshooting the set point. PD works best when frequent small changes are required. PID is applied where massive rapid load changes occur. PID reduces swinging between the process variable and set point. Figure 4-26 shows how a controller using proportional control works.

Reset Mode **The reset or integral mode** is designed to reduce the difference between the set point and a process variable by adjusting the controller's output continuously until the offset is eliminated. The reset mode responds proportionally to the size of the error, the length of time that it lasts, and the integral gain setting. Figure 4-27 shows how a controller using proportional control and reset responds to process changes.

Rate Mode **The rate or derivative mode** enhances controller output by increasing the output in relation to the changing process variable. As the

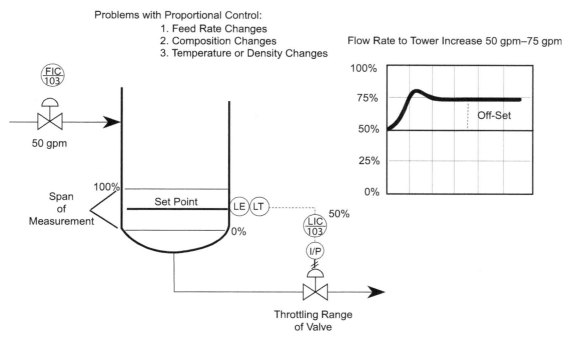

Figure 4-26 *Proportional Control Mode*

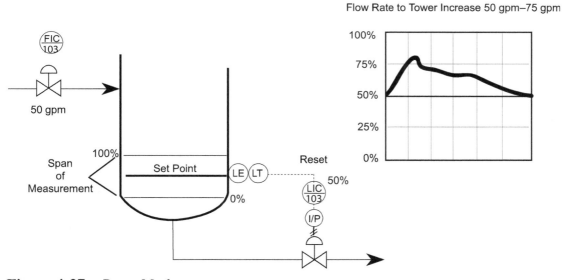

Figure 4-27 *Reset Mode*

process variable approaches the set point, the rate or derivative mode relaxes, providing a braking action that prevents overshooting the set point. The rate responds aggressively to rapid changes and passively to smaller changes in the process variable. Figure 4-28 shows how a controller using proportional control and rate responds to process changes.

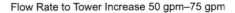

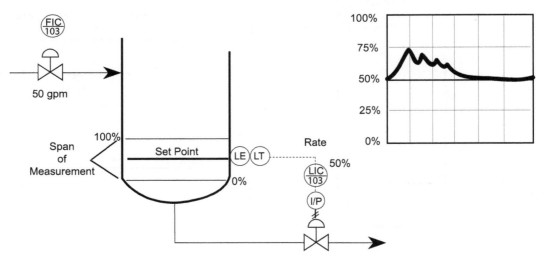

Figure 4-28 *Rate Mode*

Interlocks and Permissives

An interlock is a device designed to prevent damage to equipment and personnel. Stopping or preventing the start of certain equipment unless a preset condition has been met accomplishes this objective. There are two types of interlocks: softwire and hardwire. Softwire interlocks are contained within the logic of a programmable computer. Hardwire interlocks are a physical arrangement. The hardwire interlock usually involves electrical relays that operate independent of the control computer. In many cases, they run side by side with the computer interlocks. However, hardwire interlocks cannot be bypassed. They must be satisfied before the process they are part of can take place.

A permissive is a special type of interlock that controls a set of conditions that must be satisfied before a piece of equipment can be started. Permissives deal with start-up items, whereas hardwire interlocks deal with shut down items. A permissive is an interlock controlled by the DCS. This type of interlock will not necessarily shut down the equipment if one or more of its conditions are not met. It will, however, keep the equipment from starting up.

Summary

Automation enhances the ability of a process operator to control large and complex process networks. Automatic control is the foundation for efficient continuous flow processes. Modern advances in electronics and instrumentation have made it possible for industrial manufacturers to automate their processes. To a process operator, this means a computer has the ability to control the opening, closing, and diverting of valves, start and stop electronic equipment, measure process variables, and respond automatically.

The basic instruments used in an industrial manufacturing environment include gauges (pressure, level, temperature, and flow), DP cell, transmitter (pressure, level, temperature, and flow), recorder (pressure, level, temperature, and flow), controller (pneumatic and electronic, strip chart, panel mounted, and locally mounted), transducer, control valves, primary elements and sensors (displacer and buoyancy float, thermocouples, orifice plates, and so on), alarms, actuators, scales, and computers.

The most common type of pressure element is a bourdon tube. Bourdons come in a variety of shapes and designs. The most common are C-type, helical, and spiral. There are three commonly used pressure scales in the manufacturing environment: gauge pressure, absolute scale, and vacuum scale (inches of mercury). The absolute scale uses a "true zero" value.

Process flow measurements frequently are taken by one of the following devices: a flow transmitter (orifice plate, venturi tube, or flow nozzle), nutating disc meter, oval gear meter, rotameter, turbine flow meter, weir and level measuring device, or flume and level measuring device.

Level measurements are identified by a fixed reference point (typically zero) above or below a product. The chemical processing industry has developed very sophisticated equipment and technology to calculate accurate level readings. Correct level readings make modern manufacturing possible and profitable. Level measurements can be listed as continuous level measurement (monitors level continuously) or single-point level detection (readings are taken from single or multiple points at a vessel). Single-point detection is used to turn equipment on or off (valves, pumps, compressors, motors, alarms, and so on), and to detect high and low process levels.

Level measurement devices can be classified as direct or indirect. Direct instrumentation is in physical contact with the surface of the fluid. Direct level measurement equipment may calculate the product surface level from a specific point of reference. Indirect instrumentation incorporates pressure changes that respond proportionally to level changes.

A control loop is defined as a group of instruments that work together to control a process. These instruments typically include a transmitter coupled with a sensing device or primary element, a controller, a transducer, and a control valve. Process variables typically fall into five different groups: pressure, temperature, flow, level, and analytical variables. Control loops are specifically designed to work with a selected variable. The primary purpose of a controller is to receive a signal from a transmitter, compare this signal to a set point, and adjust the (final control element) process to stay within the range of the set point. Controllers come in three basic designs: pneumatic, electronic, and electric. Final control elements are typically classified as automated valves; however, motors or other electrical devices can be used. The final control element is the last link in the modern control loop and is the device that actually makes the change in the process.

An interlock is a device designed to prevent damage to equipment and personnel. Stopping or preventing the start of certain equipment functions unless a preset condition has been met accomplishes this. A permissive is a special type of interlock that controls a set of conditions that must be satisfied before a piece of equipment can be started. Permissives deal with start-up items, whereas hardwire interlocks deal with shut-down items. This type of interlock will not necessarily shut down the equipment if one or more of its conditions are not met; however, it will keep the equipment from starting up.

Review Questions

1. List the instruments found in the processing industry.
2. Describe the basic operation of a ΔP cell.
3. What is a rotameter used for?
4. List the basic elements of a control loop.
5. How does a process technician operate a control loop?
6. Draw a pressure control loop.
7. Draw a level control loop.
8. Draw a flow control loop.
9. Draw a temperature control loop.
10. Draw an analytical control loop.
11. How do the basic elements of a control loop work together?
12. Describe controllers and control modes.
13. List the primary elements and sensors used with temperature and pressure.
14. List the primary elements and sensors used with level and flow.
15. How are analyzers used to control product composition?
16. What is the difference between J- and K-type thermocouples?
17. Describe the following terms: pressure indicator, flow indicator, level indicator, and temperature indicators.
18. Describe computer-based control and control loops.
19. Define the term DCS.
20. Define the term PLC.

chapter 5

Statistics, Quality Tools, and Troubleshooting Techniques

LEARNING OBJECTIVES

After studying this chapter, the student will be able to:

- Describe key terms and definitions used in statistics.
- Use and describe measures of central tendency.
- Identify dependent and independent variables.
- Describe the steps used in the statistical approach.
- Describe probability.
- Describe the quality tools and techniques used by technicians.
- Explain how to use a Pareto diagram.
- Use a cause-and-effect diagram.
- Describe the purpose of a checklist.
- Design a flow diagram.
- Design a scatter diagram.
- Design a histogram.
- Fill in data on a run chart.
- Describe TPM and management tools.

Key Terms

Dependent variable—a variable that is caused or influenced by another variable.

Independent variable—a variable that causes, or influences, another variable.

Mean—the sum of the measures in a distribution divided by the number of the measures.

Measures of central tendency—descriptive measures that indicate the center of a set of values.

Measures of variation—descriptive measures that indicate the dispersion of a set of values.

Median—the middle measure in an ordered distribution.

Mode—the most frequent measure in a distribution.

Population—a group of phenomena that have something in common.

Probability—a quantitative measure of the chances for a particular outcome or outcomes.

Range—the difference between the largest and the smallest measures of a set.

Sample—a group of members of a population selected to represent that population.

Standard deviation—a measure of data variation; the square root of the variance.

Statistic—a characteristic of a sample.

Statistics—a branch of mathematics designed to draw conclusions from data that have been collected, organized, and subjected to analysis.

Variable—an observable characteristic of a phenomenon that can be measured or classified.

Variance—a measure of data variation; the mean of the squared deviation scores about the means of a distribution.

Flowchart—a picture of the activities that take place in a process.

Cause-and-effect diagram (fishbone)—a method for summarizing available knowledge about the causes of process variation.

Pareto chart—a simple bar graph with classifications along the horizontal and vertical axis. The vertical axis is usually the number of occurrences, cost, or time. The horizontal axis orders the bars from the most frequent to the least frequent.

Run chart—a graphical record of a process variable measured over time.

Control chart—a statistical tool used to determine and control process variations.

Planned experimentation—a tool used to (1) test and implement changes to a process (aimed at reducing variation) and (2) understand the causes of variation (process problems).

Histogram or frequency plot—a graphical tool used to understand variability. The chart is constructed with a block of data separated into five to twelve bars or sections from low number to high number. The vertical axis is the frequency and the horizontal axis is the "scale of characteristics." The finished chart will resemble a bell if the data are in control.

Forms for collecting data—can vary from notes jotted down on a napkin to complex forms.

Scatter plot—used to indicate relationship between two variables or pairs of data.

Improvement cycle—a four-phase system for quality improvement: (1) plan, (2) observe and analyze, (3) learn, and (4) act.

Statistical process control (SPC)—a quality system based on the principles of statistical mathematics. SPC uses charts and graphs to create a window into the process that allows a technician to consider normal variation and make changes when these limits are exceeded.

Introduction to Statistics

Statistics is a form of mathematics that refers to numbers, numerical facts, figures, or information. Statistics is a way of working with numbers to answer questions about both human and nonhuman phenomena. Statistics is also the art and science of making accurate guesses using mathematical probabilities. The purpose of statistics is to describe sets of numbers and make accurate inferences about the data based upon incomplete information. Reports of flow rate, level, pressure, temperature, and analytical variables are often called statistics. To be precise, these numbers are descriptive statistics because they are numerical data that describe phenomena. Descriptive statistics are as simple as the number of gallons a pump moves per hour or as complex as the annual report of productivity in a global chemical corporation. In this chapter, we focus on descriptive statistics: numerical and pictorial.

Examples of numerical data include (1) the **measures of central tendency** and (2) the **measures of variation.** *Measures of central tendency* include three terms: **mean, median,** and **mode.** These values tend to cluster around the middle of a set of numbers. *Measures of variation* include **standard deviation** (SD), **range,** deviation, and **variance.** *Pictorial statistics* display numerical data in graphical or pictorial information. It has been said that a picture is worth a thousand words. When data are displayed in the form of a graphic, complex and confusing information appears simple and easy to understand.

Steps in the Process

Making accurate guesses requires the application of a statistical system. The process technician must (1) gather numerical data, (2) organize them into charts, graphs, or pictorial type information, and then (3) analyze them by using tests of significance and other statistical methods. Educated groundwork using mathematical probabilities provides very accurate guesses; however, many technicians follow steps one and two and leave out step three. In this chapter, we show you how to precisely follow these procedures and finally how to use your analysis to draw an inference, or in other words an educated statistical guess, designed to solve a particular problem. This is a major feature of a modern technician's job description. While these steps may appear simple, each is grounded in the advanced statistical methods and the laws of **probability.**

Comparing Results

Let us suppose you are attempting to compare two distillation systems. The first bench-top system is a plate column and the second bench-top unit is a packed column. Each system has unique differences or operational variations. A binary mixture of propylene glycol (20%) and water is used as feedstock. The initial question is how to determine which distillation system is more efficient for commercial operation. One way might be to run both units, collect data on the results, and then statistically analyze the data to determine which areas are most and least effective. Each **variable** should be carefully controlled and repeated exactly the same way on each unit. This includes feedstock concentration, quantity, temperatures, condenser rates, and time limits. Table 5-1 shows how data were collected and then pictorially displayed, comparing a plate column with a packed column.

Graphic Displays
Graphic displays include pie charts, bar charts, histograms, **scatter plots, statistical process control** (SPC) charts, flowcharts, **run charts, Pareto charts, planned experimentation,** fishbone charts, line drawings, photographs, and pictures.

Table 5-1
Plate vs. Packed Column

	Sample (mL) (plate)	Color (plate)	Sample (mL) (packed)	Color (packed)
1	5	3	2	2
2	10	2.5	6	2
3	15	2	11	1.5
4	22	1.5	15	1
5	23	1.0	14	1
6	22	1.0	20	0.5
7	22.5	0.5	14	0.25
8	23	0.25	19.5	0
Total	142.5	11.75	101.5	8.25
Mean	17.8	1.47	12.69	1.03

Measures of Central Tendency

Measures of central tendency are described as numbers that cluster around the "middle" of a group of values. These values are called as the mean (arithmetic average, x-bar), the median (midpoint, MD), and the mode (value that occurs most often). In this section, we will collect data, graph the numerical information, analyze the graph, and attempt to understand the data. Another way to get a sense of data is to use numerical measures, certain numbers that give special insight into your values. Three types of numerical measures are useful in the analysis of variance or simple statistics: (1) measures of variation, (2) measures of central tendency, and (3) analysis of variance or ANOVA. These numerical measures can provide information about the different groupings of data. Table 5-2 graphically illustrates yearly earnings for 2004, 2005, 2006, 2007, and 2008.

Statistical Mean
A common equation used to find the mean is:

$$X\text{-bar} = \frac{\Sigma X}{n} = \frac{X_1 + X_2 + X_3 + \cdots + X_n}{n}$$

The statistical term "ΣX" means "sum of the X's." You could express your yearly earnings of the above set in a number of ways. The arithmetic mean is the sum of the measures in the set divided by the number of measures in the set. By totaling all the measures and dividing the total by the number of measures, you get $333,000 divided by 5, which equals $66,600.

Statistical Median
The median is the another measure of central tendency, which is defined as the middle value when the numbers are arranged in decreasing or increasing order. When you order the yearly earnings above, you get $61,000, $61,000, $63,000, $72,000, $76,000. The middle value is $63,000 and, therefore, $63,000 is the median.

When there is an even number of items in a set, the median is determined by averaging the two middle values. For example, if we have four values— 10, 20, 22, 26—the median would be the average of the two middle values,

2004	$61,000
2005	$61,000
2006	$63,000
2007	$72,000
2008	$76,000

Table 5-2
Yearly Earnings

137

20 and 22; thus, 21 is the median. The median is often a better indicator of central tendency than the mean, particularly when there are extreme values or outliers. For example, the following four annual salaries of a corporation are given:

CEO of Chemical Plant	$375,000
Chemical Plant Manager	$156,000
Administrator	$90,000
Clerk	$39,000

The mean of these four salaries is $660,000 ÷ 4 = $165,000. The median is the average of the middle two salaries ($246,000 ÷ 2 = $123,000). In this instance, the median appears to be a better indicator of central tendency because the CEO's salary is an extreme outlier causing the mean to lie far from the other three salaries.

Statistical Mode
Another measure of central tendency is mode. Mode is described as the value that occurs most frequently in a group of numbers. In the set of yearly earnings, the mode would be $61,000 because it appears twice and the other values appear only once.

Notation and Formulae
The mean of a **sample** is described using X-bar. Inside a **population**, the mean is typically denoted as μ. The total or sum of the measures is typically denoted as Σ. The basic formula for a sample mean is

$$X\text{-bar} = \frac{\Sigma X}{n} = \frac{X_1 + X_2 + X_3 + \cdots + X_n}{n}$$

Measures of Variation: Range, Deviation and Variance, and Standard Deviation

Range
One of the simplest measures of variation is range. Range is defined as the difference between the smallest and the largest numbers. For example, in Table 5-2, the range or spread of the data is $61,000 to $76,000; $76,000 − $61,000 = $15,000.

Deviation and Variance
Deviation can be described as the distance of the measurements away from the mean. In Table 5-3, the earnings of Technician "A" have consistently less deviation than those of Technician "B." The sum of the squared deviations of n measurements from their mean divided by $(n - 1)$ is defined as *variance*. So from the technician earnings of Table 5-3, the

Earnings of Technician A ($)	Earnings of Technician B ($)
2000	2000
2100	200
1900	4000
2010	0
1990	3900
1950	100
2050	2000
2000	3800

Table 5-3
Earnings of Employees

mean for Technician "A" is $16,000 ÷ 8 = $2000, and the deviations from the mean are

0, +100, −100, +10, −10, −50, +50, 0

The squared deviations from the mean are therefore

0, 1000, 1000, 100, 100, 2500, 2500, 0

The sum of these squared deviations from the mean equals 7200. Dividing the sum by $(n - 1)$, or $(8 - 1)$, yields 7200 ÷ 7 = 1028.57. The variance is 1028.57.

For Technician B also, the mean is $2000, and the deviations from the mean are

0, −1800, +2000, −2000, +1900, −1900, 0, +1800

The squared deviations are therefore

0, 3,240,000, 4,000,000, 4,000,000, 3,610,000, 3,610,000,
0, 3,240,000

The sum of these squared deviations equals 21,700,000. Dividing the sum by $(n - 1)$ yields 21,700,000 ÷ 7 = 3,100,000.

Although both the technicians earned the same total amount, there is significant difference in variance between the daily earnings of the two employees.

Standard Deviation

SD is described as the positive square root of the variance; thus, the SD of the earnings of Technician A is the positive square root of 1028.57, which equals 32. The SD of the daily earnings of Technician B is the positive square root of 3,100,000, or about 1760.68. The interval from one SD below the mean to one SD above the mean contains approximately 68% of the measurements.

Figure 5-1
The Normal Curve

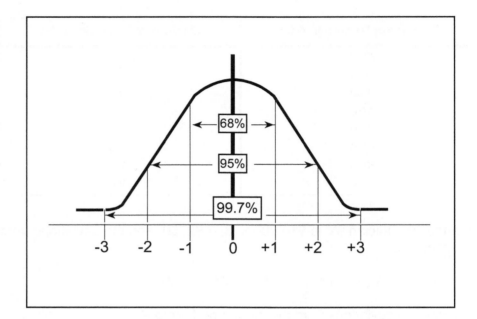

Two SD contains 95% of the measurements and three SD above and below the mean contains 99.7% of the measurements. Figure 5-1 illustrates this principle. The SD is approximately equal to the range divided by 4.

Quality Tools and Techniques

Most quality experts separate quality tools into two categories: quantitative and nonquantitative. Quantitative tools are described as acceptance sampling, reliability, experimental design, quality function deployment (QFD), failure mode and effect analysis (FMEA), Taguchi's quality engineering, and SPC: **control charts,** Pareto diagrams, process flow diagrams, scatter diagrams, histograms, cause-and-effect diagrams, and checksheets. Nonquantitative tools include ISO 9000, ISO 14000, total productive maintenance (TPM), management tools, and benchmarking.

The most popular area of quality tools falls under the SPC umbrella. SPC is composed of seven tools that every process technician should be aware of. The use and application of these tools has been proven to improve product quality.

Pareto Diagrams and Cause-and-Effect Diagrams

A Pareto diagram is a simple bar graph with classifications along the horizontal and vertical axis that ranks data classifications in descending order from left to right. The vertical axis is usually the frequency, percentage, number of occurrences, cost, or time. The horizontal axis orders the bars from the most frequent to the least frequent. These horizontal data classifications include types of field failures such as problems, causes, and types of nonconformities. The term "Pareto" takes its name from Alfredo Pareto (1848–1923) who pioneered income distribution studies. Pareto diagrams are similar to histograms; however, the horizontal scale is categorical instead of numerical. Some Pareto diagrams include a cumulative line above the bars that shows the sum of the data as it moves from left to right. Figure 5-2 illustrates how the cumulative line and the bar graphs work. Pareto diagrams provide a visual representation of the critical few characteristics that need attention. For example

- few customers—account for majority of sales and problems
- few products—account for majority of problems, scrap, and rework
- few suppliers—account for majority of rejections and cost
- few nonconformities—account for majority of complaints.

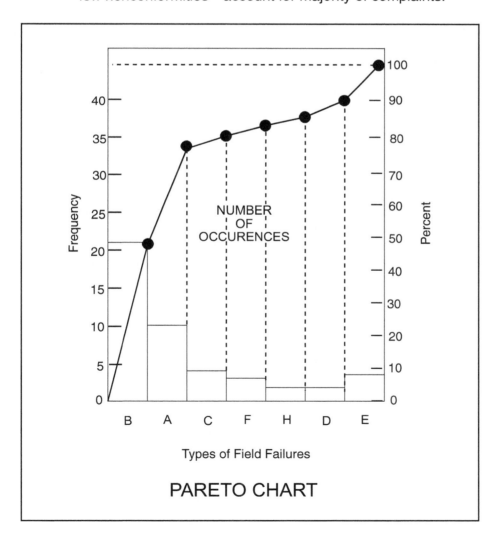

Figure 5-2
Pareto Diagram

Constructing a Pareto diagram requires six simple steps: (1) identify the method you want to use to classify the data, (2) determine if frequency or cost is to be used to rank the data, (3) collect data, (4) summarize and rank the data, (5) calculate cumulative percentage, and (6) construct the Pareto diagram and identify the critical few. The Pareto is a quality tool designed to identify a problem and measure progress.

Cause-and-Effect Diagrams—Fishbone

Another important quality tool is a cause-and-effect diagram often referred to as an Ishikawa diagram. **Cause-and-effect diagrams (fishbones)** organize the causes of variation into general categories: (1) methods, (2) materials, (3) equipment, and (4) people, and summarize available knowledge about the causes of process variation. Fishbone diagrams were developed by Dr. Kaoru Ishikawa in 1943. A cause-and-effect diagram is a graphical representation that depicts a meaningful relationship between a cause and an effect.

The first step in constructing a cause-and-effect diagram is to identify the quality problem and place it on the right side of the diagram. The second step involves the identification of the primary causes under work methods, materials, measurement, people, environment, maintenance, and management. Step three involves a team brainstorming activity to identify all minor problems. Step four is the analysis phase that determines the most likely causes. Solutions are developed, which will be used to correct the problem and improve the process. Step five is the implementation and evaluation phase where solutions are applied, revisions are made, and the process continues.

Many quality gurus believe that the Ishikawa diagram has a variety of important applications. When conducted properly, the brainstorming feature of the process is very powerful. Cause-and-effect diagrams can be used to analyze and standardize operating processes, educate and train personnel, and eliminate the conditions in a process that generate nonconforming products. Figure 5-3 illustrates what a cause-and-effect diagram looks like after the team completes part of the development process.

Checksheets and Process Flowcharts

The primary purpose of checklists is to ensure the accurate and systematic collection of operational data. Collecting data is one of the first steps found in a total quality management system. The **forms for collecting data** can range from notes jotted down on a napkin to complex checklists. Forms are very helpful in collecting and organizing raw data. In every operating system, a variety of process variables exist. Some of these variables include pressure, temperature, flow, level, and analytical variables. Most process technicians carry around small notebooks to record information collected during routine rounds. These data are transferred into logbooks, checklists, charts, graphics, and electronic environments. During each operating shift,

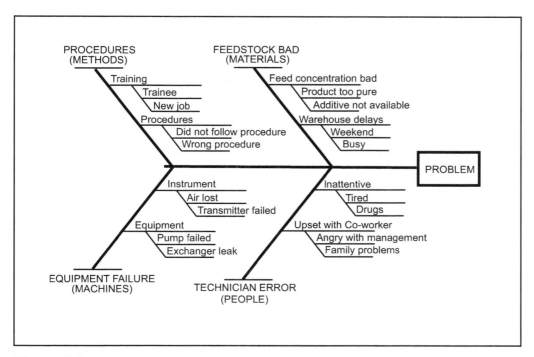

Figure 5-3 *Cause-and-Effect Diagram*

process technicians will make one or two rounds through the unit and record listed variables on a checklist. Information collected on a checklist can be used to isolate and troubleshoot complex problems. These lists can also be used for preventive maintenance and scheduled turnarounds.

Checklists should be designed to collect operating variables in a systematic order. Well-designed checklists allow a technician to collect all the variables in one area before moving into another area. Data should be organized so that they can be quickly and easily used and analyzed. New technicians will discover that filling out a checklist requires significant effort. The time required to fill out a checklist will be reduced as the trainee learns where the equipment is located. Senior technicians believe checklists are good tools for training new employees (Figure 5-4).

Process Flowcharts
A **flowchart** is a picture of the key activities that take place in a process. Flowcharts are graphical representations of how a process works. These charts are designed to provide a true picture of the current steps in a process. This information allows a technician to view the process from the outside. There are three different types of flowcharts: (1) high-level flowcharts, (2) detailed flowcharts, and (3) deployment flowcharts. High-level flowcharts map the major steps in a process. Detailed flowcharts document a step-by-step sequencing of all events in a process. Deployment flowcharts organize data in columns. Each column represents a department, an

Figure 5-4

Check-Sheet

Equipment Checklist

Date:_____ Shift No: _____ Technician:_____

EQUIPMENT	ITEM CHECKED (✓) OR RECORDED (R)	
AS-520	Proper oil level (gearbox and blower)	✓
	No unusual noise	✓
P-520 OFFSPEC	Proper oil level (gearbox)	✓
	Temperature upstream of E-520 (local)	R
	Temperature downstream of E-520 (local)	R
	No unusual noise on PUMP	✓
TK-520	Level	✓
	Area clean	✓
E-520	Steam Pressure	R
	By-Pass closed	✓
	Shell inlet/outlet lined up	✓
	TIC-520 set @ 150F	✓
	A/S on	✓
	CV-520 operating	✓
E-202	Steam Lined-up to shell @ valve and TIC 202 (200F)	✓
	Visual check on rotameter	✓
C-202	Upper Steam Tracing on	✓
	Lower Steam Tracing on	✓
	Direct Inject on	✓
	Level on Column	R
	Bottom Temperature	R
	Peroxide temperature indication (local)	R
TK-202	Level	✓
	Flow line-up from TK-520	✓
	Temperature	R
P-202A/B	Proper line-up to Ex-202-C-202	✓
	A or B pump	R
TK-530	Proper dye level in reservoir sight glass (local)	✓
	Lined-up to P-530 and P-202	✓
Over-Head System	E-204 lined-up, Ex-con lined-up	✓
	P-209 lined-up to 3-way & reflux	✓
Steam Trap	No unusual noises	✓
	Pulsing	✓

area, or a key individual. One of the common mistakes people make when flowcharting is adding too much information into the chart. Problem areas in a process can be identified during the construction of a flowchart.

Flowcharts are constructed by the following steps:
- Identify the process.
- Identify boundaries.
- Develop a high-level flowchart first to see the big picture.
- Identify nonvalue areas.
- Validate the flowchart by circulating through the organization.

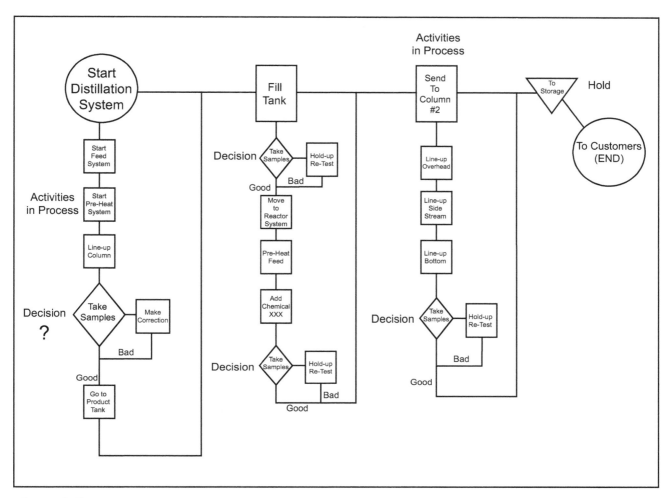

Figure 5-5 *Flowchart*

After completing the flowchart process, a modified brainstorming session is conducted to identify problems in the process. A cause-and-effect diagram is constructed to determine major and minor inputs into the process. Responsibility for each step in the process is determined during this stage of the flowchart process. Difficult or problem areas must be identified, so that a plan can be developed to handle each problem (Figure 5-5).

Scatter Diagrams and Histograms

Scatter diagrams are used to indicate relationship between two variables or pairs of data. The easiest way to determine if a cause-and-effect relationship exists between two variables is to use a scatter diagram. Variables such as flow rate and temperature can be used in a scatter diagram. In this relationship, temperature may increase or decrease as flow rate fluctuates.

Table 5-4
Scatter Diagram
Ordered Pairs
(Thomas, 2008)

Sample	Speed (mph)	Mileage (mpg)
1	25	36
2	25	36
3	30	33
4	30	30
5	35	31
6	35	29
7	40	30
8	40	29
9	45	27
10	45	28
11	50	29
12	50	27
13	55	23
14	55	24
15	60	19
16	60	20
17	65	17
18	65	18
19	70	14
20	70	16

This response would show up on a scatter diagram on the *x*- or *y*-axis. The **independent variable** is typically controllable, while the **dependent variable** is located on the opposite axis.

The steps in setting up a scatter diagram include collecting data in ordered pairs. The cause or independent variable and the effect or dependent variable are placed side-by-side in ordered pairs (*x*, *y*). Table 5-4 illustrates how this process works (see Figure 5-6).

Histograms or Frequency Plots
Histograms or frequency plots are a graphical tools used to understand variability. The chart is constructed with a block of data separated into five to twelve bars or sections from low number to high number. The vertical axis is the frequency and the horizontal axis is the "scale of characteristics." The finished chart will resemble a bell if the data are in control. Histograms describe variations in a process. A histogram can be used to graphically show the process capability of a system and illustrate the shape of the population (Figure 5-7).

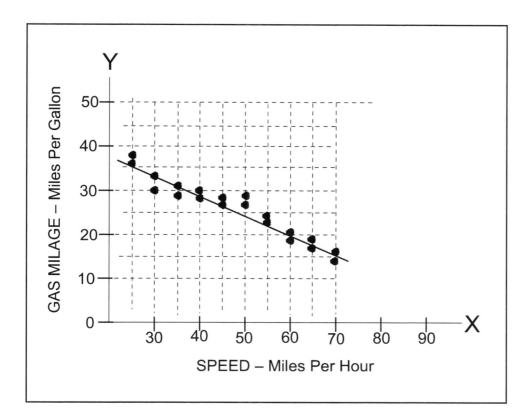

Figure 5-6
*Scatter Diagram
(Replot Data)*

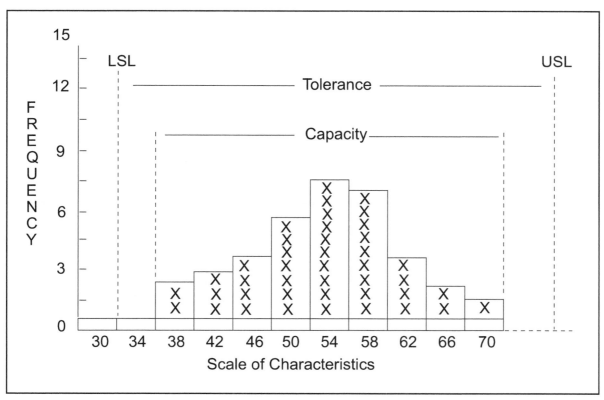

Figure 5-7 *Histograms*

Run Charts

One of the most common tools used in industry is the run chart. Run charts are very powerful tools that show a graphical record of a process variable measured over time. The following steps should be used when building a run chart:
1. Estimate the expected range of data points.
2. Develop a vertical scale for the data that uses 50–70% of the overall range so that the chart is not too narrow or too wide.
3. Plot the data over time (Figure 5-8).

Planned Experimentation

Planned experimentation is a tool used to (1) test and implement changes to a process (aimed at reducing variation) and (2) understand the causes of variation (process problems). When research is being conducted, planned experimentation provides an excellent platform. Typically these experiments are well planned and include the observational and data collection abilities of the process technicians. Figure 5-9 shows a number of small experiments; however, when these planned experiments are conducted in an operating facility, engineering technicians, supervisors, and the work team coordinate their efforts with the engineering and

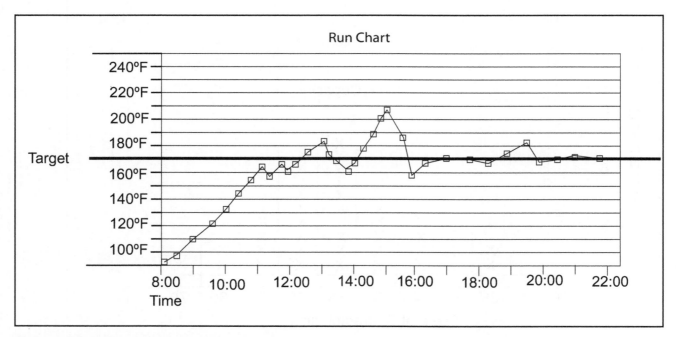

Figure 5-8 *Run Chart (Replot and Connect)*

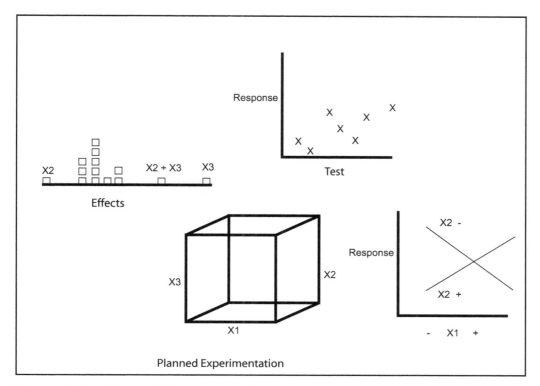

Figure 5-9 *Planned Experimentation*

chemistry departments. Planned experimentation primarily involves analytical variables and also involves other issues concerning safety, operational limits, or any variety of required data.

Summary

The term statistics refers to numbers, numerical facts, figures, or information. Reports of flow rate, level, pressure, temperature, and analytical variables are often called statistics. Making accurate guesses requires groundwork. The process technician must (1) gather data (numerical information), (2) organize them (sometimes in a pictorial type information), and then (3) analyze them (using tests of significance, and so forth). To make an educated inference that is better han chance or following a hunch, these steps need to be followed carefully. Many technicians follow steps one and two and leave out step three.

Most quality experts separate quality tools into two categories: quantitative and nonquantitative. Quantitative tools are described as acceptance sampling, reliability, experimental design, QFD, FMEA, Taguchi's quality engineering, and SPC: control charts, Pareto diagrams, process flow diagrams, scatter diagrams, histograms, cause-and-effect diagrams, and check-sheets. Nonquantitative tools include ISO 9000, ISO 14000, TPM, management tools, and benchmarking.

The most popular area of quality tools falls under the SPC umbrella. SPC is composed of seven tools that every process technician should be aware of. The use and application of these tools has been proven to improve product quality. The principles of continuous quality improvement include:
- innovation and improvement of services and products
- innovation and improvement of processes
- integration of suppliers and customers into the quality process
- using quality tools
  - SPC
  - flowcharts
  - cause-and-effect diagrams (fishbone)
  - Pareto charts
  - run charts
  - scatter diagrams
  - forms for collecting data
  - histograms or frequency plots
  - planned experimentation
  - control charts
- auditing and evaluation
- unrelenting commitment and involvement of all levels in the organization
- documentation of what you do, and doing what you say.

Companies are becoming more and more involved with customers and suppliers. Products and raw materials are tracked from inception. Documentation, quality charts, and external audits follow products and raw materials from cradle to grave. Customers are providing more information about their needs to the companies.

The first step in the **improvement cycle** is to increase current knowledge of the process. The more the team knows about the process, the more likely the changes submitted by the team will impact on quality. At the conclusion of phase one, a plan should be developed that will (1) address specific questions and (2) consider methods, resources, schedules, and people. Phase one will take a significant amount of time for the team to complete:
- plan
- observe and analyze
- learn
- act.

SPC is a quality tool based on the principles of statistical mathematics. Process technicians utilize quality tools during routine operations.

SPC charts are used to plot quality parameter points from samples taken at different times during a run. A flowchart is a picture of the key activities that take place in a process. Flowcharts describe how the process actually works. Run charts are very powerful tools that show a graphical record of a process variable measured over time. The following steps should be used when building a run chart:

- Estimate the expected range of data points.
- Develop a vertical scale for the data that uses 50–70% of the overall range so that the chart is not too narrow or too wide.
- Plot the data over time.

Review Questions

1. Describe the term measures of central tendency.
2. Define the term mean.
3. Define the term mode.
4. Define the term median.
5. Define the term range.
6. Define the term variation.
7. Describe standard deviation.
8. Give an example of how statistics is used to operate a process unit.
9. Describe the measure distribution under a normal curve.
10. What tools do you use to collect data?
11. What tools do you use to organize data?
12. What tools do you use to analyze data?
13. Describe the quality tools and techniques used by technicians.
14. Explain how to use a Pareto diagram.
15. Describe the purpose of a checklist.
16. Design a flow diagram.
17. Design a scatter diagram.
18. Design a histogram.
19. Fill in data on a run chart.
20. Explain how statistics can help in process troubleshooting.

Control Charts

Learning Objectives

After studying this chapter, the student will be able to:

- Apply the principles of quality control to a heat transfer system.
- Collect data on the heat exchanger system.
- Develop a checklist to collect data.
- Create a control chart.
- Describe patterns of instability in statistical process control (SPC).
- Apply principles of quality control to a distillation system.
- Collect data on the overhead system.
- Develop a checklist to collect data.
- Create an *x*-bar and *r*-bar control chart.
- Describe process variation.
- Describe the data collection process for a control chart.
- Explain the purpose of a control chart.
- Describe an *x*-bar chart.
- Describe a *r*-chart.
- Describe process capability.
- Describe the term "process in control."
- Describe the term "process out of control."
- Construct a control chart.
- Define the key terms associated with control charts.

Key Terms

Assignable cause—a cause of unpredictable variation that can be identified using structured techniques.

Attribute—a process characteristic that can be checked or counted.

Average—the mean or arithmetic average.

Control chart—a simple tool designed to quantitatively monitor the important characteristics of a process. It is a picture that allows us to see what changes have occurred over a period of time and to study the impact of various factors in the process that change. When properly used, it prevents us from reacting when we do not need to react and enables us to react when we need to react.

Upper control limit—the highest point established as the top specification on a control chart. $UCL_{x\text{-bar}}$ = Average x-bar + (A_2 × average r).

Lower control limit—the lowest point established as the bottom specification on a control chart. $LCL_{x\text{-bar}}$ = Average x-bar − (A_2 × average r).

***x*-Bar**—the average of a series of numbers.

Subgroup—the number of samples that are averaged together to obtain a single data point. Subgroup size or the number of samples in a subgroup is assigned by the person developing the control chart.

Range—the difference between the highest and the lowest number.

X-double bar—the average of the subgroup averages.

Control limits—the boundaries that tell you if the process is in control.

A_2—a term used on a table to identify the factors established for x-bar control limits.

D_3 and D_4—terms used on a table designed to identify factors for range control limits.

Process capability—the spread of the process, which is equal to six standard deviations when the process is in a state of statistical control.

Sample—An unbiased smaller sample of a larger population used to represent the entire population.

Standard deviation—the average of a sample, technically defined as the measure of the dispersion about the mean of a population.

Target—the desired value; also called the nominal value.

Variable—a quality characteristic that is measurable, such as pressure, temperature, flow, level, analytical, weight, length, or time.

Frequency distribution—the frequency or arrangement of data that may show repetition of numerical values.

Mean—the average of a population.

Median—the middle number in an ordered distribution, or the value that divides a series of ordered observations so that an equal distribution of numbers exists above and below.

Mode—the most frequent measure in a distribution, or the value that occurs with the greatest frequency in the data.

Assignable cause—a cause of variation or a special cause.

Attribute—a quality characteristic that varies or conforms to desired specifications.

Chance cause—a cause of variation that is small in magnitude and difficult to identify; also called random or common cause.

Introduction to Control Charts: Variation

Over the years, operations personnel have discovered that there are natural variations that occur in each equipment system. These variations have not always been understood and have inadvertently created serious operational problems. Since no two objects are identical, variability should be taken into consideration, carefully measured, and applied to customer and product specifications. Batch and continuous chemical processes exhibit variations due to equipment, materials, environment, operations personnel, and management.

The first source of variation is the equipment. This source includes individual parts wear, machine vibration, operating time on equipment, and hydraulic and electrical fluctuations. The combination of these variations can be used to calculate capability or the precision within which the equipment operates. The second source of variation is the raw materials purchased from suppliers. Variations in the feedstock will cause problems producing on-spec materials. Quality characteristics such as product purity, composition, porosity, tensile strength, ductility, thickness, and moisture content will impact overall variation in the final product. A third source of variation is the environment: humidity, temperature, pressure, radiation, electrostatic discharge, and particle size can all contribute to product variation. Experiments are conducted in clean rooms, outer space, high altitudes, under water, and other places to learn more about the effect of the environment on product variation. A fourth source of variation is the operations team. This source of variation includes the multifaceted features of human psychology. This includes procedures, methods, training, and physical and emotional well-being; the combination of all these factors contributes to process variation. It should be noted that some variability can be **attributed** to inspection, faulty equipment, and improper installation. According to Dr. W. Edwards Deming, the fifth area of variation is management. Deming believed that management was responsible for 80–85% of all quality-related problems in a chemical processing unit or refinery. Some management problems are attached to the corporate culture: how receptive the organization is to change, open communication, quality improvement, and Total Quality Management.

When natural causes of variation or **chance causes** are present in a system, the process is considered to be in a state of statistical control: stable and predictable. When variation can be attributed to an **assignable cause,** the process is classified as beyond the expected natural variation or out of control. **Control charts** are used to indicate when observed variations in quality are greater than mere chance and to describe how to respond. The control chart method graphically illustrates the variations that occur in the dispersion of a set of observations and plots the measures of central tendency.

Control Charts for Variables

SPC is a tool designed to maintain quality and reduce costs. Reducing cost is a primary goal of modern manufacturing. A second goal is to produce a quality product. Some industry studies indicate that the cost associated with correcting defective products is in excess of 25%. This includes the loss of revenue from:
- lost sales and customers
- returned products
- scrap materials
- replacement of defective products.

SPC is used by the chemical processing industry in partnership with suppliers and customers. Every chemical process has a certain amount of variability. Measuring and understanding this natural variation is a fundamental part of the quality process. SPC utilizes charts and graphs to visually represent the data. These charts are retained by the company and often sent to customers.

The goal of industry is to manufacture prime material and ensure that nothing that it produces needs to be sorted, recycled, blended, or thrown away. Quality systems are designed to keep customers happy and stay one or two steps ahead of the competitior's product. SPC enables a technician to anticipate and solve problems before they become serious. Prior to SPC, process technicians were required to follow operating directives that had rigidly controlled set points. Natural variation was not considered when making adjustments to operating conditions that were not on **target.** Early manufacturers used pass–no pass specifications that evolved into engineering specifications. This system was needed to furnish customers with the quality services and products demanded. As customers have become more involved in the manufacturing process, higher quality and lower prices have been developed. Control charts are used to reduce scrap materials, predict problems, improve quality, and reduce cost.

A control chart is a simple quantitative tool designed to monitor critical characteristics in an operating system. It provides a window into the process that graphically illustrates variations over time. When a control chart is correctly used it can clearly determine when a change is required and the magnitude of the change. Control charts are designed to:

- achieve higher effective capacity
- identify appropriate adjustments
- distinguish special from common causes of variation
- identify control problems
- achieve higher quality
- achieve lower unit cost
- establish a common language for discussing process performance
- determine if a process will meet or exceed customer expectations
- determine if a production change is needed
- provide foundation for operational decisions
- identify and establish sampling procedures.

State of Control

Assignable causes are described as **variables** that cause unpredictable variation. A process technician can use structured troubleshooting techniques to identify assignable causes. After the assignable cause has been removed and the processes return to desired specifications, a notation should be made on the control chart. After this upset, the plotted points on the process control chart should return to SPC guidelines. When this occurs, the process is considered to be in a state of control, the highest degree of uniformity an operating unit can achieve. It is, however, possible to improve the system using a structured series of quality improvement techniques.

In any process that is operating within design specifications, a natural pattern of variation exists. Identical processes will have unique characteristics that, when carefully studied, will still show signs of natural variation. This natural pattern of variation is illustrated in Figure 6-1. In the example discussed below, about 40% (6 points) of the points fall between one plus or minus one **standard deviation.** In the same example, about 47% (7 points) of the plotted points fall inside the two standard deviations and only 13% (2 points) of the plotted points fall in the third standard deviation. The variation in this example is that the 15 plotted points move back and forth across the target specifications. This natural pattern forms its own **frequency distribution.** Figure 6-1 illustrates how data points can be plotted into a normal curve.

A control chart typically has an **x-bar** or target (central) line and three standard deviations above and below the target. A single standard deviation is referred to as one sigma. Three sigma above and below x-bar takes into

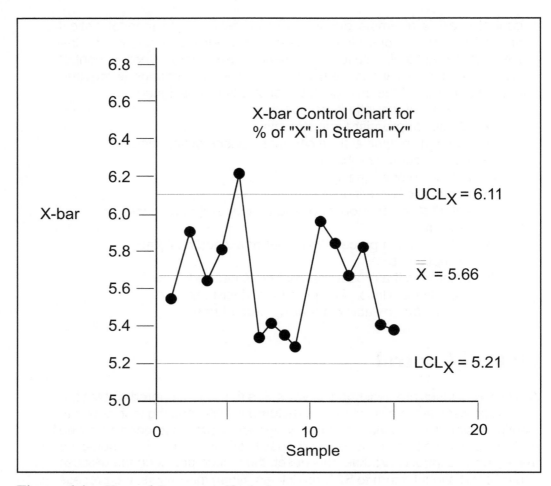

Figure 6-1 *Normal Scatter or Variation*

consideration Type 1 and Type 2 errors. Type 1 errors reflect chance cause versus the normal assignable cause. Type 1 errors (chance causes) occur at the rate of 0.27% (3 out of 1000). Type 2 errors are just the opposite. When the unit is operating within operational guidelines, only chance causes of variation are occurring. Typical variations associated with technician performance, equipment performance, or raw materials are holistically part of a stable process. Processes that are in control have a number of distinct advantages:

- product more uniform—less variation and fewer rejections
- cost of inspection reduced
- process technicians perform satisfactorily
- **process capability** within 6-sigma
- problems anticipated
- percent of product within specification
- x-bar and r-charts show customer process control.

Process Out-of-Control

When a process is out of control, it is typically due to a change in the process that is responding to an undesirable cause. These undesirable causes are easy to identify and can be helpful in getting the process back into process guidelines. A single plot point that falls outside the **upper and lower control limits** indicates that the process may be out of control. This is a red flashing signal to a process technician that something has happened and needs to be investigated. This may indicate that an assignable cause of variation is occurring. This variation may be coming from a different population or process that has a completely different set of **control limits.** Figure 6-2 shows a normal curve. In a normal curve, the plotted points from a frequency distribution are stacked for illustrative purposes to draw a normal curve for **averages.** The population **mean** is $\mu = 5.66$, and the population standard deviation for the averages is 1-sigma $= 0.15$. The frequency

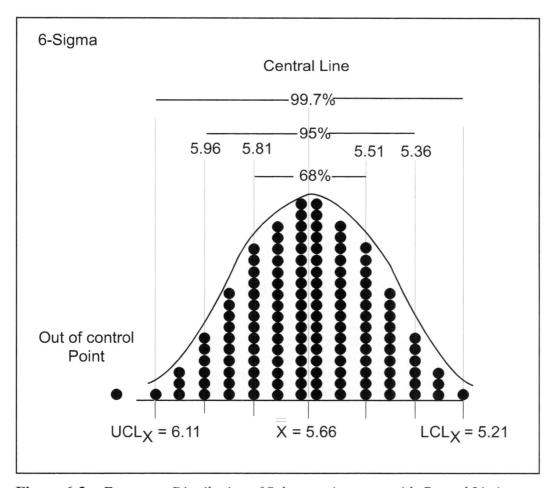

Figure 6-2 *Frequency Distribution of Subgroup Averages with Control Limits*

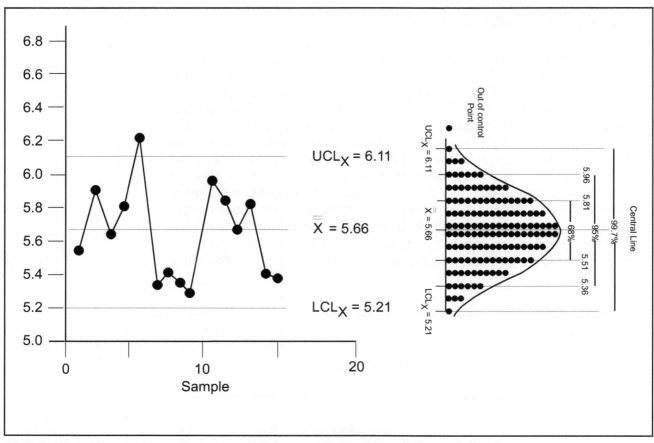

Figure 6-3   *Frequency Distribution and Statistical Process Control*

distribution is shown by the bell curve. The out-of-control point is shown outside the bell curve. This point is so far away from the 3-sigma limits (99.73%) that it is apparent that it came from another process or population (Figure 6-3 and Figure 6-4).

Example of SPC Guidelines

SPC guidelines account for normal process deviations and process upsets.
1. Seven in a row on one side of the target.
   ○ Action is usually a small set point change. Usually indicates a process shift. Using the average of the last two results, adjust according to the product directive.
2. Three results in a row or three out of four results above yellow (+1 sigma) or below yellow (−1 sigma).
   ○ The three out of four case scenario keeps one isolated plot point from upsetting the process. Using the average of the last two results, make adjustments according to the product directive.

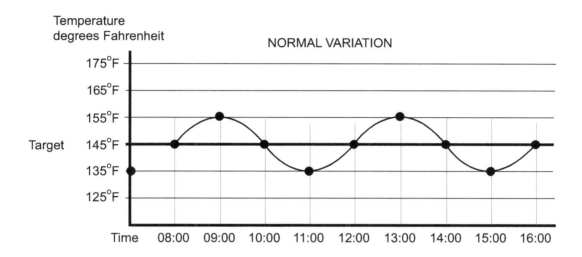

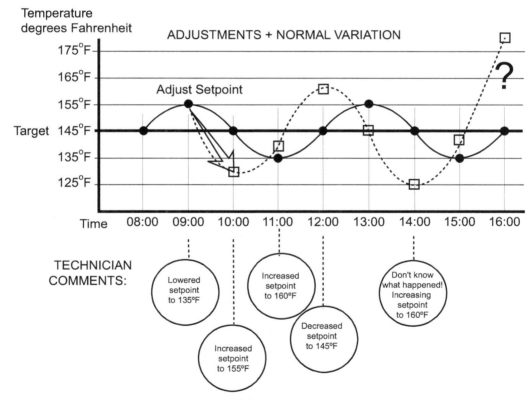

Figure 6-4 *Quality Control Before SPC*

3. Two results in a row or two out of three results above orange (+2 sigma) or below orange (−2 sigma).
   o Using the average of the last two results, make adjustments according to the product directive.
4. One result above/below red (+3 or −3 sigma).
   o A red plot point requires immediate evaluation. Check process trends to see if a significant step change occurred. If a shift in the normal process is identified, make adjustments according to the product directive. If the change does not look reasonable (no visible change in operating conditions), resample and send a green tag to the lab. If the green tag result is in control, disregard the questionable result and follow normal procedures. If the resample confirms the previous result, take the appropriate action and resample after 30 minutes. If the green tag result is above quality limits, divert to off test until the process is back in control.
5. One result crosses four sigma lines.
   o A four sigma jump requires immediate investigation. Check process trends to see if a significant step change occurred. If a shift in the normal process is identified, make (+3 or −3 sigma) product directive adjustments. If the change does not look reasonable (no visible change in operating conditions) resample and send a green tag to the lab. If the green tag result is in control, disregard the questionable result and follow normal procedures. If the resample confirms the previous result, take the appropriate action and resample after 30 minutes. If the green tag result is above quality limits, divert to off test until the process is back in control.

Description of Heat Transfer System

Industry uses a variety of methods and techniques to transfer heat energy between two separate streams. In this example, a pump and heat exchanger are used to heat up the material. A *centrifugal pump* is a device designed to accelerate the flow of liquids using the principles of centrifugal force. A typical centrifugal pump operates with a motor, coupling, gearbox, outer casing, impeller, suction eye, and discharge outlet. A *heat exchanger* is a device used to transfer heat energy between two different process streams. A typical heat exchanger has a shell, tubes, an inlet and an outlet head, a tube inlet and outlet, and a shell inlet and outlet.

Quality tools will be applied to the pressure differential (Δp) across the heat exchanger and the exit temperature of the product stream. Figure 6-5 illustrates what this system looks like and the variables that are being

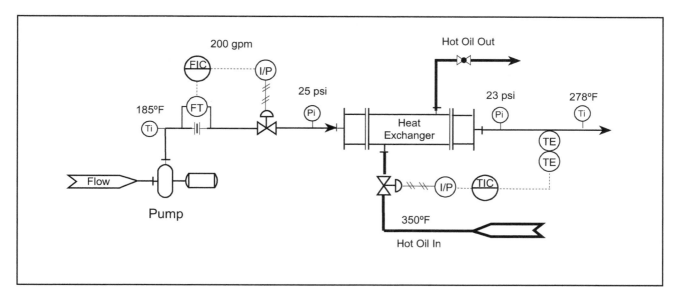

Figure 6-5   *Heat Transfer System*

monitored. Hot oil is used as the heating medium running through the shell side of the exchanger. Heat energy is transferred through the tube and into the cooler feedstock. Heat exchangers utilize conductive and convective heat transfer methods. Temperatures are closely monitored on the tube-side outlet at 278°F. It is important for the charge to be at this temperature when it enters the distillation column. A narrow operating **range** has been established for temperature control of the feed. Hot oil temperatures are controlled at 350°F. The pressure differential across the exchanger is 2 psi. The inlet feed temperature is set at 185°F.

Data Collection

Data collection is an important part of a process technician's job. At the beginning of each shift, technicians frequently make rounds through the unit to ensure that the equipment and the systems are operating properly. A number of process variables are important to this process and are recorded on checklists. Historical data allow a process technician to view a piece of equipment's performance over an extended period of time. This is particularly important with heat exchangers that tend to foul or plug up over time. Fouling creates nonuniform heating cycles that tend to aggravate the plugging problem. The differential measurement across the exchanger measures the effects of fouling over a long period as flow rates decrease, inlet pressures increase, and outlet flows and pressures decline. Data collection is one of the more important quality tools used by process technicians. Table 6-1 shows a historical collection of data on an exchanger that is showing signs of fouling.

Table 6-1
Fouling Table

Sample	Inlet (psi)	Outlet (psi)	Δp	Flow (gph)	Outlet Temperature (°F)
1	25	23	2	200	278
2	25	23	2	200	278
3	25.5	23	2.5	195	278
4	25	22.5	2.5	195	278.5
5	25.5	22.5	3.0	190	278.5
6	26	22.5	3.5	190	277
7	26	22.5	3.5	185	277
8	26	22	4	185	278.5
9	26	22	4	180	278.5
10	26	21	5	180	282
11	27	21	6	175	282
12	28.5	21.5	7	175	279
13	30	21.5	8.5	160	279
14	30	21	9	160	320
15	32	20	12	120	320
16	32	19	13	120	288

You will notice that as the heat exchanger begins to foul, the flow rate through the device decreases. When the temperature control loop attempts to adjust, the reading becomes erratic as the plugging becomes more severe. It is difficult to see the changes within a single month. Over a period of months, a historical profile is developed. The data are much easier to read and to analyze after a variety of time periods have been observed.

Scatter Plots and Histograms

The use of scatter plots and histograms provides additional information for a process technician to use. Scatter diagrams are used to indicate relationships between two variables or pairs of data. The easiest way to determine if a cause-and-effect relationship exists between two variables is to use a scatter diagram. Variables such as flow, pressure, and temperature can be used in a scatter diagram. In this relationship, pressure differential may increase or decrease as flow rate fluctuates through the tubes in the exchanger. This response will show up on a scatter diagram on either the x- or the y-axis. The independent **variable** is typically controllable, while the dependent variable is located on the opposite axis.

The steps in setting up a scatter diagram include collecting data in ordered pairs. The cause or independent variable and the effect or dependent variable are placed side-by-side in ordered pairs (x, y). Table 6-2 illustrates how this process works.

Sample	Flow	Δ*p*	Sample	Flow	Δ*p*
1.	200	2	9.	180	4
2.	200	2	10.	180	5
3.	195	2.5	11.	175	6
4.	195	2.5	12.	175	7
5.	190	3	13.	160	8.5
6.	190	3.5	14.	160	9
7.	185	3.5	15.	120	12
8.	185	4	16.	120	13

Table 6-2
*Scatter Diagram
Ordered Pairs*

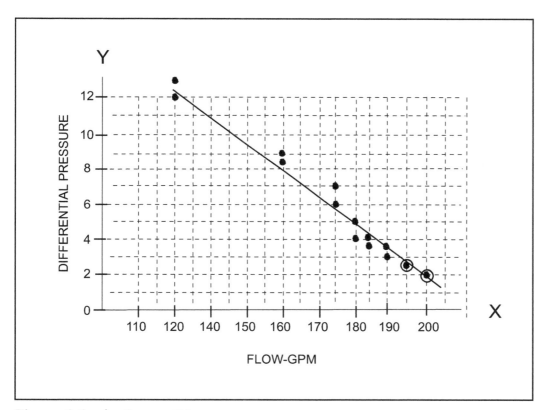

Figure 6-6   Δ*p Scatter Diagram*

Histograms or Frequency Plots

Histograms or frequency plots constitute a graphical tool used to understand variability. The chart is constructed with a block of data separated into five to twelve bars or sections from low number to high number. The vertical axis is the frequency and the horizontal axis is the "scale of characteristics." The finished chart will resemble a bell if the data are in control. Histograms describe variation in a process. A histogram can be used to

165

Figure 6-7
Histograms of Δp and Flow Rate

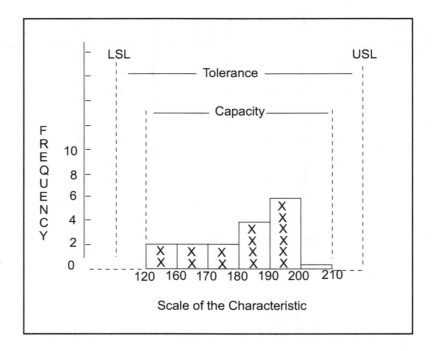

graphically show the process capability of a system and to illustrate the shape of the population (Figure 6-7).

Statistical Process Control

Another valuable tool that can be used with this heat transfer system is SPC. Defective equipment can lead to loss in product quality. Detecting and responding to fouling data is an important part of successfully operating the heat transfer system. According to SPC, each process or system has its own unique characteristics. To develop a control chart, the process needs to be in a steady state; otherwise the chart control limits will be too wide (Table 6-3).

x-chart

$$UCL = x\text{-bar} + (A_2 \times r)\ 199 + (1.023 \times 5.8) = 204.93$$

$$LCL = x\text{-bar} - (A_2 \times r)\ 199 - (1.023 \times 5.8) = 193.06$$

r-chart

$$UCL = r \times D_4 = 5.8 \times 2.575 = 14.935$$

$$LCL = r \times D_3 = 5.8 \times 0 = 0$$

Figure 6-8 shows how these data can be included on a moving *x*-bar chart and *r*-bar chart. While the *x*-bar and moving *x*-bar chart look similar, the

Sample	Date	Flow (gpm)			Σx	x-bar	r
		x_1	x_2	x_3			
1.	5-1-05	205	200	195	600	200	10
2.	5-8-05	200	203	197	600	200	6
3.	5-15-05	196	199	194	589	196	5
4.	5-22-05	193	197	204	594	198	11
5.	5-29-05	191	194	196	581	194	3
6.	6-5-05	196	197	200	593	198	4
7.	6-12-05	189	196	194	579	193	7
8.	6-19-05	193	199	201	593	198	8
9.	6-26-05	199	202	201	602	201	2
10.	7-3-05	205	202	201	608	203	4
11.	7-10-05	199	201	203	603	201	4
12.	7-17-05	200	203	206	609	203	6
Average						199	5.8

Table 6-3
*Heat Exchanger
Flow Rate*

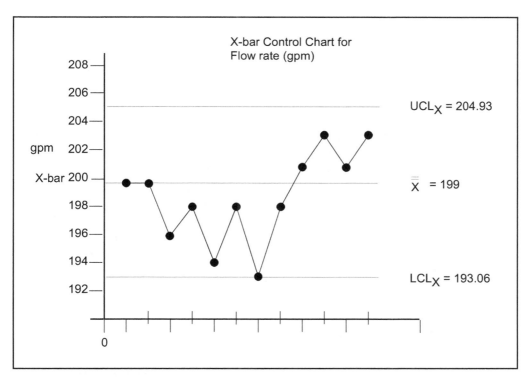

Figure 6-8   *Flow Rate Control Chart—x-Bar*

Figure 6-9
*Flow Rate Control
Chart—Range*

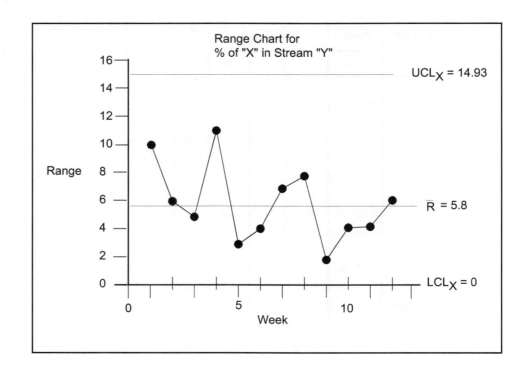

difference is in the use of averages to calculate the individual plot point (Figure 6-9).

Detecting and Responding to Trends

One of the more important skills needed to operate an equipment system is the ability to detect and respond to trends. Control charts provide an excellent window into the process and can accurately predict future trends. The rules of SPC enable standardization of the response each technician has to various trends and process upsets. Instability occurs when abnormal variations in temperature, flow, pressure, or composition occur. Recognizing patterns of instability early enables a process technician to correct the situation. Patterns in the data include level changes, runs, trends, and cycles. Level changes are characterized by a sudden shift above or below the centerline. A run is described as 7 out of 8 or 12 out of 14 data points on one side of the centerline. The principles of mathematics indicate that this type of data is not a good example of normal variation. A trend shows an upward or downward movement of data on the control chart. Cycles are patterns in the data that repeat (Figure 6-10).

Applications of Quality Tools

A variety of quality tools can be applied to the various processes found in the chemical processing industry. Important quality tools include data

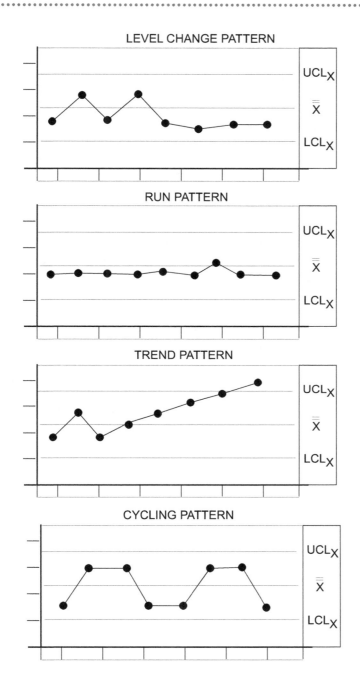

Figure 6-10
Patterns of Instability

collection methods, checklists, Pareto charts, histograms, scatter plots, flow charts, run charts, and control charts. Quality tools can be applied to process variables such as flow, level, temperature, pressure, and composition. Quality tool applications have been connected to pumps, compressors, heat exchangers, cooling towers, boilers, furnaces, reactors, and distillation systems.

Statistical Process Control and Distillation Systems

The principles of SPC can easily be applied to a distillation system. Distillation is a common process that requires SPC for product optimization. Inside a distillation column vapors move up and liquids flow downward. This process varies depending on the design of the column: plate or packed. Larger hydrocarbon fractions collect at the bottom of the column, while lighter components move up. Product purity or concentration in the overhead, side, and bottom streams are the important quality factors. These numerical values are typically measured in percent and carefully controlled to customer specifications. A product that is too pure may not meet customer specifications. Distillation systems are a complex network of variables that respond quickly to each other.

The purpose of SPC in distillation systems is to decrease deviation from product specifications. Prior to industry acceptance of SPC, customer specifications were either exceeded or not met at all. Over the past thirty years, SPC has proven its value to the chemical processing industry. The quality movement has recently lost a number of key experts who championed the importance of the application of statistical methods. Competition in the global economy requires dedicated adherence to the principles of quality control.

Distillation Systems

Distillation is often described as the separation of components in a mixture by boiling point. A distillation column is often characterized as plate or packed, with an enriching or rectifying section, feed section, and stripping section. A distillation system, however, includes numerous assortments of equipment and systems that support the process. A distillation system includes feed system, preheat system, distillation process, and product storage. The quality system provides the mathematical foundation that standardizes plant operations.

The feed system is composed of a variety of equipment systems, including a feed tank, valves, piping, instruments, and pumps. Inside the feed system, the composition of the feedstock is carefully monitored. Flow rates, pressures, temperature, and levels are carefully maintained. Figure 6-11 shows the basic equipment found in a feed system. Compressed air or nitrogen systems are also used in the feed system. Nitrogen is considered to be an inert gas that provides protection from fire or explosion. Compressed air is used to open, close, or throttle control valves. Basic instrument systems used include indicators, control loops, recorders, and analyzers.

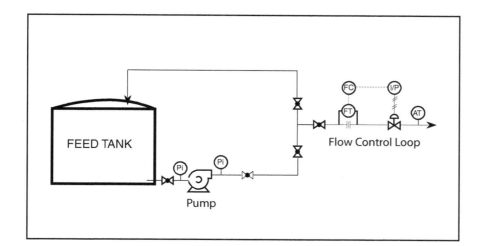

Figure 6-11
Distillation Feed System

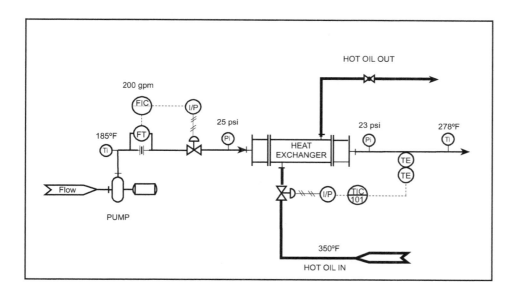

Figure 6-12
Preheat System

Before feed can be sent to a distillation column it needs to be preheated to a temperature range that will allow the distillation process to occur. The chemical processing industry uses fired heaters, heat exchangers, or a combination of these devices depending on the quantity being transferred. Heating the feed initiates the distillation process as the various components in the mixture respond differently. Figure 6-12 illustrates what a simple preheat system looks like. As the feed is heated, pressure increases inside the pipe due to the lighter components attempting to escape from the liquid. Heat increases molecular movement within the liquid. A temperature control loop maintains unit specifications.

After the feed has been blended and preheated, it is sent to the distillation column. In a plate column, the heated feed enters on the feed tray. The part of the column above the feed line is called the rectifying or enriching section. The part of the distillation column below the feed line is referred to as the stripping section. As the heated feed enters the column, part of it vaporizes and moves up the column. The heavier part of the mixture flows down the column through devices called downcomers.

Each tray in the column forms a liquid seal that provides good vapor–liquid contact. Theoretically, each tray in the column would have a different molecular structure, ranging from heavier components at the bottom to lighter components at the top. The lighter fractions exert a higher vapor pressure. Typically, the temperatures are higher at the bottom of the column and lower at the top. This is referred to as temperature gradient.

There are two scientific principles that must be balanced on a distillation column: energy and mass. Heat is returned into the system using a kettle reboiler. The reboiler is connected to the bottom of the distillation column. A reboiler is a heat-transfer device designed to add energy to the bottom fractions. The heavier bottoms product is still rich in lighter components that need help breaking free from the larger molecules. As the heated fluid passes through the reboiler, the lighter components vaporize and flow into

Figure 6-13
Distillation Column

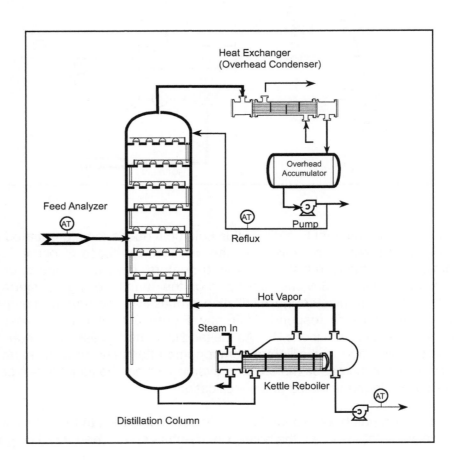

the column under the bottom tray. The space below the bottom tray allows the liquid to free-roll and boil.

As feed flows into the column through a single feed source the rate is carefully monitored. The various fractions in the mixture can flow out of the overhead, side, or bottom. The old saying "what flows into the column must flow out" is still true. The overhead system is specifically designed to condense the lighter fractions and send them to product storage and back to the top tray in the column. The cooled, condensed product that flows back to the column is called reflux. It is specifically designed to control product purity and control temperature at the top of the column. The overhead system includes a condenser, accumulator, pump, and piping. Figure 6-13 shows a basic distillation column.

In Figure 6-13, the bottoms product is sent to the tank farm for storage. The tank farm has a number of product and off-spec tanks. The overhead stream is sent to the tank farm in preparation for being tested and, if approved, shipped to the customer.

Variation in Binary Distillation Systems

In a binary mixture, two components are separated from the feedstock. In this example, our distillation system has a 40/60 mixture of butane and pentane. Three butane analyzers are used to monitor the concentration of butane in the various streams. Butane should be found in the following concentrations:

Overhead	98.5% butane, 1.5% pentane
Bottom	8% butane, 92% pentane
Feedstock	40% butane

Variations in product composition are carefully monitored and controlled. A variety of modern control features are incorporated into the system, including temperature, pressure, level, and flow control loops. Product variation is the enemy of any chemical process. When product specifications are not met, customers will take their business elsewhere.

Data Collection

Data collection includes feed composition, overhead purity, and bottoms purity. Three analyzers are placed in each of these streams to enable each process technician on shift to monitor and record the process. Table 6-4 contains information collected from the overhead line.

Table 6-4

Distillation-Overhead Purity

Sample	Date	Flow (gpm)			Σx	x-bar	r
		x_1	x_2	x_3			
1.	5-1-05	98.6	98.7	99	296	98.76	0.4
2.	5-2-05	99.1	98.7	98.5	296	98.76	0.6
3.	5-3-05	98.3	98.6	99.1	296	98.66	0.8
4.	5-4-05	99	98.7	98.2	295	98.63	0.8
5.	5-5-05	97.8	97.5	98	293	97.77	0.5
6.	5-6-05	98.3	98	98.8	295	98.37	0.8
7.	5-7-05	96.9	97.5	98.1	292	97.5	1.2
8.	5-8-05	97.8	98.4	98.6	295	98.26	0.8
9.	5-9-05	99	99.1	98.7	297	98.93	0.7
10.	5-10-05	97.9	98.1	98.6	295	98.2	0.7
11.	5-11-05	98.1	98.5	99.2	296	98.6	1.1
12.	5-12-05	98.3	98.8	99	296	98.7	0.7
Average						98.43	0.758

x-chart

$$\text{UCL} = x\text{-bar} + (A_2 \times r)\ 98.43 + (1.023 \times .758) = 99.2$$

$$\text{LCL} = x\text{-bar} - (A_2 \times r)\ 98.43 - (1.023 \times .758) = 97.66$$

r-chart

$$\text{UCL} = r \times D_4 = .758 \times 2.575 = 1.95$$

$$\text{LCL} = r \times D_3 = .758 \times 0 = 0$$

Control Charts

Controlling the overhead butane concentration is an important quality aspect of process control on a distillation system. This is accomplished by carefully monitoring the analyzer on the overhead reflux stream and recording the lab results from each shift. The three daily results are averaged and included on the control chart. The range is calculated by subtracting the low reading from the highest. Using the equations for the x-bar chart and the r-bar chart, the upper and lower control limits can be calculated. The average of the three daily results and the range between the variables enable us to develop the control chart (Figure 6-14, Figure 6-15, Figure 6-16, and Figure 6-17).

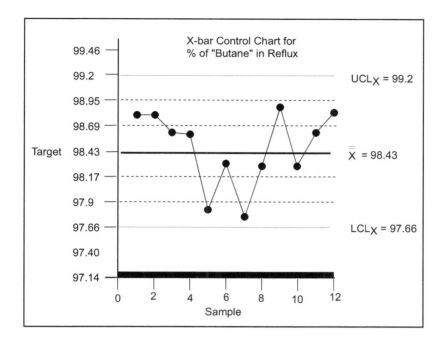

Figure 6-14
Butane Overhead Control Chart x-Bar Chart

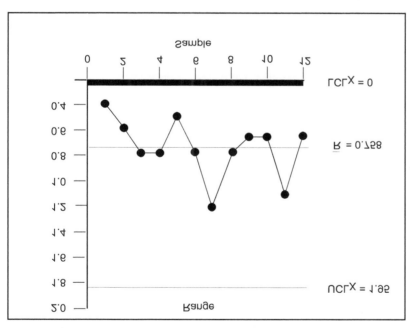

Figure 6-15
Butane Overhead Control Chart x-Bar Chart

Variables in the Distillation System

A number of variables are controlled in a distillation process, including pressure, temperature, level, flow, and analytical variables. The composition of the overhead stream is only one of a large number of variables. The three analyzers on the unit are primary targets for a quality system. Table 6-5 shows the typical variables found in this system.

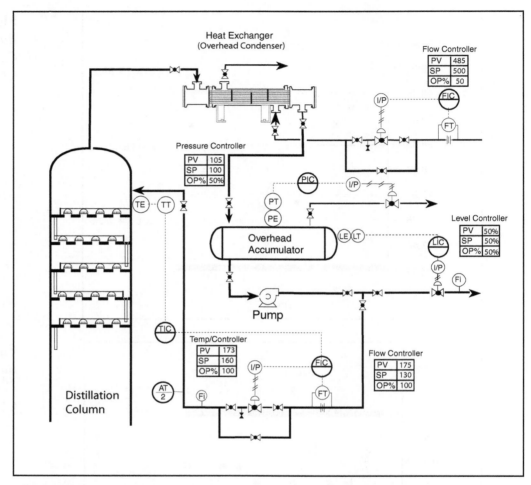

Figure 6-16 *Overhead System*

Summary

The purpose of a control chart is to:
- define natural variability
- produce a more consistent product
- characterize product or process
- anticipate problems
- provide a standardized communication tool for identifying problems, seeing the impact of solutions, communicating with customers and suppliers, and making decisions based upon facts rather than opinions.

The steps to creating a control chart include:
- selecting a process characteristic of interest
- deciding who will collect the data, which method of sampling and testing will be used, how data will be recorded, and who will be responsible for the control chart

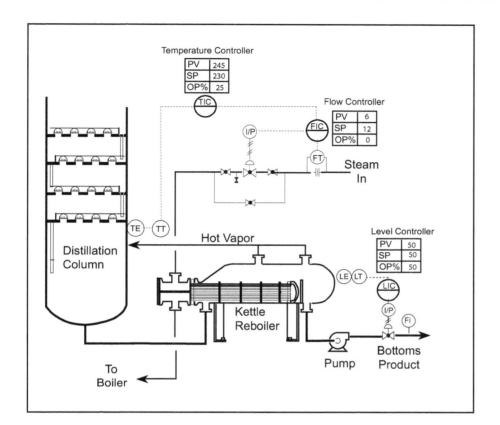

Figure 6-17
Bottom System

Tag#	Description	Set Point
AT-1	Butane feed analyzer	40%
AT-2	Butane bottom analyzer	8%
AT-3	Butane overhead analyzer	98.5%
TIC-101	Preheater hot oil	180.6 °F
TIC-102	Column bottom tray #1	220 °F
TIC-103	Column top tray #10	160 °F
FIC-101	Unit feed	225 gpm
FIC-102	Reflux flow rate	143 gpm
FIC-103	Bottom flow rate	135 gpm
FIC-104	Steam to reboiler	14 mlb/h
FIC-105	Cooling water to overhead condenser	525 gpm
FI-102B	Overhead flow to tank farm	89 gpm
LIC-101	Feed tank level	50%
LIC-102	Overhead accumulator level	50%
LIC-103	Reboiler level	50%
LIC	All product tank levels	50%

Table 6-5
Distillation—Typical Process Variables

- setting up a **sample** plan
- deciding on a **subgroup** size
- drawing the *x*-bar and the *r*-chart axes
- collecting data and plotting it on the graph
- calculating an average *x*-bar and range
- plotting the values for *x*-bar and *r*-bar on the control chart
- calculating your control limits
- plotting the control limits on the control charts.

The questions to ask yourself when deciding on a characteristic to follow are as follows:
- Is the characteristic important?
- Does it reflect quality?
- Is it measurable?
- Will it return immediate results?

Industry uses a variety of methods and techniques to transfer heat energy between two separate streams. In this chapter, a pump and a heat exchanger are used to heat up a process liquid. A heat exchanger is a device used to transfer heat energy between two different process streams. Quality tools can be applied to the pressure differential (Δp) across the heat exchanger and the exiting temperature of the product stream. Heat exchangers utilize conductive and convective heat transfer methods.

Data collection is an important part of a process technician's job. At the beginning of each shift, technicians frequently make rounds through the unit to ensure that the equipment and systems are operating properly. Historical data enable a process technician to view a piece of equipment's performance over an extended period of time. This is particularly important with heat exchangers that tend to foul or plug up over time. Fouling creates nonuniform heating cycles that tend to aggravate the plugging problem.

Scatter diagrams are used to indicate relationships between two variables or pairs of data. The easiest way to determine if a cause-and-effect relationship exists between two variables is to use a scatter diagram. The independent variable is typically controllable, while the dependent variable is located on the opposite axis. The steps in setting up a scatter diagram include collecting data in ordered pairs. The cause or independent variable and the effect or dependent variable are placed side-by-side in ordered pairs (*x, y*).

Histograms or frequency plots constitute a graphical tool used to understand variability. The chart is constructed with a block of data separated into five to twelve bars or sections from low number to high number. The vertical axis is the frequency and the horizontal axis is the "scale of characteristics."

Another valuable tool that can be used with this heat transfer system is SPC. Detecting and responding to fouling data is an important part of successfully operating the heat transfer system. According to SPC, each process or system has its own unique characteristics. The following equations are used to develop SPC charts:

x-chart	**r-chart**
UCL = x-bar + (A$_2$ × r) LCL = x-bar − (A$_2$ × r)	UCL = r × D$_4$ LCL = r × D$_3$

One of the more important skills needed to operate an equipment system is the ability to detect and respond to trends. Control charts provide an excellent window into the process and can accurately predict future trends. Instability occurs when abnormal variations in temperature, flow, pressure, or composition occur. Recognizing patterns of instability early enables a process technician to correct the situation. Patterns in the data include level changes, runs, trends, and cycles.

The purpose of SPC in distillation systems is to decrease deviation from product specifications. Prior to industry acceptance of SPC, customer specifications were either exceeded or not met at all. The principles of SPC can easily be applied to a distillation system. Distillation is a common process that requires SPC for product optimization. Product purity or concentration in the overhead, side, and bottom streams are the important quality factors. These numerical values are typically measured in percent and carefully controlled to customer specifications. A product that is too pure may not meet customer specifications. Distillation systems are a complex network of variables that respond quickly to each other. Over the past thirty years, SPC has proven its value to the chemical processing industry. Competition in the global economy requires dedicated adherence to the principles of quality control.

A distillation system includes feed system, preheat system, distillation process, and product storage. The quality system provides the mathematical foundation that standardizes plant operations.

Review Questions

1. Describe natural variability in a process.

2. What are the basic elements of a control chart?

3. What is an "assignable cause"?

4. What is the primary purpose of a control chart?

5. What is a control chart and why do we want to use it?

6. List seven benefits of using a control chart.

7. What is a subgroup?

8. Describe the term "process in control."

9. Describe the term "process out of control."

10. Draw and describe the four patterns of instability listed in this chapter.

11. How do control charts provide a window into the process?

12. Describe how to set up a scatter plot diagram.

13. List the four equations used to develop x-bar and r-bar charts.

14. Draw a sketch of a simple heat transfer process.

15. What role does data collection play in the quality system?

16. Describe histograms or frequency plots.

17. Describe the purpose of SPC in a distillation system.

18. Describe the variables controlled in a distillation process.

19. Describe the four basic systems found in a distillation system.

20. Write the SPC equations used to develop an x-bar chart.

21. Write the equations used to calculate range.

chapter 7

Introduction to Process Troubleshooting

LEARNING OBJECTIVES

After studying this chapter, the student will be able to:

- Describe the various troubleshooting models.
- Describe how different variables affect each other.
- Explain how problems with process equipment affect other systems.
- Analyze process problems and provide solutions.
- Troubleshoot specific operational scenarios.
- Describe various instrumentation used to troubleshoot process problems.
- Distinguish between primary and secondary problems.
- Collect, organize, and analyze data.
- Respond to alarms and control systems that are outside operational guidelines.
- Compare troubleshooting methods and models.

Key Terms

Process variables—instruments used to detect process variables provide clues that can be used to complete the big picture.

Failed equipment—a term used to describe equipment that has broken down, ruptured, or is no longer responding to its design specifications.

Troubleshooting models—tools used to teach troubleshooting techniques. Basic models include distillation, reaction, absorption and stripping, or a combination of these three.

Primary operational problems—a term used to describe the first problem that created a process break-down.

Secondary operational problems—are created or responded to a primary operational problem that has created a process upset and a trickle down effect.

Troubleshooting methods—include educational, instrumental, experiential, and scientific methods.

Process flow diagram—a diagram used in troubleshooting to quickly identify the primary flow path and the control instrumentation being used in the process.

Introduction to Process Troubleshooting

Troubleshooting the operation of process equipment requires a good understanding of basic components and equipment operation. Equipment used in modern manufacturing is run twenty-four hours a day, seven days a week, and fifty-two weeks a year. Routine maintenance is performed on this equipment during scheduled maintenance.

Redundancy is a process that uses backup systems for critical equipment. For example, pumps and compressors typically have two or three backups. Process technicians should attempt to obtain as much information as possible on the equipment found in their units. Much of this information can be found in technical manuals or the operating manuals. Manufacturer information is typically included in engineering specifications, drawings, and equipment descriptions.

Data collection, organization, and analysis can also be used to troubleshoot process problems. Check sheets are used to collect large quantities of data. These quantitative data can be organized into graphics or trends to plot process variation or changes. Data analysis utilizes a variety of quality techniques to put all of the parts in place.

Process Troubleshooting Methods

Troubleshooting Methods
There are a number of **troubleshooting methods** that can be used with these models. Methods vary depending on individual education faculty, consultants, and industry. The basic approach to most methods includes the development of a good educational foundation.

Method One: Educational
- Basic knowledge of the equipment and technology
- Understand the math, physics, and chemistry associated with the equipment
- Study equipment arrangements in systems
- Study process control instrumentation
- Operate equipment in complex arrangements
- Troubleshoot process problems

Troubleshooting is a process that requires a wide array of skills and techniques. Modern control instrumentation includes indicators, alarms, transmitters, controllers, control valves, transducers, analyzers, and interlocks. The primary goal is to control variables such as temperature, pressure, flow, level, or analytical variables. It is possible to control large, complex processes from a single room. In these types of systems, process set points and **process variables** on controllers should clearly reflect each other. Process problems are quickly identified when these two do not line up. For example, if the flow rate is set at 200 gpm and the process variable is 175 gpm, a 25 gpm difference exists. This could indicate a serious problem.

Method Two: Instrumental
- Basic understanding of process control instrumentation
- Basic understanding of the unit process flow plan
- Advanced training in controller operation—PLC & DCS
- Troubleshooting process problems.

Method Three: Experiential
- Experience in operating specific equipment and system
- Familiarity with past problems and solutions
- Ability to think outside the box
- Ability to think critically—identify and challenge assumptions
- Ability to evaluate, monitor, measure, and test alternatives
- Ability to troubleshoot process problems.

Method Four: Scientific
- Grounded in principles of mathematics, physics, and chemistry
- Understands theory-based operations
- Understands equipment design and operation
- Views the problem from the outside in
- Utilizes outside information and expertise and reflective thinking
- Generates alternatives, brainstorming, and rank alternatives
- Troubleshoots process problems.
1. In Figure 7-1, what will happen if P-104 fails and cannot be restarted?
   a. Steam flow will:
      o increase,
      o decrease, or
      o stay the same.

Figure 7-1
Basic
Troubleshooting 1

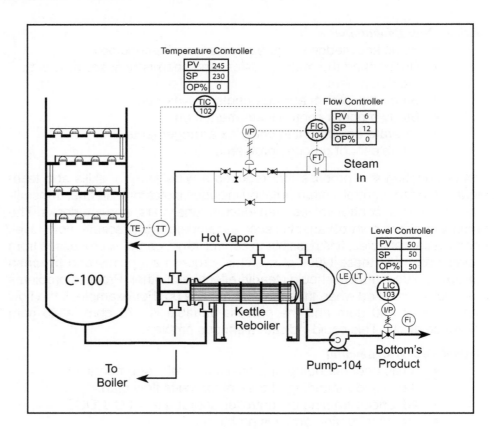

b. Bottom's flow rate will:
   ○ increase,
   ○ decrease, or
   ○ stay the same.
c. Bottom's temperature will:
   ○ increase,
   ○ decrease, or
   ○ stay the same.
d. Bottom's level will:
   ○ increase,
   ○ decrease, or
   ○ stay the same.

The questions that are developed for troubleshooting scenarios can vary from equipment failure to instrument failure. Each of these failures provides good experience to the new technician. Collecting and organizing these scenarios is a difficult process that takes time and effort. In Figure 7-1, the steam flow will increase, the bottom's flow rate will decrease, the bottom's temperature will decrease, and the bottoms level will increase. Why?

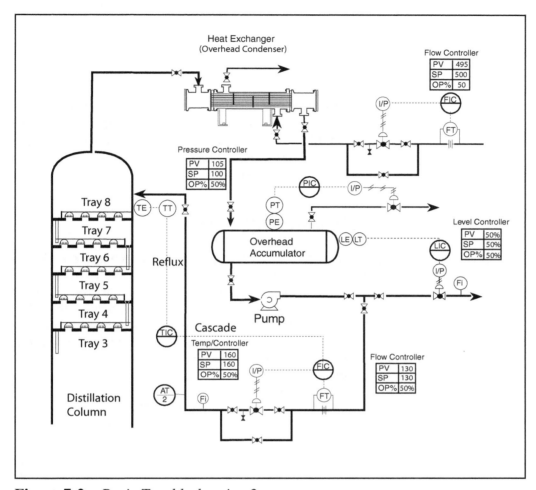

Figure 7-2 *Basic Troubleshooting 2*

2. In Figure 7-2, what will happen if the reflux control valve fails to close?
   a. Column pressure will:
      o increase,
      o decrease, or
      o stay the same.
   b. Reflux flow rate will:
      o increase,
      o decrease, or
      o stay the same.
   c. Column top temperature will:
      o increase,
      o decrease, or
      o stay the same.

d. Level in D-100 will:
- increase,
- decrease, or
- stay the same.

In Figure 7-2, the column pressure will stay the same, the reflux rate will increase, the column temperature will go down, and the level in D-100 will decrease. Technicians should study these problems and ask why a specific control loop responds a certain way. Controllers in AUTO, MAN, and cascade respond differently in operation.

Process Troubleshooting Models

One of the highest levels a process technician can achieve is the ability to clearly see the process and to sequentially break down, identify, and resolve process problems. Process troubleshooting has traditionally been considered the area of senior technicians. However, some people believe that successful techniques can be taught to all technicians. Experience has proven over time to be the best teacher on equipment that is manually operated; however, new computer technology provides advanced control instrumentation that can be used to quickly and methodically track down process problems. It is well known that a single problem can have a cascading effect on all surrounding equipment and instrumentation. This phenomenon is commonly associated with primary problems and secondary problems.

Troubleshooting models are attached to equipment and systems presently being taught in community colleges and universities. Some of these models include the reaction model, absorption and stripping model, separation model, and distillation model. These models are completely outfitted with alarms, analyzers, interlocks, permissives, video trends, recorders, and control instrumentation. Process problems can be simulated using these models. College curriculum includes the use of advanced computer and computer system software that closely simulates console operations. Software companies such as Advanced Training Resources (ATR®) lead the way in the development of this type of software and computer systems. Some college training systems have modern control instrumentation mounted on operational pilot units. Students using these types of systems receive a true hands-on experience.

The ten models used to teach process troubleshooting are as follows:
1. Pump and tank model
2. Compressor model
3. Heat exchanger model
4. Cooling tower model
5. Boiler model
6. Furnace model

7. Distillation model
8. Reaction model
9. Separation model
10. Combination of the above models.

These ten models provide the hardware or framework from which the various troubleshooting methodologies are applied. Each model has a complete set of process control instrumentation and equipment arrangements. A complete range of troubleshooting scenarios has been developed by some educators and is typically included with these models.

Simple Reactor Model

The reactor model provides a framework to develop a series of troubleshooting scenarios. Educators use this model to teach one part of a much larger process. Problems are developed from simple to complex as students learn one section and move to the next. It is possible to create a multiprocess plant that can be used within the walls of the classroom. Figure 7-3 illustrates the basic components of a simple reactor model.

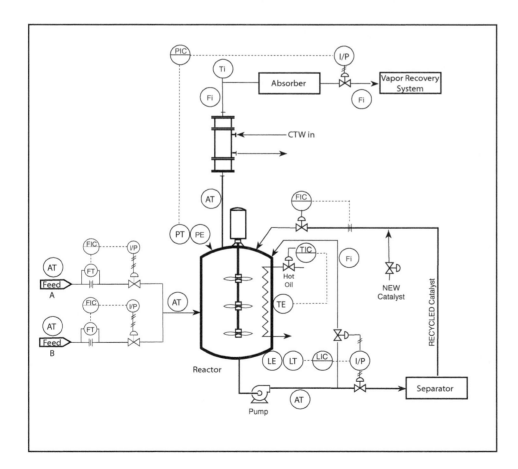

Figure 7-3
Reactor Model

Troubleshooting the reactor system requires the student to become familiar with the typical operation of the unit. As equipment and instrumentation fail, the student sees the cascading effect a single problem can have on the unit. Since a single problem can create a series of other problems, students must identify the primary problem that started the system failures. The stirred reactor combines reactants and catalyst at controlled temperatures, pressures, and flows. The liquid catalyst is designed to speed up the reaction between reactants A and B without becoming part of the actual reaction. The temperature of the reactor is controlled by a hot oil system. A feed ratio is used to control the amount of reactant A to reactant B. The reaction requires heat to continue and is classified as endothermic. During the reaction, some of the catalyst will vaporize and flow through the overhead line. If the hot oil system fails, pressure and vapor production will increase. The reaction rate is also linked to agitation and catalyst flow.

Distillation Model

A variety of distillation models can be used with the various troubleshooting methods. A distillation system includes a well-defined feed system, overhead system, bottom system, and column system overview. Plate columns and packed columns offer a different variety of process troubleshooting scenarios. In Figure 7-4, a binary mixture is heated and sent to an eight-tray distillation column. The feed system uses flow control, temperature control, and primary variable indicators, including composition. Changes can be made to one or more of these variables and observations recorded by individual learners. The software from a number of companies supports these changes and responses. Actual problems on the feed system can be paper-based or electronic. These same problems can be plugged into the overhead and bottom system.

Modern Process Control

Process control instrumentation is used to control each variable in the column. The level in the bottom of the column and the overhead accumulator needs to be controlled at 50%. The thermosyphon reboiler maintains the energy balance on the column at a set temperature. The bottom and top temperatures form a gradient. Flow to the column and reflux lines enables the system to operate in automatic. Process technicians can monitor the response of these systems from the control console. Pressure is held at specified values on the overhead accumulator. This has an effect on the entire distillation column, even though pressures and temperatures will vary slightly up and down the column.

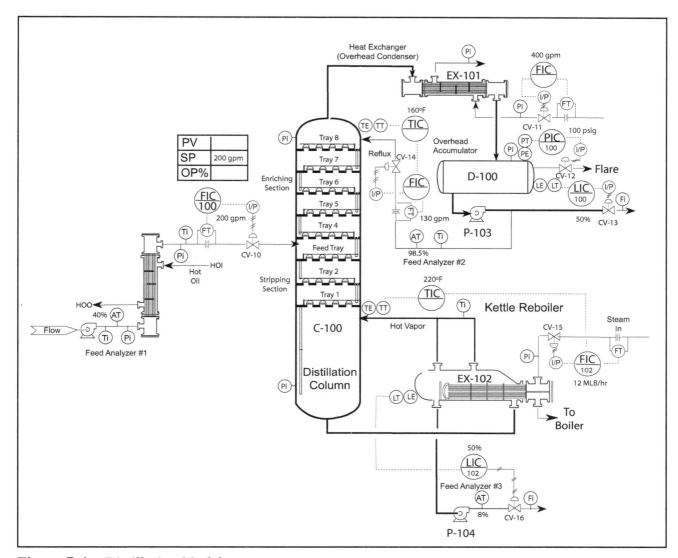

Figure 7-4 *Distillation Model*

Figure 7-5 shows how all of these control loops would be located in the system. The various types of instruments and control loops allow process technicians to operate larger and more complex processes.

Summary

The ability to clearly see the process and to sequentially break down, identify, and resolve process problems is the highest level of technician learning. Troubleshooting has traditionally been considered the area of senior technicians. However, some people believe that successful techniques can

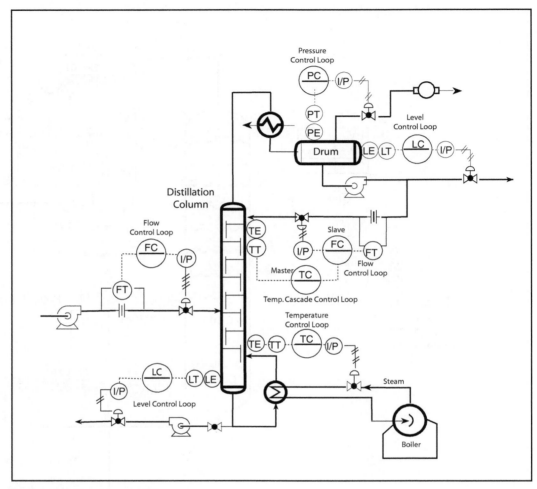

Figure 7-5 *Control Loops*

be taught to all technicians. Experience is a proven teacher; however, new computer technology can be used to quickly and methodically track down process problems. A single problem can have a cascading effect on all surrounding equipment and instrumentation.

Data collection, organization, and analysis can also be used to troubleshoot process problems. Check sheets are used to collect large quantities of data. These quantitative data can be organized into graphics or trends to plot process variation or changes. Data analysis utilizes a variety of quality techniques to put all of the parts in place.

Troubleshooting models are attached to equipment and systems presently being taught in community colleges and universities. Some of these models include the reaction model, absorption and stripping model, separation model, and distillation model. These models are completely outfitted with alarms, analyzers, interlocks, permissives, video trends, recorders, and control instrumentation. Process problems can be simulated using these models. The college curriculum includes the use of advanced computer and computer system software that closely simulates console operations. Some college training systems have modern control instrumentation mounted on operational pilot units. Students using these types of systems receive a true hands-on experience.

The ten models used to teach process troubleshooting are as follows:
1. Distillation model
2. Reaction model
3. Separation model
4. Pump and tank model
5. Compressor model
6. Heat exchanger model
7. Cooling tower model
8. Boiler model
9. Furnace model
10. Combination of any of the above models.

The ten models provide the hardware or framework from which the various troubleshooting methodologies are applied. Each model has a complete set of process control instrumentation and equipment arrangements. A complete range of troubleshooting scenarios has been developed by some educators and is typically included with these models. There are a number of troubleshooting methods that can be used with these models. Methods vary depending on individual education faculty, consultants, and industry. These methods include the following:
- Educational
- Instrumental
- Experiential
- Scientific.

Troubleshooting is a process that requires a wide array of skills and techniques. Modern control instrumentation includes indicators, alarms, transmitters, controllers, control valves, transducers, analyzers, and interlocks. The primary goal is to control variables such as temperature, pressure, flow, level, or analytical variables. It is possible to control large, complex processes from a single room.

Review Questions

1. List the various types of troubleshooting models.

2. Compare and contrast troubleshooting methods with models.

3. Describe how control loops are used in process troubleshooting.

4. Compare and contrast primary and secondary problems in an operating system.

5. How are checklists used to troubleshoot problems?

6. List the various types of instruments used to troubleshoot.

chapter 8

Pump Model

LEARNING OBJECTIVES

After studying this chapter, the student will be able to:

- Describe the various components of the pump system.
- Describe the principles of fluid flow in the pump system.
- Draw a simple block flow diagram of the pump system.
- List the safety aspects associated with the operation of the pump system.
- List the operational specifications of the pump system: pressures, flow rates, etc.
- Solve various troubleshooting scenarios associated with pump system operations.
- Identify common problems associated with pump operations.
- Operate the pump system.

Key Terms

Agitator—a set of blades attached to the end of a rotating shaft designed to mix or blend components into a homogenous mixture.

Cavitation—the formation and collapse of gas pockets around the impellers during pump operation; results from insufficient suction head (or height) at the inlet to the pump.

Centrifugal pump—a dynamic pump that accelerates fluid in a circular motion. Commonly used in automatic control with fluid flow and level control.

Discharge head—the resistance or pressure on the outlet side of a pump.

Dynamic—class of equipment such as pumps and compressors that convert kinetic energy to pressure; can be described as axial or centrifugal.

Head—described as pressure (at suction) $\times$ 2.31 $\div$ specific gravity. One psi is equal to 2.31 ft. of head.

Impeller—a device attached to the shaft of a centrifugal pump that imparts velocity and pressure to a liquid.

Mechanical seal—provides a leak-tight seal on a pump; consists of one stationary sealing element usually made of carbon, and one that rotates with the shaft.

Net positive suction head (NPSH)—the head (pressure) in feet of liquid necessary to push the required amount of liquid into the impeller of a dynamic pump without causing cavitation.

Pressure-relief valve—used to relieve excessive pressure on a pump system.

Priming the pump—a procedure used to fill the pump with liquid and remove air.

Slip—the percentage of fluid that leaks or slips past the internal clearances of a pump over a given time.

Specific gravity—the ratio of the density of a solid or a liquid to that of water or the ratio of the density of a gas to that of air.

Vapor lock—condition in which a pump loses liquid prime and the impellers rotate in vapor.

Valve lineup—a term used to describe opening and closing of a series of valves to provide fluid flow to a specific point or tank before starting a pump.

Pump start-up—includes arranging the valves so that fluid can flow to the correct place, checking the equipment, liquid levels, pump suction and discharge pressures, and other process variables associated with the system prior to starting the pump system.

Fluid flow—characterized by fluid particle movements such as laminar and turbulent. Fluids assume the shape of the container they occupy. A fluid can be classified as a liquid or a gas. When a liquid is in motion, it will remain in motion until it reaches its own level or is stopped.

Fluid pressure—the pressure exerted by a confined fluid. Fluid pressure is exerted equally and perpendicularly to all surfaces confining it.

Bernoulli's principle—states that in a closed process with a constant flow rate:
- Changes in fluid velocity (kinetic energy) decrease or increase pressure.
- Changes in kinetic energy and pressure energy correspond to pipe size changes.
- Changes in pipe diameter cause velocity changes.
- Changes in pressure energy, kinetic energy, or fluid velocity are related to pipe diameter.

Pascal's law—pressure in a fluid is transmitted equally in all directions. The basic principles are:
- Pressure in a fluid is transmitted equally in all directions.
- Molecules in liquid move freely.
- Molecules are placed close to each other in a liquid.

Introduction to Pump Operations

The pump system is designed to provide a constant **fluid flow** rate to the distillation system for the separation of butane, pentane, and liquid catalyst. Variations in the pumping process will cause other variables in the system to go into alarm. For this reason, pump-202 (P-202) has a back-up system that can be placed online quickly. The pump system takes suction from tank-202, which has a specially designed mixing system to ensure that the feedstock is blended properly. The feedstock is comprised of three components:
- liquid catalyst—designed to help separate the butane and pentane and to form a new product formed in the stirred reactor system and separated in the separator system. 15 gal. −6.6%
- butane—85 gal. −37.7%
- pentane—125 gal. −55.5%
- total feed rate—225 gpm.

Chemical plants and refineries use **centrifugal pumps** to move liquids. P-202 is classified as a **dynamic** pump that accelerates fluid through centrifugal force, a center-seeking rotation that utilizes an **impeller** and volute design. Centrifugal pumps are used widely in chemical processing plants and refineries. The primary principle used by centrifugal pumps is centrifugal force. As liquid enters the suction eye of a centrifugal pump, it encounters the spinning impeller. The liquid is propelled in a circular rotation that forces it outward and into the volute. Centrifugal force and volute design convert velocity energy to pressure. As the liquid leaves the volute, it slows down, building pressure. Some designs include devices such as diffusers that deflect and slow down the liquid, increasing pressure. Figure 8-1 shows the major components of the pump system. To operate and troubleshoot this system, a process technician needs to clearly understand the process and be able to sketch it from memory.

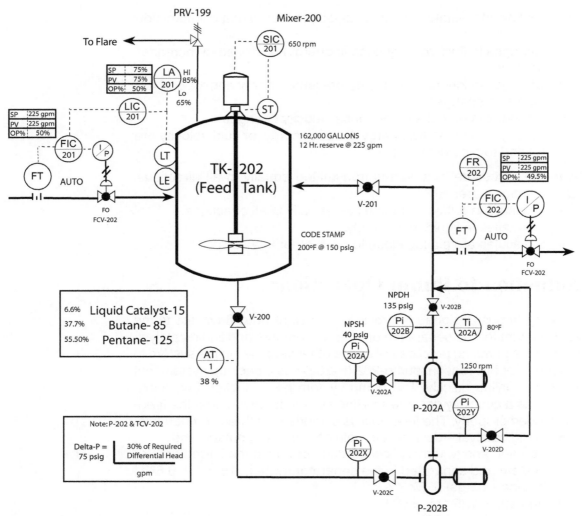

Figure 8-1 *Pump-202 System*

Flow rates are controlled at 225 gpm; however, the system has the ability to operate from 0 gpm to over 500 gpm. A relationship exists between the pump and the control valve, which is unique to modern process control. A positive displacement (PD) pump is typically not used in a flow control system without special relief valve control systems. If the control valve closes on a PD pump, internal pressure will rapidly increase and damage various components in the system. Internal **slip** in a PD pump is very low since, by design, it moves a specific amount of liquid on each rotation or stroke. Pumps are available in a variety of shapes and designs. Figure 8-2 displays the pump family tree.

Modern process control utilizes the centrifugal pump and control valve because of the principle of internal slip. Internal slip is defined as the percentage of fluid that leaks or slips past the internal clearances of a pump

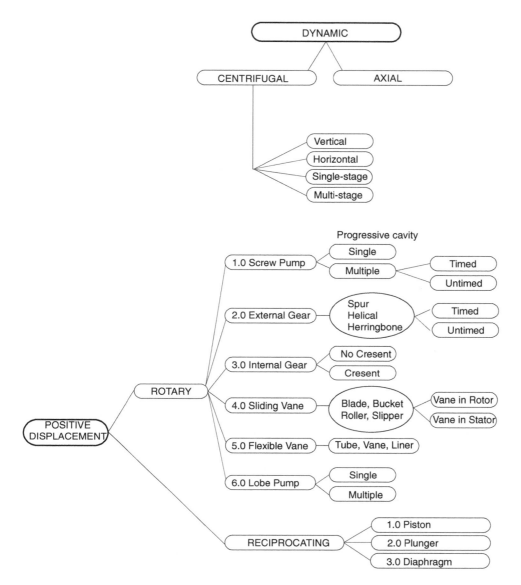

Figure 8-2
Pump Family Tree

over a given time. Slip can also be defined as the difference between how much liquid a pump can move and how much it actually does. A centrifugal pump can have as much as 100% slip when the discharge valve is closed. This principle allows a control valve to regulate or throttle fluid flow without damaging the pump or valve. Figure 8-3 shows how the liquid reacts or responds to valve stem movement on the control valve. A number of factors are linked to the operation of a centrifugal pump, which makes it a complicated operating system when all the factors are considered together. Many process technicians do not understand how these various factors impact the operation of the pump system. Some of these factors include NPSH, NPDH, viscosity, motor size, **specific gravity,** type of liquid being pumped,

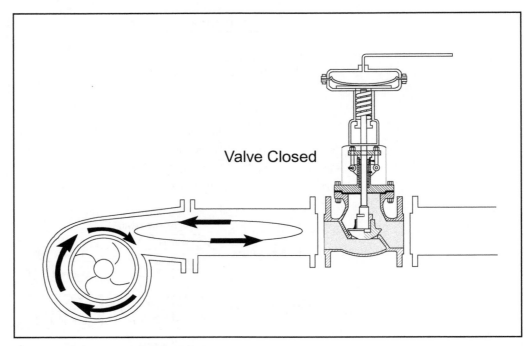

Valve Closed

Figure 8-3 *Internal Slip*

cavitation, vapor lock, air trapped in the system, rotation of the pump, and control valve operation in relation to the pump.

During slip, liquid enters the suction eye and is accelerated outward against the casing and through the volute. Initial velocity or V-1 is very high, while pressure P-1 is relatively low. As liquid accelerates through the pump, it expands through the volute into a wider opening called the discharge. At this point, velocity slows down as pressure begins to build. At the discharge point, V-2 slows as P-2 increases. This process transfers the liquid through the piping, where fluid variations called laminar and turbulent flows occur. As the control valve moves through its operating range, pressure variations move up and down. When the valve is completely closed, pressure will reach its highest point and internal slip will occur. Control valves are designed to operate through 30% of the required differential **head** that takes into account the difference between the suction and the discharge pressure.

Operating a centrifugal pump system is a dynamic process affected by ever changing variables in the suction and discharge system. Temperature variations, level changes in the tank, restrictions in equipment, and pressure increases in the distillation system will also affect the operation of the pump. Technicians need to constantly monitor the conditions of the pump system to ensure steady operations. Modern control features make this job a little easier by showing graphs and electronic video trends, which continuously monitor process variations.

Modern industrial process plants are connected by a complex network of pipes, valves, pumps, and tanks. Centrifugal and PD pumps are used to transfer fluids from place to place inside and outside the plant. The combination of pumps and pipes closely resembles the way the human heart pumps fluid into arteries and veins. Fluids assume the shape of the container they occupy. A fluid can be classified as a liquid or a gas. When a liquid is in motion, it will remain in motion until it reaches its own level or is stopped. Fluid flow is a critical process utilized in the day-to-day operation of all plants.

A Swiss scientist, Daniel Bernoulli, developed a key scientific principle for fluid flow. **Bernoulli's principle** states that in a closed process with a constant flow rate:

- Changes in fluid velocity (kinetic energy) decrease or increase pressure.
- Changes in kinetic energy and pressure energy correspond to pipe size changes.
- Changes in pipe diameter cause velocity changes.
- Changes in pressure energy, kinetic energy, or fluid velocity are related to pipe diameter.

Industry uses four different ways to express the heaviness of a fluid:

1. density
2. specific gravity
3. baume
4. API.

The density of a fluid is defined as the mass of a substance per unit volume. Density measurements are used to determine heaviness. For example, 1 gal. of:

1. water = 8.33 lb.
2. crude oil = 7.20 lb.
3. gasoline = 6.15 lb.

Another common term used by industry to describe the flow characteristics of a substance is viscosity. Viscosity is defined as the resistance of a fluid to flow. Figure 8-4 illustrates characteristics associated with the ability of a fluid to resist flow. Viscosity and density have a strong impact on fluid flow characteristics.

Specific gravity is defined as the comparison of a fluid (liquid or gas) to the density of water or air. It is a common mistake of technicians to confuse specific gravity with density. This is easy to understand because specific gravity is a method for determining the heaviness of a fluid. Density is the heaviness of a substance, while specific gravity compares this heaviness to a standard and then calculates a new ratio. Most hydrocarbons have specific gravity below 1.0.

Figure 8-4
Viscosity

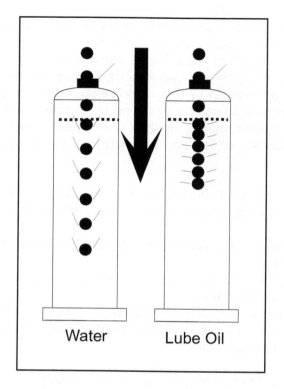

Water   Lube Oil

Key points to remember:
1. The specific gravity of water is 8.33 lb./gal. ÷ 8.33 = 1.0.
2. The specific gravity of gasoline is 6.15 lb./gal. ÷ 8.33 = 0.738.
3. The density of 1 gal. of water is 8.33 lb./gal.
4. The density of air is 0.08 lb./ft.^{3}.
5. Density is calculated by weighing unit volumes of a fluid at 60°F.

Baume gravity is the standard measurement used by industrial manufacturers to measure nonhydrocarbon heaviness. The American Petroleum Institute applies API gravity standards to measure the heaviness of a hydrocarbon. A specially designed hydrometer marked in units API is used to determine the heaviness or density of a hydrocarbon. High API readings indicate low fluid gravity. Figure 8-5 illustrates what laminar and turbulent flows look like inside a pipe.

Two major classifications of fluid flow are laminar and turbulent. Laminar flow or streamline flow moves through a system in thin cylindrical sheets of liquid flowing inside one another. This type of flow has little if any turbulence in it. Laminar flow usually exists at lower flow rates. As flows increase, the laminar flow pattern breaks into turbulent flow. Turbulent flow is the random movement or mixing of fluids. Once turbulent flow is initiated, molecular activity speeds up until the fluid is uniformly turbulent.

Turbulent flow allows molecules of fluid to mix more readily and absorb heat. Laminar flow promotes the development of static film, which acts as an insulator. Turbulent flow decreases the thickness of static film.

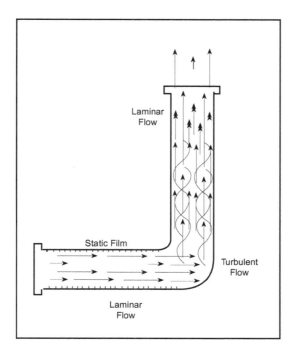

Figure 8-5
Laminar and Turbulent Flow

Fluid Flow Measurement

Gases have flow characteristics similar to liquids; they mix easily and are affected by temperature, pressure, and turbulent and streamlined flow patterns. Higher temperature causes increase of molecular motion in liquids, gases, or vapors, and the motion speeds up. Pressure increases cause molecules to move closer and shift the boiling point upwards. The effect is more drastic in gases and vapors than in liquids.

Fluids flow through a series of pipes, valves, pumps, and vessels. Knowing and controlling the flow rate of a particular process stream is critical to the operation of an operating unit. Continuous chemical reactions require precise measurements to ensure that all of the reactants (raw materials) are combined in the proper proportions to form the final products. Feed rates and product rates must be accurately controlled for economic reasons.

Principles of Liquid Pressure

- Liquid pressure is directly proportional to its density.
- Liquid pressure is proportional to the height of the liquid.
- Liquid pressure is exerted in a perpendicular direction on the walls of a vessel.
- Liquid pressure is exerted equally in all directions.
- Liquid pressure at the base of a tank is not affected by the size or the shape of the tank.
- Liquid pressure transmits applied force equally, without loss, inside an enclosed container.

Figure 8-6

psia-psig-psiv-psid

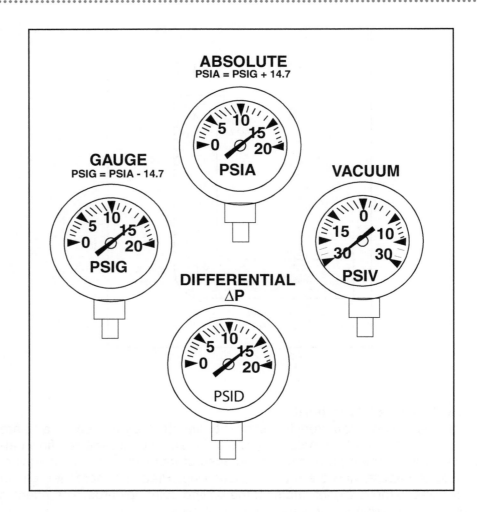

Three different types of pressure gauges can be found in industrial environments: (1) absolute (psia), (2) gauge (psig), and (3) vacuum (psiv). Absolute pressure is equal to gauge pressure plus local atmospheric pressure of 14.7. Gauge pressure is equal to the absolute pressure minus the local atmospheric pressure of 14.7. Vacuum is typically measured in inches of mercury. Any pressure below atmospheric pressure 14.7 is referred to as vacuum. Figure 8-6 illustrates the four basic types of pressure gauges found in the chemical processing industry.

Equipment Descriptions—Block Flow Diagrams (Safety)

Tank-202 Feed Tank

The feed tank is equipped with a variable speed **agitator** and motor that is designed to blend butane, pentane, and liquid catalyst before they are heated and sent to the distillation column for separation. The tank has a total

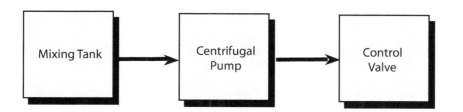

Figure 8-7
Block Flow Diagram of the Pump System

capacity of 162,000 gal. or 12 hours' reserve. Mixer-200 is typically run at 650 rpm during normal operation. A safety relief system is set at 150 psig and is designed to prevent any damage to the vessel. The relief system is connected to the flare header and only opens during emergency situations. Resonance time for the feed inside the tank is set at 30 min. and is designed to ensure adequate blending. If Mixer-200 fails, or if the agitation falls below design specifications, inadequate blending will occur. This will result in increased pressure and a lower reading on the butane analyzer. The increased pressure will cause the safety relief valve to lift and release the flare header. An analyzer is connected to the discharge outlet of TK-202 and is designed specifically to measure the amount of butane and liquid catalyst in the system. Block flow diagrams are often used to simplify the process. Figure 8-7 shows the basic components of the pump system.

Analyzer-1 A butane analyzer is located on the discharge line from TK-202 and is specifically designed to measure butane and liquid catalyst concentration. Only the instrument department is qualified to troubleshoot and calibrate the analyzer. Analyzer Transmitter-1 is designed to measure the concentration of butane in the feed stream. Butane concentrations typically range from 36% to 38%.

Mixer-200 Mixer-200 is a variable speed agitator and motor designed to ensure complete mixing of the butane, pentane, and liquid catalyst. It is designed to operate at 650 rpm. The agitator has one lower set of blades located at the end of the rotating shaft. A **mechanical seal** and bearings are located where the shaft first penetrates the shell of the tank. The tank is composed of stainless steel and provides protection from the elements while having an operating temperature of 200°F at 150 psig.

Pump(s)-202A/B

P-202 is a horizontally mounted centrifugal pump designed to operate at 225 gpm. Suction pressure is maintained at 40 psig, and the discharge pressure typically runs at 135 psig. The differential pressure across the pump is 75 psig. This differential is necessary for the system to operate properly. The pressure is obtained as the control valve moves through 30% of its operating range on the discharge side of the pump. A back-up pump is provided to ensure continuous flow to the system. P-202 is designed to operate continuously twenty-four hours a day, seven days a week.

Figure 8-8
*Speed-to-Torque
Conversion*

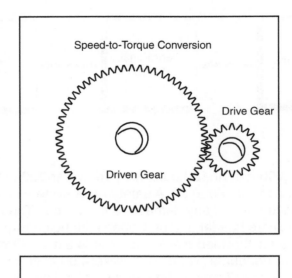

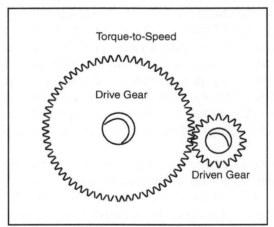

Gear Boxes & Power Transmission

Every six weeks the pumps are alternated for preventive maintenance. The motors on P-202A/B are not variable in speed and are designed to operate at full capacity, 1250 rpm. A protective cover is in place over the rotating parts of the pump. The motor is connected to a flexible coupling, which is connected to a gear box where a speed-to-torque conversion occurs. Gear boxes are common devices used by the chemical processing industry. These devices are often part of the pump and provide the connection with the coupling and motor or driver. Figure 8-8 illustrates the two basic features of this device: speed to torque and torque to speed.

Flow-Control Loop 202   The flow-control loop will be covered in more detail in the instrument section; however, it is an integral part of the pump system and must be included in the equipment descriptions. Flow-control valve 202 works in harmony with P-202A/B.

Instrumentation Systems

Temperature
Ti-202A (temperature indicator)—design specifications typically run around 80°F.

Flow
FIC-201 (flow-indicating controller)—located on the inlet side of TK-202 and is designed to control the flow into the system. Due to variations in the supply side, these rates can vary from 225 to 150 gpm. Because of these variations, a 50% level is maintained in TK-202.

FIC-202 (flow-indicating controller)—the primary flow-control system to C-202, which is designed to control flow at 225 gpm.

FR-202 (flow recorder)—provides an electronic video trend of flow to C-202.

Level
LIC-201 (level-indicating controller)—the master controller on TK-202, cascaded to FIC-201. It is designed to control the level on TK-202 at 75%.

LA-201 (level alarm hi/lo)—alerts operational staff to problems associated with the level in tank-202.

Speed
SIC-201 (speed-indicating controller)—a variable speed controller designed to regulate the rotations per minute of the agitator. Mixer-200 rotates at 650 rpm.

Analyzer
AT-1 (analyzer transmitter 1)—one of four transmitters located on the various systems. It is specifically designed to measure the concentration of butane flowing out of TK-202. Samples sent to the lab indicate concentration of liquid catalyst and pentane.

Pressure
Pi-202A—pressure indicator located on the suction side of P-202A.

Pi-202B—pressure indicator located on the discharge side of P-202A. The differential across P-202 is 75 psig and is used in engineering calculations for flow rate.

Pi-202C—pressure indicator located on the suction side of P-202B.

Pi-202D—pressure indicator located on the discharge side of P-202B.

PRV-199 (**pressure-relief valve**)—is designed to open at 150 psig to the flare header and reset once system pressure drops below set point.

The Pump System

The basic components of a centrifugal P-202 include the suction and discharge, impeller and rotating shaft, casing and volute, wear rings, seals and bearings, pressure indicators, and a start-stop switch. The outer casing is designed to enhance fluid flow. It contains the fluid, forms an ever-increasing cavity called the volute, surrounds the pumping chamber that contains the impeller, and is the largest part of the pump. The inlet to the pump is often referred to as the suction eye. The spinning impeller and stationary suction eye form a seamless part of the pump. A set of wear rings (seals) keeps high-pressure liquid in the pump from returning back into the low-pressure suction line. Air is vented from the pumping chamber through a high-point bleed valve. A bleed valve on the suction line allows liquid to be drained from the system. Liquid enters the center of the impeller and is spun toward the outside of the volute. The inlet line is designed to run when primed (full of liquid). The outlet line receives the discharge from the pump volute. Because the discharge line is wider than the inlet volute, fluid velocity slows, creating pressure. The impeller is a circular device attached to the shaft on P-202 with a closed design. The closed design works well with the pentane, butane, and liquid catalyst feed and can generate higher pressure than semi-open or open impeller designs. The impeller resembles a circular device that can be connected to the rotating shaft. The closed-impeller design has two circular plates that sandwich a series of vanes that curve toward the left. This curved-vane impeller design was invented by the Englishman John Appold in 1851. It was a significant improvement over the 1600s straight-vane design of French inventor Denis Papin. The curved-vane design significantly enhanced the liquid acceleration aspects related to centrifugal force.

The pump system includes a feed tank, piping and valves, the pump, and control valve. Process technicians view the pump from a split approach: the suction side and the discharge side. Process variables constantly affect each side and cause different things to occur. One of the most important parts in the P-202 system is the motor. The motor is classified as a driver that has a fixed coupling attached to the rotating shaft of the pump. The motor on P-202 is designed to operate at a fixed speed measured in rpm. Figure 8-9 shows the basic components of a centrifugal pump, including the gear box. Figures 8-9B, C, and D show two different impeller designs: open and closed. Refer to the equipment text for information about the semi-open impeller design. It is important to note that the impeller is connected to the rotating shaft.

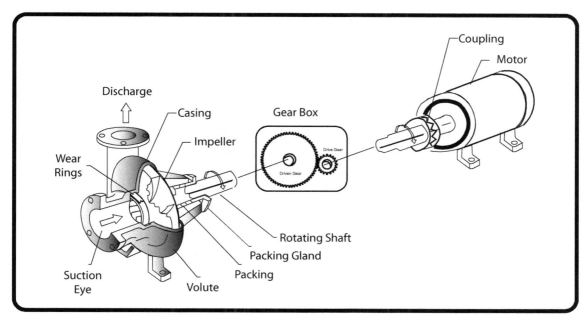

Figure 8-9A   *Basic Components of a Centrifugal Pump*

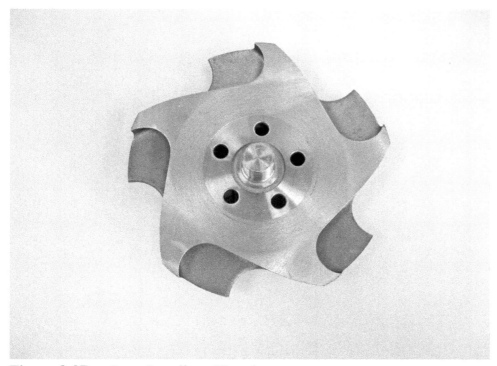

Figure 8-9B   *Open Impeller—View 1*

Figure 8-9C
*Open Impeller—
View 2*

Figure 8-9D
*Closed-Impeller
Design*

Operational Specifications (SOP, SPEC Sheet, and Checklist)

The operating procedures associated with the pump system include the following:

Action	Notes
1. Establish 75% level on TK-202 and set LIC-201 to AUTO. 2. Set FIC-201 at 225 gpm and cascade to LIC-201. 3. Lineup P-202A, open V-200, V-202A, V-202B, V-201. 4. Ensure P-202B is in the closed position and V-202C and V-202D are closed. 5. Ensure FIC-202 is set at zero and closed. 6. Start P-202A and monitor Pi-202A (40 psig) and Pi-202B (135 psig). Establish flow. 7. Set SIC-201 to 650 rpm and put in AUTO and blend for 30 min. 8. Sample TK-202 and send to lab. 9. Verify AT-1 is reading 36–38%. 10. Set FIC-202 to 225 gpm and put in AUTO. 11. Monitor system pressures. 12. Cross-check process variables with SPEC sheet.	

Specification Sheet and Checklist

Level

1.	LIC-201	75%	AUTO	Feed tank level
2.	LA-201	85%	High	Feed tank level high
3.	LA-201	65%	Low	Feed tank level low

Flow

4.	FIC-201	225 gpm	CASC	Feed flow
5.	FIC-202	225 gpm	AUTO	Feed flow to C-202
6.	FR-202	—	—	Feed flow to C-202

Analytical

7.	AT-1	38%	—	Butane-liquid catalyst
8.	SIC-201	650 rpm	AUTO	Mixer-200

Pressure

9.	Pi-202A	40 psig	—	Suction-P-202A
10.	Pi-202B	135 psig	—	Discharge-P-202A
11.	Pi-202C	40 psig	—	Suction-P-202B
12.	Pi-202D	135 psig	—	Discharge-P-202B
13.	PRV-199	150 psig	—	Tank-202

Temperature

14.	Ti-202A	125°F	AUTO	Hot-water return

Common Pump Problems and Solutions

During operation, a single feed pump is used to transfer feed while the other is used as a back-up. As long as the procedure is followed correctly, the pumps can be switched without causing serious fluctuations in the system pressure and flow. As with most industrial equipment, the pumps will occasionally fail due to bearing or seal damage, motor failure, coupling failure, power outage, impeller problems, or a variety of other problems. For these reasons, process technicians frequently inspect and monitor the equipment. When the pumps are switched, maintenance performs a more detailed inspection and repair if necessary.

The most common problems associated with pump operation include improper lineup and lack of understanding of how a centrifugal pump operates. Centrifugal pumps are used frequently in automatic process control. This means that a throttling type valve is installed in the system and working in conjunction with the pump. Centrifugal pumps require a certain amount of pressure on the suction side to push the liquid into the suction eye. The natural drawing action of the impeller creates a low-pressure phenomenon inside the impeller. Problems associated with centrifugal pumps include:

- cavitation
- vapor lock
- improper lineup
- high discharge pressure variations or NPDH
- variations in suction pressure or NPSH
- feed composition changes
- gear box problems
- seals and bearings problem
- broken suction and discharge gauges
- breaker trips on motor
- motor problems
- gasket leaks
- seal flush tubing plug ups.

Cavitation is defined as the formation and collapse of air pockets inside the pumping chamber. It can also be described as boiling, a process which can be very violent, with rapid pressure increases and decreases. Cavitation can damage the impeller, shaft, casing, or wear rings. This phenomenon can break the pump loose from the piping or foundation and sounds like marbles being agitated in a large blender.

Cavitation can be prevented by simply increasing the NPSH or pinching down on the discharge valve. It appears to be caused when the pump outruns the liquid entering the suction eye, forming a serious vacuum and reducing the boiling point of the liquid to a point where it violently expands and then collapses as the pressure builds.

Vapor lock occurs when the discharge pressure exceeds the ability of the pump to transfer liquid. When a pump is in vapor lock, it sounds very quiet since it is not transferring liquid and the impellers are spinning in air. Suction and discharge gauges appear to be working; however, liquid is not being transferred. Operating a centrifugal pump is not as easy as simply turning a switch on or off and letting it run. Sometimes a technician needs to carefully consider all the variables in a system. This includes monitoring both suction and discharge gauges and ensuring liquid is being transferred to the destination tank. Vapor lock can be fixed by shutting down the pump, bleeding off the trapped air, and decreasing the pressure in the destination tank.

Improper lineup is a serious problem that involves mistakes made in opening and closing valves from source to destination tanks. New technicians frequently make errors that contaminate product or put the product in the wrong place. Adequate training is critical during this process, and new technicians should carefully walk the line and inspect each lineup to ensure that it has been made correctly. Errors in lineup can cost the company money and may result in the loss of employment for the new technician.

Troubleshooting Scenario 1

Figure 8-10 displays the pump system and presents a series of what-if-scenarios. Refer to the figure and select the best answers.

Troubleshooting Scenario 2

Figure 8-11 displays the pump system and presents a series of what-if-scenarios. Refer to the figure and select the best answers.

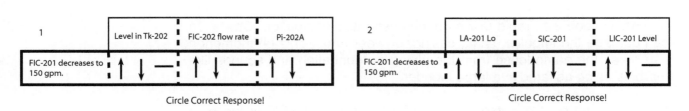

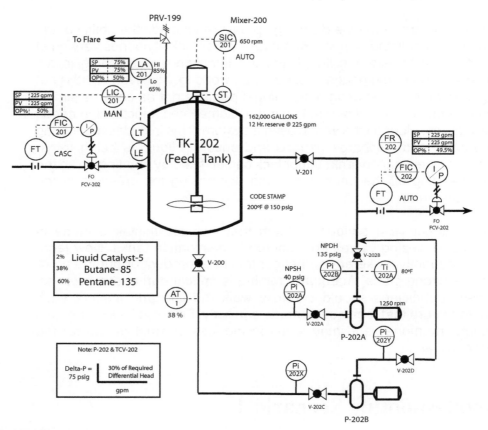

Figure 8-10 *Troubleshooting Scenario #1*

Troubleshooting Scenario 3

Figure 8-12 displays the pump system and presents a series of what-if-scenarios. These scenarios are designed to give first-hand experience in the operation of the pump system with a large safety net and an instructor to ensure correct instruction and a variety of opinions. The opinions expressed in this text present a simplistic approach. Refer to the figure and select the best answers.

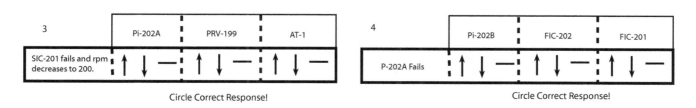

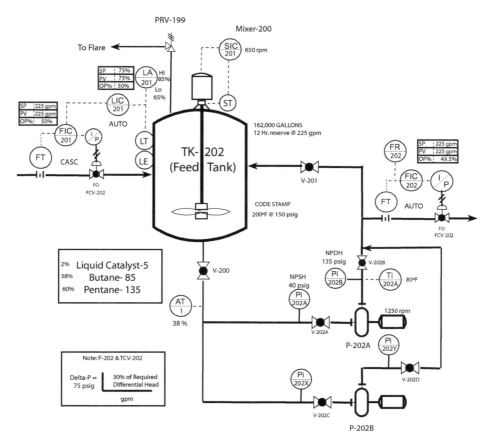

Figure 8-11   *Troubleshooting Scenario #2*

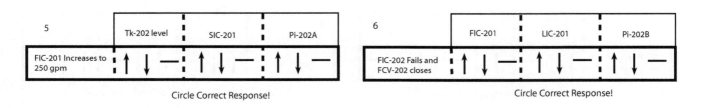

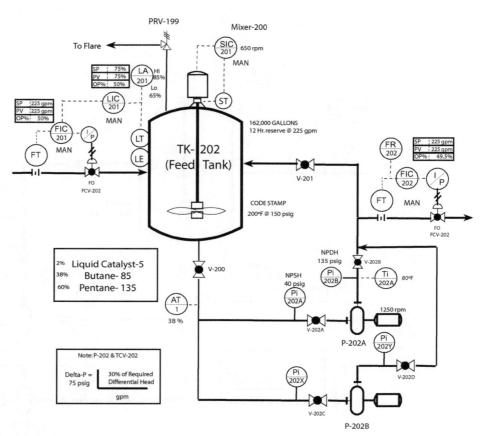

Figure 8-12 *Troubleshooting Scenario #3*

Summary

Operating the pump system requires a technician to study and understand the basic components and the standard operating procedures associated with the pump system. Preparing for a career in the chemical processing industry requires students to be familiar with problems typically encountered on a pump system. Feed is controlled at 225 gpm in TK-202, which has a total capacity of 162,000 gal.; however, the system is typically maintained at 75%, with a low alarm set at 65% and a high alarm set at 85%. Mixer-200 is mounted on the top of TK-202 and blends 38% butane, 60% pentane, and 2% liquid catalyst. The mixer rotates at 650 rpm. If agitation is lost, pressure will rapidly build in the tank, causing the butane concentration to be low. An analyzer is located at the base of the tank and measures the butane-liquid catalyst concentration. A centrifugal pump, P-202, has a back-up or redundant pump to ensure continuous flow to the distillation system. At start-up, P-202 is lined up to TK-202, which allows recirculation to occur. TK-202 is a stainless steel tank designed to handle pressures in excess of 150 psig at 200°F. P-202 is driven by a motor that operates at 1250 rpm. Suction pressure is maintained at 40 psig on the inlet side or suction eye, and discharge pressures are measured at 135 psig when the system is lined up to the distillation system. Flow-indicating controller 202 regulates flow at 225 gpm, and a flow recorder electronically collects and displays a video trend of flow to the system. A pressure-relief valve is located at the top of the tank and is set at 150 psig.

Review Questions

1. Describe how NPSH and NPDH affect the operation of pump-202.

2. Describe the basic components of pump-202.

3. Explain the basic steps required to start up P-202.

4. Explain how a centrifugal pump builds pressure for liquid transfer.

5. Explain how changes in NPSH affect the operation of P-202.

6. Contrast NPDH and NPSH and describe how each one affects the operation of the pump system.

7. Calculate how many gallons would be left in tank-202 if FIC-201 failed completely.

8. Describe what happens when agitation in tank-202 decreases or stops.

9. Explain how leaving a controller in manual affects the way it responds to process changes.

10. Calculate how much feed passes through TK-202 during a normal 24-hour period.

chapter 9

Compressor Model

LEARNING OBJECTIVES

After studying this chapter, the student will be able to:

- Explain the principles of compression in a multistage centrifugal compressor.
- Identify and describe the basic components of a compressor system.
- Draw a simple block flow diagram of a compressor system.
- Describe the various safety aspects associated with working on compressed air systems.
- Describe how instrument air is used in process control systems.
- Identify common problems associated with compressor operations.
- Operate the compressor system.
- Solve various troubleshooting scenarios associated with compressed air systems.

Key Terms

Centrifugal compressor—uses centrifugal force to accelerate gas and convert energy to pressure.

Compression ratio—the ratio of discharge pressure (psia) to suction pressure (psia). Multistage compressors use a compression ratio in the 3–4 range, with the same approximate compression ratio in each stage. For example, if the desired pressure is 1500 psia, a four-stage compressor with 3.2 compression in each stage might be used. The pressure at the discharge of each stage would be: 1st = 47 psia, 2nd = 150 psia, 3rd = 480 psia, 4th = 1536 psia.

Desiccant dryer—used to remove moisture from compressor gases as they are passed over a chemical desiccant, which absorbs water.

Receiver—a compressed-gas storage tank that is a key part of the compressor system. The receiver provides the volume required for compressed systems to discharge into and operate.

Impeller design—a compressor has a variety of impeller designs, including open backward-bladed impeller, open radial-bladed impeller, and closed backward-bladed impeller.

Suction vane tip—a part of the compressor impeller vane that first comes into contact with air.

Discharge vane—a part of the compressor impeller vane that comes into contact with air as it exits the impeller.

Pressure control—a variety of different methods are used to control the pressure on industrial systems. Some smaller systems include a pressure regulator on the receiver. Other systems utilize a pressure control loop, which measures the pressure at the industrial air header and adjusts an automatic valve that allows air from the discharge side to flow into the suction line.

Introduction to Compressor Operations (Complete System)

The compression of gases and vapors in the process industry is very important, particularly in systems used for modern process control. Compressed air systems provide the medium used to open or close pneumatically operated valves. Compressor-100 is a multistage centrifugal device designed to provide clean, dry air for instruments and control devices. Operating and maintaining this system is an important part of a technician's job. In areas where high humidity exists the compression process produces liquid that accumulates in the **receiver** and needs to be drained off periodically. Figure 9-1 shows a simple air system that has been incorporated into the operation of the multivariable plant.

The basic components of compressor-100 include the compressor, receiver, dryer, instruments, piping, valves, and a **pressure control** system. The instrument air header provides clean, dry instrument air to the entire plant. The instrument air header operates at 100 psig and feeds a series of pressure reducing valves, which are located in key areas where the pressure needs to be reduced for operating purposes, typically 20 psig. The compressed air system takes suction

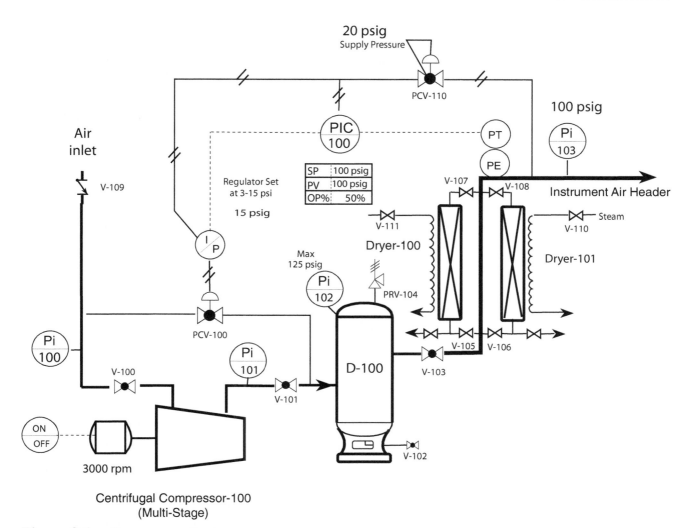

Figure 9-1   *Compressor System*

from a high point located inside the plant. The air is compressed as it enters D-100. A pressure relief valve is located on top of the vessel and vents to the atmosphere when the pressure exceeds 125 psig. Liquid is bled off the bottom of D-100 once each shift. After the air is compressed, it is sent to a series of dryers filled with desiccant. The desiccant removes moisture from the compressed air. Only one dryer at a time is ever placed in service. Dryer-101 is equipped with a copper coil steam tracing designed to heat the dryer, vaporize the moisture, and dry out the desiccant so that it can be reused. PIC-100 controls the pressure on the system by venting excess pressure back into the suction line. PCV-110 is a back-pressure regulator designed to reduce the pressure of 100 psig found at the instrument air header to a usable 20 psig for 3–15 psig valve actuator operation. Figure 9-2 is a simplified block flow diagram of the air system.

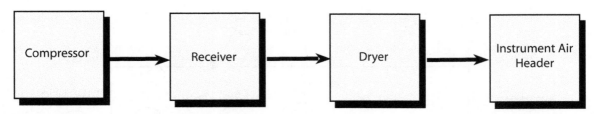

Figure 9-2 *Compressor Block Flow Diagram*

The principles of compression include the following:
- Compression decreases volume.
- Gases and vapors are compressible.
- Liquids and solids are typically considered not to be compressible.
- Compression generates heat.
- Compression moves gas molecules close together.
- Compressed gases will resume their original shape when the pressure is released.
- Compressed gases produce or generate heat because of molecular friction.
- The smaller the volume the higher the pressure.
- Force ÷ area = pressure.
- Gas volume varies with temperature and pressure.

A **centrifugal compressor** is very similar to a centrifugal pump; however, the blade design is uniquely different, as illustrated in Figure 9-3. The material is thinner and designed for operation with gases and vapors.

Pressure is an important variable that must be carefully monitored and controlled in an industrial environment. Pressure is often defined as the force per unit area or, in other words, the amount of pressure exerted by a fluid on the equipment in which it is contained. In physics, the term pressure usually is applied to a fluid (gas or liquid). Pressure is measured in pounds per square inch (psi) in the English system and kilograms per square centimeter or Pascals in the metric system. Increases in liquid pressure are immediately distributed throughout the fluid medium in a closed system. This process illustrates the noncompressible nature of a liquid.

Two of the most common types of pressure are atmospheric and hydrostatic. Atmospheric pressure is the force exerted on the earth by the weight of the gases that surround it. At sea level, atmospheric pressure is about 14.7 psi (1.3 kPa). This pressure decreases with altitude because of the reduced height (weight) of gas.

Hydrostatic pressure is the pressure exerted on a contained liquid and is determined by the depth of the liquid. Even a novice swimmer is familiar with pressure differences between the surface of water and bottom of the pool. This pressure difference is what causes your ears to "pop" as you swim to the bottom of a 10-ft. swimming pool (hydrostatic) or drive over a high mountain range in Colorado (atmospheric). Figure 9-4 shows how

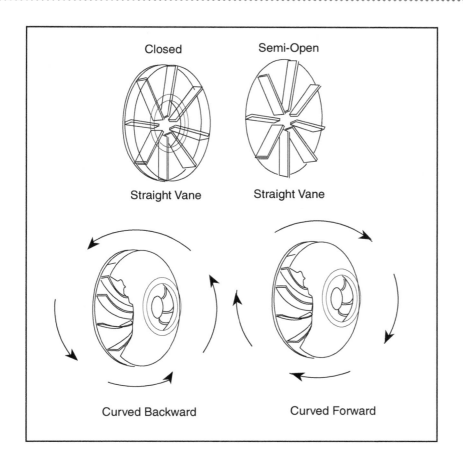

Figure 9-3
Impeller Design for Compressor-100

Closed

Semi-Open

Straight Vane

Straight Vane

Curved Backward

Curved Forward

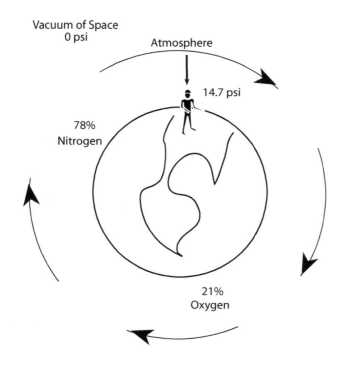

Figure 9-4
Pressure Effects

Vacuum of Space
0 psi

Atmosphere

14.7 psi

78%
Nitrogen

21%
Oxygen

pressure varies with height. The fluid that surrounds the earth is composed primarily of 78% nitrogen and 21% oxygen. This fluid has weight that can be expressed as 14.7 psi at sea level. A perfect vacuum exists in space. The pressure on the top of Mt. Everest is less than that at sea level.

When a fluid (gas or liquid) is at rest, pressure is transmitted equally to all parts. Fluids act this way because the molecules move about freely. The distance between the molecules depends on whether they are in a solid, liquid, or gaseous state. Molecules in gases are much further apart than they are in solids. A French scientist named Blaise Pascal discovered that pressure in fluids is transmitted equally to all distances and in all directions. From this discovery, Pascal formulated laws that describe the effects of pressure within a liquid. These laws have many practical applications in hydraulics and presses.

An Irish scientist named Robert Boyle developed laws that describe how the volume of a gas changes when the pressure changes. The higher the gas pressure the closer the gas molecules are and the smaller the volume they occupy. Under ordinary conditions, gas volumes decrease by half when the pressure doubles. Liquids and solids also respond to pressure increases but in much smaller proportions than gases. Liquids and solids are generally considered incompressible.

Pressure is defined as force or weight per unit area (force ÷ area = pressure). The term "pressure" is typically applied to gases or liquids. Pressure is measured in psi. Atmospheric pressure is produced by the weight of the atmosphere as it presses down on an object resting on the surface of the earth. Pressure is directly proportional to height; the higher the atmosphere, gas or liquid, the greater the pressure. At sea level, atmospheric pressure equals 14.7 psi. The boiling point of a substance is (1) the temperature at which the vapor pressure exceeds atmospheric pressure, (2) bubbles become visible in the liquid, and (3) vaporization begins. Vapor pressure is the weight of a liquid's vapor. Molecular motion in water vapor:
- produces pressure
- increases as temperature is added to the liquid.

The vapor pressure of a substance can be directly linked to the strength of the molecular bonds of a substance. The stronger the bonds or molecular attraction the lower the vapor pressure. If a substance has a low vapor pressure, it will have a high boiling point. For example, gold changes from a solid to a liquid state at 1064°C and will boil when the temperature reaches 2807°C. Water changes from a solid to a liquid state at 0°C and will boil when the temperature reaches 100°C.

Liquids do not need to reach their boiling points in order to begin the process of evaporation. For example, a pan of water placed outside on a hot summer day (98°F) will evaporate over time. The sun will increase the molecular activity of the water vapor, and some of the molecules will escape into the atmosphere. Wind currents will enhance the process of evaporation by sweeping away water molecules that will be replaced by other water molecules. Pressure directly impacts the boiling point of a substance. As the pressure increases:

- The boiling point increases.
- The escape of molecules from the surface of the liquid is reduced proportionally.
- The gas or vapor molecules are forced closer together.
- The vapor phase above a liquid could be forced back into solution.

This is an important fact for a process technician to understand. A change in pressure will shift the boiling points of raw materials and products. Pressure problems are common in industrial manufacturing environments and must be controlled. Atmospheric pressure is 14.7 psi, and any pressure below this is referred to as a vacuum. Vacuum impacts the boiling point of a substance in a manner opposite that of positive pressure. Vacuum systems:

- lower the boiling point of a substance
- enhance molecular escape of liquid
- reduce energy costs
- reduce molecular damage by overheating
- reduce equipment damage.

Blaise Pascal was a French scientist who discovered that pressure in a fluid is transmitted equally in all directions. Pascal successfully described the effects of pressure in a liquid and established the scientific foundation for hydraulics. Key facts for process technicians:

- Pressure in a fluid is transmitted equally in all directions.
- Molecules in liquids move freely.
- Molecules are close together in a liquid.

Robert Boyle was an Irish scientist who developed the law that describes how the volume of a gas responds to pressure changes. Key facts for process technicians:

- Pressure decreases volume and moves gas molecules closer together.
- The higher the pressure the smaller the volume.
- Gas volume decreases by half when pressure doubles.

The pressure a liquid exerts on a container is determined by the height and the weight of the fluid. The pressure exerted by a 20-ft. column of mercury would be more than that of a 20-ft. column of water. The specific gravity of mercury (Hg) is much higher than water. Liquids are typically considered to be noncompressible even though a 10% decrease in volume can be observed when a pressure of 65,000 psi is applied. Gases behave much differently than liquids. Gases are very compressible. The volume of a gas is determined by the shape of the vessel containing it, temperature, and pressure. Operators utilize these three factors in the control and storage of gases.

Dalton's law states that the total pressure of a gas mixture is the sum of the pressures of the individual gases: $P_{total} = P_1 + P_2 + P_3. \ldots$

The ideal gas law calculates the pressure, temperature, volume, or moles of any ideal gas: $PV = nRT$.

P = pressure of the gas

V = volume

n = moles of gas

T = temperature in Kelvin

R = ideal gas constant, 0.08206 L atm/mol K

The combined gas law

$$\frac{P_1 V_1}{T_1} = \frac{P_2 V_2}{T_2}$$

calculates changes in a gaseous substance from one condition to another.

Equipment Descriptions

Air-Compressor
Compressor-100 is a horizontally mounted, multistage compressor. It is considered small by industrial standards. It has an operating pressure of 100 psig, although the system can achieve much higher pressures. Centrifugal compressors are considered invaluable in the operation of the chemical processing industry. Compressor-100 was selected because of its low initial installation cost, low operation and maintenance cost, simple new piping installations, interchangeable drivers, large volume capacity per unit plot area, and long service life. Compressor-100 is designed to deliver much higher flow rates than positive displacement compressors.

Drum
D-100 is an integral part of the compressor system, providing the volume required for efficient operation. Maximum operating pressures have been fixed at 125 psig; however, the drum is rated for pressures exceeding 500 psig. A pressure relief valve 104 is located at the top of the drum and is designed to vent safely to the atmosphere. Pressure indicator 102 provides a visual indication to process technicians of system pressure. A low point bleeder valve 102 is located at the base of the vertically mounted drum.

Dryer
Dryers 100 and 101 are filled with desiccant and are designed to remove excess moisture from the system that can be harmful to the process instruments used in the chemical processing industry.

Instrument Systems

Temperature
Ti-100 temperature indicator is located on the discharge side of compressor-100. Temperature range is 90–110°F.

Pressure
PIC-100 pressure indicating controller is used to control system pressure at 100 psig.

Pi-100 pressure indicator is located on the suction side of the compressor. During compressor operation, pressure readings slightly below atmospheric may be achieved.

Pi-101 pressure indicator is located on the discharge line between compressor-100 and D-100. Pressure readings are typically 105–110 psig.

Pi-102 pressure indicator is located on the top of D-100. Pressure range is 95–105 psig.

Pi-103 pressure indicator is located on the instrument air header. Pressure range is 95–105 psig.

The Compressor System

During operation, air enters the compressor at the suction inlet and is accelerated by moving impellers. Compressor-100 has one moving element (the drive shaft) and an impeller. In the compressor, the impeller discharges into a circular, narrow chamber called the diffuser. This narrow opening completely surrounds the impellers. As back-pressure builds in the compression chamber, gas velocity is accelerated through the diffuser assembly and into a circular volute. As high-velocity gas moves through the diffuser and into the volute, kinetic energy is converted into pressure as gas speed slows in the ever-widening volute before exiting the discharge port.

Because compressor performance is linked to the compressibility of the gas or air it is moving, centrifugal compressors are more sensitive to density characteristics than positive displacement compressors. Compressor-100 is a multistage device that compresses the air in the first stage impeller and then discharges it into the second stage impeller. System pressure is controlled around 100 psig in D-100. Figure 9-5 shows the basic components of a multistage centrifugal compressor.

Compression ratio is defined as the ratio of discharge pressure (psia) to suction pressure (psia). Frequently, the desired discharge pressure is very high, over 100 times the inlet pressure. When a gas is compressed, the temperature of the gas increases. If a gas were compressed in one stage to a pressure 100 times that of the inlet pressure, the gas temperature would be

Figure 9-5
Multistage Centrifugal Compressor

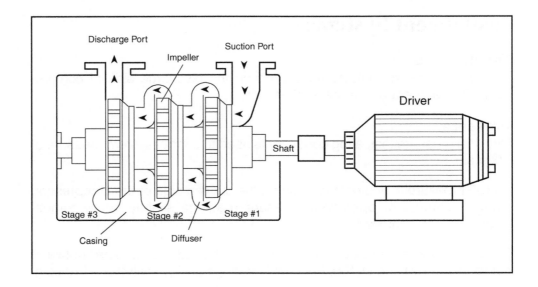

extremely high. Multistage compressors, with cooling between stages, are used to develop high pressures to allow for the heat of compression. The compression ratio normally runs in the 3–4 range, with the same approximate compression ratio in each stage. For example, if the desired discharge pressure is 1500 psia, a four-stage compressor with 3.2 compression in each stage might be used. The pressure at the discharge of each stage would be: 1st = 47 psia, 2nd = 150 psia, 3rd = 480 psia, 4th = 1536 psia.

The part of the impeller vane that comes into contact with the air first is called the **suction vane tip.** The part of the impeller vane that comes into contact with the air at the end is called the **discharge vane.** The driver is an electric motor that operates at 3000 rpm. Compressor-100 utilizes a closed backward bladed impeller.

As the desiccant in the dryers becomes saturated with water, pressure increases across the body of the dryer. Dryers can be switched during operation by opening V-105 and V-107 and allowing the pressure to move across the body of the dryer. When this is complete, V-108 and V-106 are closed, isolating the dryer. A continuous set of copper coils is wrapped around the shell of the dryer. Low-pressure steam is slowly admitted into the copper coils by opening V-110 for dryer-101 or V-111 on dryer-100. During this operation, the lower leg bleeder valve is opened, V-110 or V-111. This allows the steam to heat up the desiccant while nitrogen slowly flows from top to bottom, encouraging the liquid to drain out the bottom of the dryer. This process takes about 12 hours before the desiccant is regenerated and ready to be placed by in service.

The compressor system is a vital part of modern process control. Compressors come in two basic designs, dynamic and positive displacement. Figure 9-5 illustrates the various members of the compressor family. Minor problems are occasionally experienced with compressor systems. These problems are usually the result of dirt, adjustment problems, liquid in the receiver, or inexperience in operating the system. A number of safety issues should be addressed prior to operating a compressor system. Some of these safety issues include:

- noise hazards
- high-pressure hoses blowing loose
- hazards associated with compressed gas systems
- hazards associated with rotating equipment
- mixing air and hydrocarbons into flammable or explosive concentrations
- avoiding high-pressure releases: eyes, ears, nose, skin.

Experienced technicians can quickly fix compressor problems by making proper adjustments, cleaning the equipment, ensuring lubrication or auxiliary systems are maintained according to guidelines, temperatures, and pressures. Figure 9-6 shows the compressor family tree, which includes

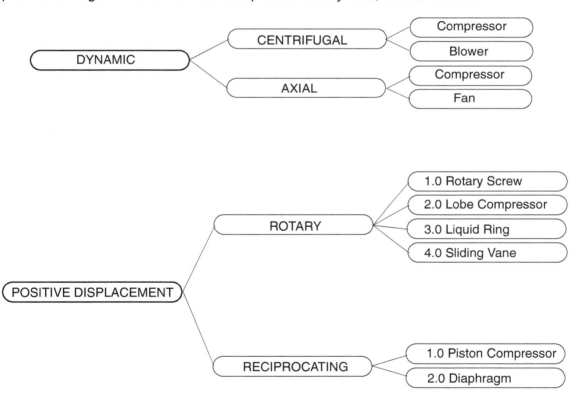

Figure 9-6 *Compressor Family Tree*

rotary and reciprocating positive displacement compressors and the dynamic family.

Operational Specifications (SOP, SPEC Sheet, and Checklist)

The operating procedures associated with the compressor system include:

Action	Notes
1. Ensure the area is clear of debris and all auxiliary systems are lined up and operational. 2. Set PIC 100 to 50 psig and raise in 25 psi increments until desired operational conditions are achieved. 3. Drain D-100 and close bleeder valve. 4. Open V-100, V-101, V-103, V-105, V-107. 5. Set back pressure regulator PCV-110 to 20 psig. 6. Turn on compressor-100 and closely monitor all process variables. 7. Ensure PCV-100 is operating properly and place PIC-100 into AUTO. 8. Complete checklist of compressor system. 9. Cross-check process variables with SPEC sheet.	

Specification Sheet and Checklist

Pressure

1.	PIC-100	100 psig	AUTO	Instrument air header
2.	Pi-103	100 psig	—	Instrument air header

Troubleshooting Scenario 1

Figure 9-7 displays the compressor system and presents a series of what-if-scenarios. Refer to the figure and select the best answers.

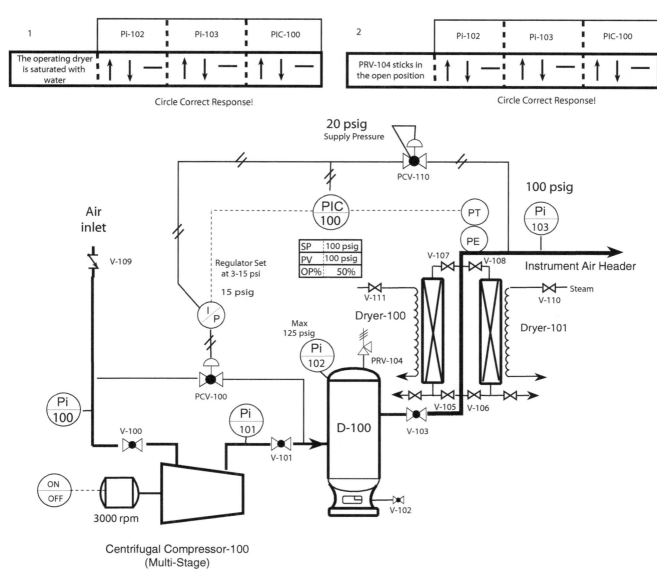

Figure 9-7   *Compressor Troubleshooting Scenario 1*

Troubleshooting Scenario 2

Figure 9-8 displays the compressor system and presents a series of what-if-scenarios. Refer to the figure and select the best answers.

Troubleshooting Scenario 3

Figure 9-9 displays the compressor system and presents a series of what-if-scenarios. Look at each variable and determine how it would affect the others. Refer to the figure and select the best answers.

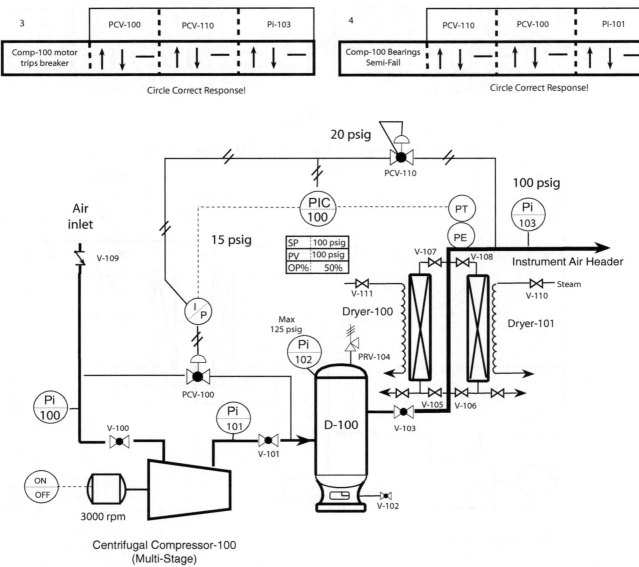

Figure 9-8 *Compressor Troubleshooting Scenario 2*

Summary

The compression of gases and vapors in the process industry is very important, particularly in systems used for modern process control. Compressed air systems provide the medium used to open or close pneumatically operated valves. Compressor-100 is a multistage centrifugal device designed to provide clean, dry air for instruments and control devices. Operating and maintaining this system is an important part of a technician's job. In areas where high humidity exists, the compression process produces liquid that accumulates in the receiver and needs to be drained off periodically.

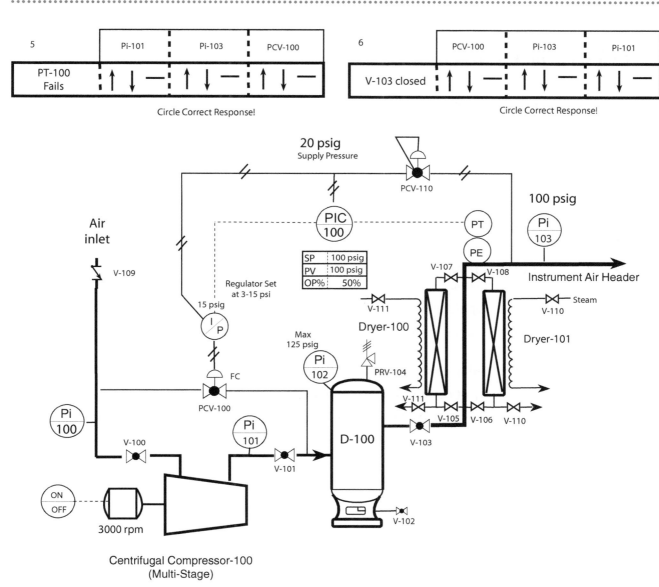

Figure 9-9 *Compressor Troubleshooting Scenario 3*

The basic components of compressor-100 include the compressor, receiver, dryer, instruments, piping, valves, and a pressure control system. The instrument air header provides clean, dry air to the entire plant. The instrument air header operates at 100 psig and feeds a series of pressure-reducing valves that are located in key areas where the pressure needs to be reduced for operating purposes, typically 20 psig.

Review Questions

1. Describe the operation and purpose of the centrifugal compressor system.

2. List the basic components of a centrifugal compressor.

3. List the basic components of the compressor system.

4. Explain the primary purpose of the receiver.

5. Describe how to operate the compressor system.

6. Explain how to regenerate the **desiccant dryer.**

7. Draw and describe the pressure control system used in compressor-100.

8. Describe how the pressure on the instrument air header is reduced and used to operate other automatic valves.

9. List the basic principles of compression and explain how these operate industrial valves.

10. List the primary advantages associated with the installation and operation of centrifugal compressor systems.

11. List the basic **impeller designs** used in centrifugal compressors.

12. Describe the theory of compression ratio in a centrifugal compressor.

13. Describe the names used for the impeller vane that touches the air first and last.

14. Explain how the impeller vanes, diffuser, volute, and drive shaft operate together.

Heat Exchanger Model

LEARNING OBJECTIVES

After studying this chapter, the student will be able to:

- Apply the principles of heat transfer to problems found in this chapter.
- Identify and describe the basic components of a heat exchanger system.
- Draw a simple block flow diagram (BFD) of a heat exchanger system.
- List and draw the two control loops used in the heat exchanger system.
- Identify the simple instrument used by a technician to operate the heat transfer system.
- Explain the primary purpose of each control loop and explain how the two control loops work and complement each other.
- Operate the heat exchanger system.
- Identify the common problems associated with heat exchanger systems.
- Solve various troubleshooting scenarios associated with heat exchanger system.
- Compare and contrast the set point, process variable, and valve operating percentage.

Key Terms

Principles of heat transfer—include radiant, conductive, and convective heat transfer. The primary methods of heat transfer in a heat exchanger are conductive and convective. Conductive heat transfer is characterized by energy transfer through a solid. Convective heat transfer is characterized by energy transfer through the moving fluid. Conductive heat transfers through the tubes and into the flowing liquid, and convective heat transfers once it enters the flowing liquid. Radiant heat transfer is not typically applied in heat exchanger operation.

Fouling—buildup on the internal surfaces of devices such as cooling towers and heat exchangers, resulting in reduced heat transfer and plugging. Fouling also reduces fluid flow. In most cases, fouling takes place in the tube side of the exchanger; however, it is possible that it could occur in the shell if there is a large number of support baffles.

Differential pressure and temperature $\Delta P/\Delta T$—key factors associated with the operation of heat exchangers. ΔP is the difference between inlet and outlet pressures; ΔT is the difference between the inlet and outlet temperatures.

Baffles—evenly spaced partitions in a shell and tube heat exchanger that support the tubes, prevent vibration, control fluid velocity and direction, increase turbulent flow, and reduce hot spots.

Channel head—a device mounted on the inlet side of a shell and tube heat exchanger that is used to channel tube-side flow in a multipass heat exchanger.

Condenser—a shell and tube heat exchanger used to cool and condense hot vapors.

Fixed head—a term applied to a shell and tube heat exchanger that has the tube sheet firmly attached to the shell.

Floating head—a term applied to a tube sheet on a heat exchanger that is not firmly attached to the shell on the return head and is designed to expand (float) inside the shell as temperature rises.

Laminar flow—streamline flow that is more or less unbroken; layers of liquid flowing in a parallel path.

Multipass heat exchanger—a type of shell and tube heat exchanger that channels the tube-side flow across the tube bundle (heating source) more than once.

Parallel flow—refers to the movement of two flow streams in the same direction; for example, tube-side flow and shell-side flow in a heat exchanger; also called concurrent.

Shell and tube heat exchanger—a heat exchanger that has a cylindrical shell surrounding a tube bundle.

Tube sheet—a flat plate to which the ends of the tubes in a heat exchanger are fixed by rolling, welding, or both.

Turbulent flow—random movement or mixing in swirls and eddies of a fluid.

Heat exchanger control system—includes the pump and control valves and heat exchangers. In this system, flow control loop-202 (FIC-202) controls flow through the shell side of Exchanger-202

(Ex-202) and Exchanger-203 (Ex-203). Temperature control loop-100 (TIC-100) controls the flow of heated oil through the tube side of Ex-203 and measures the temperature of feed through the shell side of Ex-203.

Tube growth—a process of thermal expansion where the tubes expand lengthwise.

Temperature—the hotness or coldness of a substance.

Heat—a form of energy caused by increased molecular activity. A basic principle of heat states that it can not be created or destroyed, only transferred from one substance to another.

Heat transfer—Heat is transmitted through:
- *Conduction*—heat energy is transferred through a solid object. Example: heat exchanger.
- *Convection*—requires fluid currents to transfer heat from a heat source. Example: the convection section of furnace or economizer section of boiler.
- *Radiation*—the transfer of energy through space by the means of electromagnetic waves. Example: the sun.

Introduction to Heat Exchanger Operation (Complete System)

Heat exchangers are commonly used to heat or cool process flows in the chemical processing industry. A **shell and tube heat exchanger** has a cylindrical shell that surrounds a tube bundle. Fluid flow through the exchanger is referred to as tube-side flow or shell-side flow. A series of **baffles** support the tubes, direct flow, decrease tube vibration, increase velocity, create pressure drops, and protect the tubes. Shell and tube heat exchangers can be classified as (1) **fixed head,** single-pass, (2) fixed head, multipass, (3) **floating head,** multipass, and (4) floating head, multipass, U-tube operation. The term "fixed head" refers to how the **tube sheet** is bolted between two flanges: one on the **channel head** and one on the shell. Floating head refers to how the tube return is designed. There are a variety of applications where the tube return head is designed to expand or float. This enables the device to operate at **temperature** differentials above 200°F. Fixed head heat exchangers are limited to temperature differentials below 200°F. Thermal expansion prevents a fixed head heat exchanger from exceeding this temperature differential. The basic components of a shell and tube heat exchanger include:
- shell inlet
- shell outlet
- tube inlet
- tube outlet
- channel head

- shell
- tubes
- baffles.

Ex-202 and Ex-203 are designed to heat the feed as it moves at 225 gpm to the distillation column. Feed enters the shell side of Ex-202 at 80°F and is heated up to 115°F before entering the shell side of Ex-203. When it exits the exchanger, the temperature of the feed mixture is at 180°F. During operation, Ex-202 and Ex-203 have one tube sheet that is fixed while the return head "floats" inside the shell. The floating end allows for **tube growth** or expansion length-wise and is not attached to the shell.

The shell nozzle arrangement is different on Ex-202 and Ex-203, indicating different internal designs. On Ex-202, a longitudinal baffle is in place, while Ex-203 has a series of segmental baffles. The heating medium on both exchangers comes from the tubes. Ex-202 takes advantage of the heat energy in the liquid pentane as it flows from the kettle reboiler. By transferring heat to the feed, pentane cools enough to be safely stored in the primary bottom tank. Ex-203 utilizes a hot oil system specifically designed to heat the process flow up to 180°F.

Heat exchangers are characterized typically by a shell inlet and outlet and a tube inlet and outlet. In a shell and tube exchanger, the tube-side flow is separated from the shell-side flow. Heat energy flows from hot to cold areas and through conductive and convective **heat transfer** methods. Figure 10-1 is a review of the basic components found in a shell and tube heat exchanger.

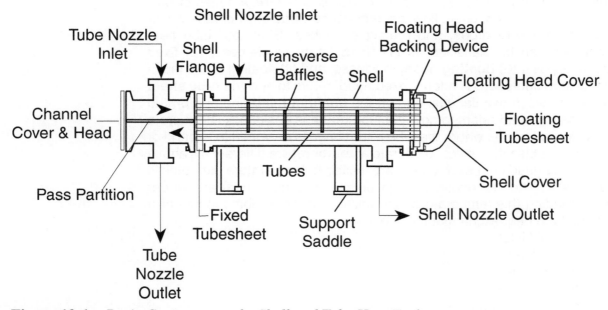

Figure 10-1 *Basic Components of a Shell and Tube Heat Exchanger*

Fluid flow through the heat exchanger system is initiated by three centrifugal pumps: feed pump 202, bottoms pump 205, and the hot oil pump. Differences in baffle arrangement produce a variety of fluid flow patterns, mostly turbulent. Fluid flow in heat exchangers is often described as cross-flow, counter-flow, and parallel-flow. Baffles are designed to support the tubes and direct and redirect fluid flow. This process enhances the heat transfer process.

Ex-202 and Ex-203 are characterized as shell and tube, multipass, floating head heat exchangers. Floating head heat exchangers are designed for high-temperature differentials above 200°F. During operation, one tube sheet is fixed and the other "floats" inside the shell. The floating end is not attached to the shell and is free to expand. Engineering specifications take into account the thermal tube expansion or tube growth. The tube sheet on the channel head is fixed on both Ex-202 and Ex-203. The floating head creaks as it slowly expands during start-up. The exchangers in this system are classified as two-pass devices. The shell side of Ex-202 is connected in a series that flows to the shell side of Ex-203. Tube-side flow on Ex-202 occurs from the bottom of the distillation column, while the tube side of Ex-203 is connected to the hot oil system. The heat exchanger system described in this chapter is illustrated by Figure 10-2.

Heat is a form of energy caused by increased molecular activity. A basic principle of heat states that it can not be created or destroyed, only transferred from one substance to another. Heat energy moves from hot to cold areas, transferring energy in the process. This process will continue until the heat energy has been equally distributed. A stone thrown into a still pool of water sends ripples out in all directions. Heat energy moves in a similar pattern. Heat is measured in energy units called British thermal units (BTUs). A BTU is the amount of heat needed to raise the temperature of 1 pound of water by 1°C. Another common unit used in industrial manufacturing is the calorie. One calorie is roughly equal to the heat energy required to raise the temperature of 1 gram of water by 1°C. The effects of absorbed heat are:
- molecular activity—increases
- change of state—solid, liquid, and gas
- chemical change—matches
- energy movement—hot to cold
- heat transfer—radiant
- heat transfer—conductive and convective
- electrical transfer—thermocouple
- volume—increases
- temperature—increases.

Heat and heat transfer can be described in the following terms:
- Conductive heat transfer—heat energy is transferred through a solid object. Example: tubes in a heat exchanger.

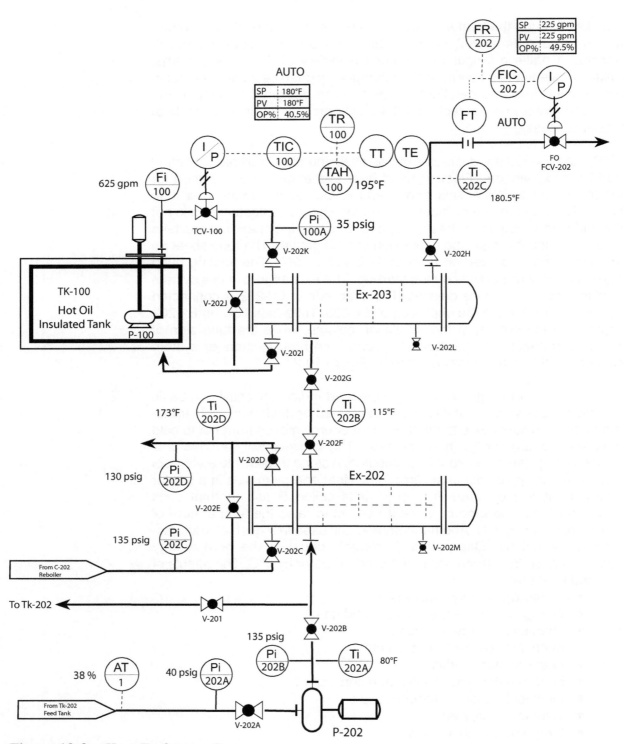

Figure 10-2 *Heat Exchanger System*

- Convective heat transfer—requires fluid currents to transfer heat from a heat source. Example: convection section of furnace.
- Radiant heat transfer—the transfer of energy through space by the means of electromagnetic waves. Example: the sun.
- Evaporation—a form of convective heat transfer. Example: cooling tower.
- Sensible heat—heat that can be sensed or measured; increase or decrease in temperature. Example: thermometer.
- Latent heat—hidden heat; does not cause a temperature change.
- Latent heat of fusion—heat required to melt a substance. Heat removed to freeze a substance.
- Latent heat of vaporization—heat required to change a liquid to a gas.
- Latent heat of condensation—heat removed to condense a gas.
- Specific heat—the required BTUs needed to raise the temperature of 1 pound of a specific substance by 1°F.

Temperature

By measuring the hotness or coldness of a substance, we determine temperature. Process operators use a variety of temperature systems.

The most common four temperatures are as follows:

		Water boils	Water freezes
1.	Kelvin (K)	373 K	273 K
2.	Celsius (C)	100°C	0°C
3.	Fahrenheit (F)	212°F	32°F
4.	Rankin (R)	672°R	492°R

Conversion Formulas
1. K = C + 273
2. C = (F − 32) ÷ 1.8
3. F = 1.8°C + 32
4. R = F + 460

Temperature and Heat Key Points
- Heat is a form of energy caused by increased molecular activity; it cannot be created or destroyed, only transferred from one substance to another.
- The hotness or coldness of a substance will determine the temperature.

Figure 10-3
Temperature Scales

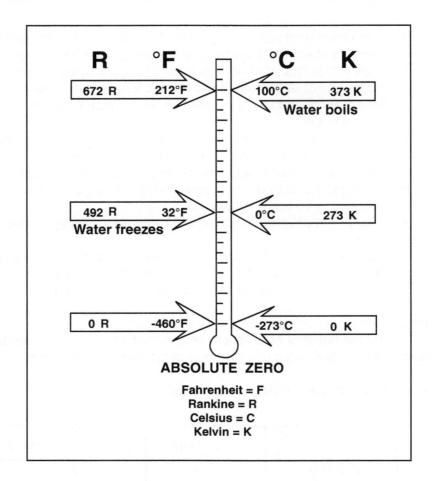

• Heat is measured in BTUs and temperature is measured by K, C, F, or R.
• Temperature and heat are not the same.

The chemical processing industry utilizes a variety of temperature scales. The most common are illustrated on Figure 10-3.

Equipment Descriptions (Block Flow and Safety)

Preheater 202
Ex-202 is a horizontally mounted, shell and tube, floating head, two-pass system. A ternary mixture of butane, pentane, and liquid catalyst enters the shell side at 80°F at 225 gpm and encounters a longitudinal baffle that forces the fluid to run countercurrent to the tube-side flow. A series of smaller vertical baffles allow the fluid to cascade from the longitudinal center plate to the outer shell and back until it reaches the return head and

passes across the exchanger again and into the outlet. The outlet temperature of the shell-side flow is typically 115°F, resulting in a 35°F differential. The operating pressure on the feed to the column is 135 psig.

The tube side of Ex-202 receives 152 gpm at 223°F as it fills the tubes from the bottom to the top. The exiting temperature of the liquid pentane is 173°F at 130 psig. The bottoms flow specifications are 92% pentane and 8% butane. The differential pressure across the tubes is 5 psi and the differential temperature is 50°F.

Preheater 203

Ex-203 is also a horizontally mounted, shell and tube, floating head, two-pass system. The primary difference between Ex-202 and Ex-203 is the application of a hot oil system that has been incorporated into the transfer of heat via the tube inlet and outlet of Ex-203. Like Ex-202, the shell-side parallel connection has a flow rate of 225 gpm at 115°F and 135 psig. The operating pressure on column-202 (C-202) is 100 psig, so the pressure at pump 202's discharge is monitored closely. The flow rate through Ex-202 and Ex-203 is controlled by FIC-202, which is located on the shell discharge between Ex-203 and C-202. The temperature differential across the shell side of Ex-203 is 65.5°F. This is the difference between the shell outlet at 180.5°F and the shell inlet at 115°F.

The tube-side flow is regulated at 180°F by TIC-100 and the flow of hot oil from TK-100. TIC-100 measures the temperature of the exiting shell-side flow on Ex-203 and uses this variable to compare to the set point.

Pump 202A/B

Pump 202 is a horizontally mounted centrifugal pump designed to operate at 225 gpm. Suction pressure is maintained at 40 psig, and the discharge pressure typically runs at 135 psig. The differential pressure across the pump is 75 psig. This differential is necessary in order to operate the system properly. The pressure is obtained as the control valve moves through 30% of its operating range on the discharge side of the pump.

Flow Control Loop-202 (FIC-202)

The FIC-202 is designed to work with the pump and the backpressure available in the control valve. Figure 10-4 illustrates the basic components of FIC-202. The basic components of FIC-202 are:
- orifice plate
- flow transmitter
- flow indicator controller
- transducer
- flow control valve 202
- flow recorder 202.

Figure 10-4
FIC-202 Control Loop

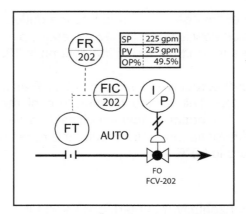

An automatic process controls the centrifugal pump and control valve to form a unique relationship as the control valve begins to throttle flow. This throttling effect does not hurt the pump or valve, as the principle of internal slip protects the system. This is not the case when a positive displacement (PD) pump is used. PD pumps are not typically used for flow control that requires throttling since they will damage the equipment. With a centrifugal pump the control valve operates through a 30% value designed to allow it to regulate flow from 0 gpm to the desired process set point.

Temperature Control Loop-100 (TIC-100)
The TIC-100 utilizes a hot oil system to heat up the feed to C-202. The set point for the system is 180°F at 225 gpm at 135 psig. Figure 10-5 illustrates TIC-100. The basic components of TIC-100 include:
* temperature recorder 100
* temperature element—thermocouple in thermowell
* temperature transmitter
* temperature indicating controller
* transducer
* temperature control valve.

Tank-100
The hot oil system is designed to provide heat to various operations that require a stable heat source. The primary use of the hot oil system is to provide a heat source to a series of heat exchangers, which are used to preheat feed in a number of distillation systems. The tank operating capacity is 20,000 gal. The tank is constructed of stainless steel and insulated to conserve and retain heat energy. The system is run under atmospheric pressure; however, the tank is sealed and caution signs are posted on the vessel. A level and pressure relief system is attached to the tank.

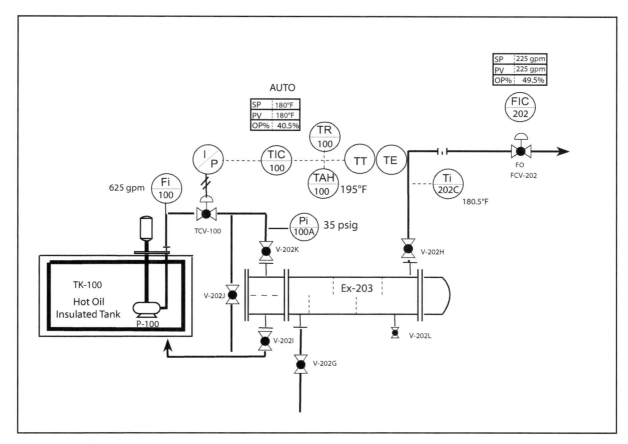

Figure 10-5 *Temperature Control Loop-100*

Process technicians are responsible for monitoring and controlling all of these variables.

Instrumentation Systems

The instrumentation system must be applied correctly with an eye to safety when incorporating it into a heat exchanger system. It is possible to turn a heat exchanger into a bomb. Since liquids are considered to be noncompressible, higher temperatures will cause pressure to build and possibly explode and catch fire. In this system, flow control valve 202 is designed to fail in the open position, ensuring that the hot oil system does not damage Ex-203.

Temperature

TIC-100 (temperature indicating controller)—located on the tube inlet for Ex-203.

TR-100 (temperature recorder)—provides a video trend of variations in the feed entering C-202.

Ti-202A—temperature indicator located on the discharge of pump 202. Runs at 80°F.

Ti-202B—temperature indicator located on the shell outlet line from Ex-202. Runs at 115°F.

Ti-202C—temperature indicator located on the shell outlet line from Ex-203. Runs at 180.5°F.

Ti-202D—temperature indicator located on the tube outlet line from Ex-202. Runs at 173°F.

Pressure

Pi-202A—pressure indicator located on the suction side of P-202. Runs at 40 psig.

Pi-202B—pressure indicator located on the discharge side of P-202. Runs at 135 psig.

Pi-202C—pressure indicator located on the tube inlet of Ex-202. Runs at 135 psig.

Pi-202D—pressure indicator located on the tube outlet of Ex-202. Runs at 130 psig.

Pi-100A—pressure indicator located on the tube inlet of Ex-203. Runs at 35 psig.

Flow

FIC-202—flow-indicating controller located on the shell outlet of Ex-203. Set Point 225 gpm.

Analytical

AT-1 (analytical transmitter)—located on the suction side of P-202.

Measures the percentage of butane in the feed.

The Heat Exchanger System

The primary purpose of the heat exchanger system is to transfer heat to the feed. This is accomplished in a two-step process where the liquid is heated up from 80 to 115°F in the first exchanger and from 115 to 180°F in the second heat exchanger. Heat exchangers can be used to heat or cool a substance. The chemical processing industry utilizes a variety of unique approaches in the process. Before operating a heat exchanger system, a technician spends significant time to memorize and learn the various types of heat exchangers, and how they can be arranged in different systems.

The safety aspects associated with the operation of a heat exchanger system include all of the following:

- chemical hazards associated with spills and leaks (see chemical list and MSDS)
- hazards associated with burns
- hazards associated with fires
- hazards associated with explosions and BLEVE (boiling liquid expanding vapor explosions)
- confined space entry; larger exchangers with tube bundles removed
- equipment failure: tube leak, gasket leak, shell puncture, or leak
- error with valve line up resulting in explosion or fire
- pump failure resulting in overheating in heat exchanger
- gauge failures
- sampling, purging, or venting the shell
- exceeding pressure and temperature ratings on heat exchanger code stamp for tubes and shell
- utilizing incompatible materials with chemicals
- working with hot materials under pressure.

In looking at the above list, a large number of items can be listed under "operator error." Unfortunately, a high number of safety incidents can be attributed to mistakes made by process technicians or engineering. A process hazard analysis should always be performed prior to allowing technicians to operate a heat exchanger system. Proper training is also critical for new technicians who are assigned to heat exchanger systems. Some serious industrial accidents have been linked to a lack of training for new employees.

Operational Specifications (SOP, SPEC Sheet, and Checklist)

The operating procedures associated with the heat exchanger system include the following:

Action	Notes
1. Verify pump recirculation to TK-202	This includes looking at pump suction and discharge pressures and level in TK-202
2. Lineup heat exchanger system to C-202	Open V-202F, V-202G, V-202H

Action	Notes
3. Open V-202C and V-202D	This will admit flow into the lower tube-side of Ex-202. Flow comes from the kettle reboiler
4. Set FIC-202 to 75 gpm and set in AUTO	Ensure C-202 has been prepared prior to this step
5. Close V-201 and ensure the feed tank system is operating properly and the analyzer (AT-202A) is between 36% and 38%	
6. Set FIC-202 to 225 gpm	
7. Inspect hot oil system and ensure it is operating properly	
8. Open V-202J	
9. Set TIC-100 to 180°F and put in AUTO	
10. Open V-202K and V-202L. Close V-202J	This will admit flow into the tube-side of Ex-203
11. Monitor and ensure that both the control systems are operating properly	This includes the flow and temperature recorder
12. Place the system on-spec when operating temperatures approach set point	
13. Cross-check process variables with SPEC sheet	

Specification Sheet and Checklist

Flow

1.	FIC-202	225 gpm	AUTO	Feed flow
2.	FR-202	225 gpm	AUTO	Feed flow to C-202

Analytical

3.	AT-1	38%	Butane–liquid catalyst	—

Pressure

4.	Pi-202A	40 psig	Suction P-202A	—
5.	Pi-202B	135 psig	Discharge P-202A	—
6.	Pi-202C	40 psig	Suction P-202B	—
7.	Pi-202D	130 psig	Discharge P-202B	—
8.	Pi-100A	35 psig	Tube inlet Ex-203	—

Temperature

9.	TIC-100	180°F	AUTO	Tube inlet Ex-203 hot oil system
10.	Ti-202A	80°F	Gauge	Discharge P-202
11.	Ti-202B	115°F	Gauge	Discharge shell-side Ex-202
12.	Ti-202D	173°F	Gauge	Tube outlet Ex-202
13.	TR-100	180°F	—	TR hot oil system
14.	TAH-100	195°F	—	High-temperature alarm

Common Heat Exchanger Problems and Solutions

Heat exchanger systems are simple in design; however, since so many different arrangements are available they can be a little tricky. Common heat exchanger problems include all of those listed in the safety aspects section of this chapter. Solutions to each of these problems will depend on the individual training procedure used at the company you work for. This chapter will cover some of the typical problems and responses used in the industry.

Troubleshooting Scenario 1

Refer to the figure and select the best answers. Figure 10-6 displays the heat exchanger system and presents a series of what-if scenarios. Opening and closing the correct valves are critical in the operation of a heat exchanger. It is possible to close the wrong valve and turn the device into a bomb. For this reason, new technicians are carefully trained in the operation of a heat transfer system. This reduces injuries and increases productivity.

Troubleshooting Scenario 2

Refer to the figure and select the best answers. Figure 10-7 displays the heat exchanger system and presents a series of what-if scenarios. Consider all of the possible scenarios that could appear during the operation of this type of equipment.

Troubleshooting Scenario 3

Refer to the figure and select the best answers. Figure 10-8 displays the heat exchanger system and presents a series of what-if scenarios.

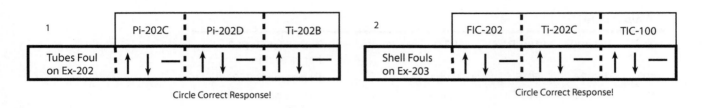

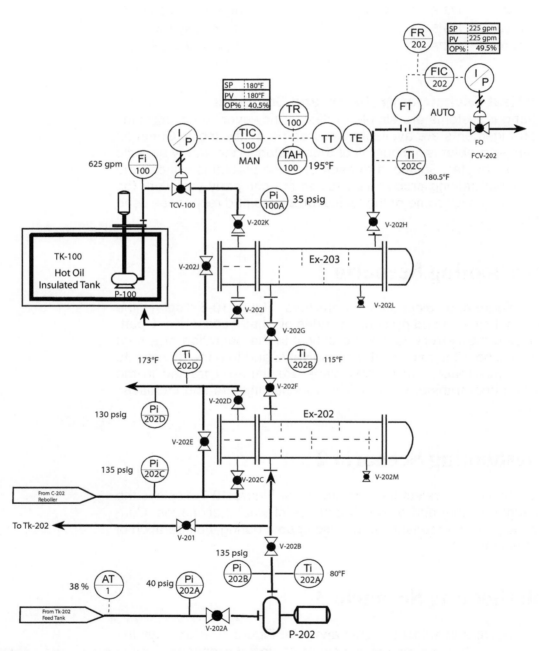

Figure 10-6 *Heat Exchanger System—Troubleshooting Scenario 1*

3	TIC-100 PV			Fi-100			FIC-202 PV		
P-100 Fails (Trips Off)	↑	↓	—	↑	↓	—	↑	↓	—

Circle Correct Response!

4	Ti-202B			Pi-202D			TIC-100 OP%		
C-202 Reboiler flow fails (P-208)	↑	↓	—	↑	↓	—	↑	↓	—

Circle Correct Response!

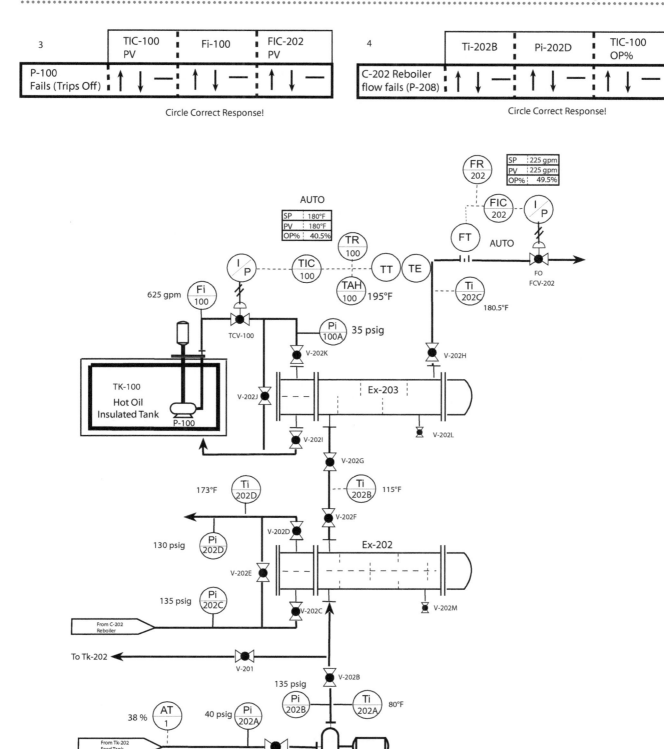

Figure 10-7 *Heat Exchanger System—Troubleshooting Scenario 2*

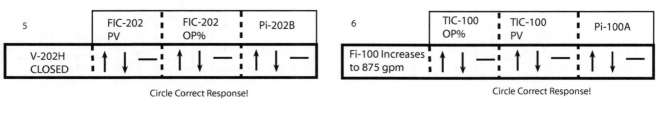

5	FIC-202 PV	FIC-202 OP%	Pi-202B
V-202H CLOSED	↑ ↓ —	↑ ↓ —	↑ ↓ —

Circle Correct Response!

6	TIC-100 OP%	TIC-100 PV	Pi-100A
Fi-100 Increases to 875 gpm	↑ ↓ —	↑ ↓ —	↑ ↓ —

Circle Correct Response!

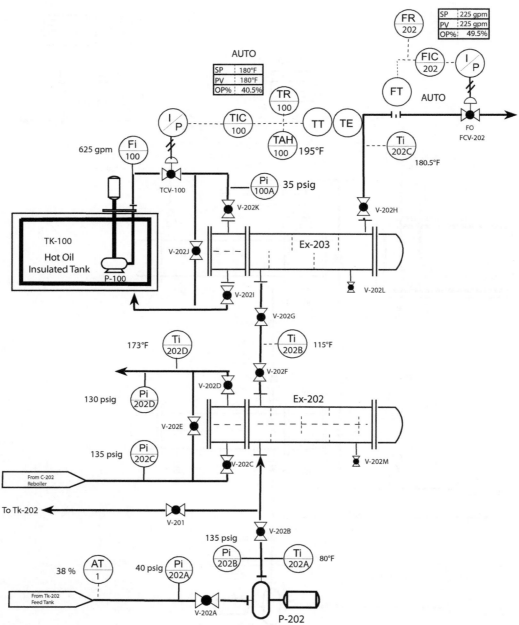

Figure 10-8 *Heat Exchanger System—Troubleshooting Scenario 3*

Summary

The primary purpose of the Ex-202 and Ex-203 heat exchanger systems is to transfer heat to the feed. This is accomplished in a two-step process in which the liquid is heated up from 80 to 115°F in Ex-202 and from 115 to 180°F in Ex-203. It should be noted that the heat transfer system is a part or component of a much larger system. In order to operate effectively, three centrifugal pumps are needed: one on the hot oil, one on the bottoms product, and one primary feed pump. Working together, these pumping systems move liquid through the two heat exchanger systems. The direction of heat transfer for each system includes:

- **Tube-side Ex-202**—receives heat energy through conductive and convective heat transfer as P-208 discharges into the lower tube inlet at 152 gpm at 223°F. The tube-side flow is much hotter than the entering shell-side flow at 80°F. When liquid exits the exchanger on the tube side, it is at 173°F. The difference between the tube inlet temperature and the tube outlet temperature is 223°F − 173°F = 50°F. A slight pressure loss is also encountered as the liquid flows across the tubes, 135 psig − 130 psig, a difference of 5 psig.
- **Shell-side Ex-202**—flow encounters heat energy radiating from the tubes. As the feed passes over the tubes, the temperature increases from 80 to 115°F. The corresponding ΔT across Ex-202 is equal to 115°F − 80°F = 35°F. In a heat exchanger, heat energy cannot be contained inside the tube as it transfers its energy to the liquid flowing into the shell side of Ex-202. The two temperatures are quite different: 80°F on the entering shell-side flow and 223°F on the tube-side flow, a temperature difference of 143°F. Heat energy transfers from hot to cold areas. The design of the floating head heat exchanger can tolerate this large temperature difference. Since the fluid flow is moving much faster on the shell side, 225 gpm vs. 152 gpm on the tube, the heat transfer between the two flows is 115°F − 80°F = 35°F.
- **Tube-side Ex-203**—receives heat energy through conductive and convective heat transfer as the hot oil system flows through the tubes. The flow rate on the hot oil system is 625 gpm. P-100 is submerged inside the hot oil system and discharges at 35 psig into the upper tube inlet on Ex-203. A TIC-100 controls the flow of hot oil through the tubes. The control loop takes a measurement of the flow exiting the shell side on Ex-203 and adjusts the hot oil flow rate to ensure that the shell-side temperature remains at 180°F. The hot oil system is designed to be maintained at 280°F.
- **Shell-side Ex-203**—operates at 225 gpm at 115°F on the inlet side and at 180°F on the outlet side. The temperature

differential is 65°F between the shell inlet and outlet. The temperature on the shell inlet, 115°F, and 300°F on the tube inlet results in a differential of 185°F.

When a technician studies the thermodynamics associated with heat transfer in a heat exchanger system, it can become very complicated. These differences are typically monitored between the shell and tube inlet and outlet flows. How much heat energy is transferred is carefully controlled in most cases. Another area that is critical to heat exchanger operation is pressure, specifically the difference between the inlet and outlet pressures. If the pressure changes on the inlet and outlet of the tubes or shells on the heat exchanger, it is a sure indication that **fouling** or plugging is occurring. When the tubes begin to foul, heat transfer is diminished.

Review Questions

1. Describe the purpose of FIC-202 and how it works.

2. Describe the purpose of TIC-100 and how it works.

3. What is the controlled flow rate of P-202?

4. What is the flow rate of P-100 to Ex-203?

5. What is the initial feed temperature to C-202?

6. What is the temperature differential across the shell side of Ex-202?

7. What is the temperature differential across the tube side of Ex-202?

8. What is the temperature differential across the shell side of Ex-203?

9. What is the temperature differential across the tube side of Ex-203?

10. Describe the purpose and operation of AT-1.

11. Describe the primary difference between Ex-202 and Ex-203.

12. Draw a process flow diagram (PFD) for the heat exchanger system. List all instruments and control systems.

13. Define the term "tube growth."

14. Explain the various methods of heat transfer used in the heat exchanger system.

15. Define the term "fouling."

16. Describe the correct procedure for starting up the heat exchanger system.

17. What temperature differentials are associated with fixed head and floating head heat exchangers?

18. Identify the instruments associated with Ex-202.

19. Identify the instruments associated with Ex-203.

20. Compare and contrast the various pump systems and operating circuits on the heat exchanger system.

Cooling Tower Model

LEARNING OBJECTIVES

After studying this chapter, the student will be able to:

- Describe the various components of the cooling tower system.
- Describe the principles of heat transfer in the cooling tower system.
- Describe the relationship between heat exchangers and cooling towers.
- Draw a simple block-flow diagram (BFD) of the cooling tower model.
- List the safety aspects associated with the cooling tower system.
- Describe the cooling tower instrumentation systems.
- List the operational specifications for the cooling tower system.
- Solve various troubleshooting scenarios associated with cooling tower systems.
- Identify common problems associated with cooling tower operation.
- Operate the cooling tower system.

Key Terms

Cooling towers—evaporative coolers specifically designed to cool water or other mediums to the ambient wet-bulb air temperature.

Air intake louvers—slats located on the sides of the cooling tower to direct airflow into the cooling tower.

Approach to the tower—the temperature difference between water leaving the cooling tower and the wet-bulb temperature of air entering the tower.

Basin heaters—designed to keep water from freezing during the winter months.

Water basin—concrete storage compartment or catch basin located at the bottom of the cooling tower.

Blowdown or draw-off—a process designed to control the level of suspended or dissolved solids in a cooling tower by removing a certain amount of water from the basin and replacing it with fresh makeup water.

Leaching—the loss of wood preservative chemicals in the supporting structure of the cooling tower as water washes or flows over the exposed components.

Capacity—the amount of water a cooling tower can cool.

Cell—the smallest subdivision of a cooling tower that can function as an independent unit. Some cooling tower systems have multiple cells.

Plume—the water-saturated exhaust (fog) exiting the cooling tower.

Windage or drift—small water droplets that are carried out of the cooling tower by flowing air.

Cooling range—the temperature difference between the hot and cold water in a cooling tower.

Drift eliminators—devices used in a cooling tower to keep water from blowing out.

Drift loss—entrained water lost from a cooling tower in the exiting air; also called windage loss.

Dry-bulb temperature (DBT)—the air temperature as measured without taking relative humidity into account.

Evaporative coolers—a term often used to describe a cooling tower.

Evaporate—to turn to vapor; evaporation removes heat energy from hot water.

Fill—plastic or wood surfaces that direct airflow and provide for contact of water and air in a cooling tower (see Splash bar).

Biocides and algaecides—prevent biological growths from interfering with water circulation.

Cooling tower types—classified as induced (requires fan), forced (requires fan), atmospheric, and natural draft.

Induced draft—cooling tower-302 (CTW-302) is a mechanical-draft cooling tower that has a single cell and utilizes a fan located on the top of the tower to slowly draw air into the system and out of the system.

Plenum—an open area sandwiched between the fill in the center of an induced draft cooling tower.

Relative humidity—a measurement of how much water the air has absorbed at a given temperature.

Scale—the result of suspended solids adhering to internal surfaces of equipment in the form of deposits.

Splash bar—a device used in a cooling tower to direct the flow of falling water and increase surface area for air–water contact.

Water distribution system—consists of a deep pan with holes equipped with nozzles that distribute water across the fill using gravity. Some systems utilize a pipe and spray nozzle design.

Wet-bulb temperature (WBT)—the air temperature as measured by a thermometer that takes into account the relative humidity.

Psychrometry—described as the study of cooling by evaporation.

Introduction to Cooling Tower Operation (Complete System)

Cooling towers are classified by how they produce airflow and how they produce airflow in relation to the downward flow of water. A cooling tower can produce airflow mechanically or naturally. Airflow can enter the cooling tower and cross the downward flow of water or run counter to the downward flow of water.

The world's tallest cooling tower is located at the Niederaussem Power Station, standing 200 m. Wet cooling towers are operated by using the scientific principle of evaporation. Examples of wet cooling towers include (1) **induced draft** cross-flow, (2) forced draft counter-flow, (3) atmospheric flow, and (4) hyperbolic or chimney flow. Dry cooling towers operate using the scientific principle of convection by heat transmission through pipes or tubes or surfaces that separate the working fluid from the ambient air. Hybrids of wet and dry cooling towers can be found in operation at many systems in the chemical processing industry.

Cooling tower-302 (CTW-302) is classified as an induced draft (or draw-through), cross-flow, single cell device that is primarily designed to control the temperature on condenser EX-204. A cooling tower is often referred to as a heat rejection device designed to extract excess heat from the returning water and expel it into the atmosphere. Figure 11-1 shows what the cooling tower system looks like. This type of heat transfer relies on the principle of evaporation. When this process occurs, the heat from water is absorbed by the air stream, which raises the **relative humidity** to near 100%. These heated currents are quickly dissipated by the wind. CTW-302

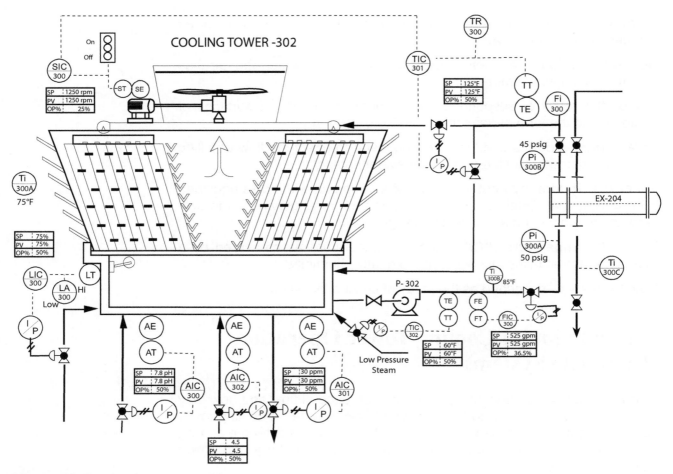

Figure 11-1 *Cooling Tower-302*

is an evaporative heat rejection device that can significantly reduce water temperatures.

CTW-302 is an enclosed structure with a system of air louvers designed to direct airflow across the **fill.** When warm water enters the top of the cooling tower, the **water distribution system** carefully sprays or directs fluid flow over a labyrinth-like honeycomb, splashboards, or fill. The purpose of the fill is to allow the hot water to spread out over the surface of the boards. The fill provides a vastly expanded surface area interface that enhances air–liquid contact. As evaporation occurs, air becomes saturated with water and is carried out of the cooling tower system. CTW-302 uses a fan to draw air into the fill and across the fill. The fan is located on the top of the cooling tower and slowly draws air into the system and rapidly discharges it at the exit point. The cooled water continues to drop through and over the fill until it enters the basin. A typical cooling tower is a heat transfer device designed to cool water so it can be reused in industrial applications.

Equipment Descriptions (Block Flow and Safety)

Water Pump

Pump 302 is a vertically mounted centrifugal pump designed to operate at 525 gpm at 85°F. A backup pump is used when problems occur with the primary pumping system. The suction pressure should be carefully monitored. A strainer is located on the suction side of the pump to ensure that no debris enters the pump. The strainer needs to be blown down weekly. Pump motor temperature and vibrations are checked at each shift. The pump discharge is typically 50 psig. Figure 11-2 is a review of the basic components of a cooling tower.

Heat Exchanger

Ex-204 is a horizontally mounted, shell and tube, multipass heat exchanger designed to condense vaporized butane into the liquid state. The pressure is 115 psig on the shell side and 50 psig on the tube side. The tube side contains cooling water from the cooling tower system. The shell side contains vapor or liquid butane at 115 psig. The heat exchanger is comprised of stainless steel and is rated at 225 psig at 250°F on the tubes and 225 psig at 300°F on the shell. The inlet pressure on the tube side is run at 50 psig and 45 psig on the tube outlet. The tube delta pressure is 5 psig.

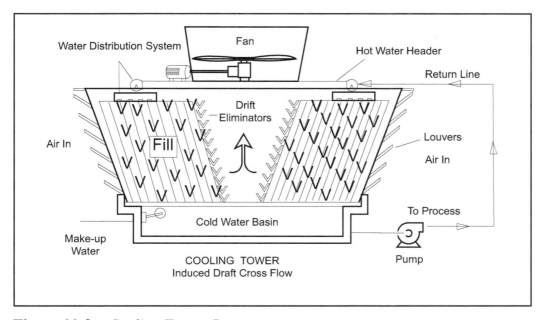

Figure 11-2 *Cooling Tower Components*

Induced Draft Fan

The fan is mounted on the top of the water distribution pipe in the center of the cooling tower. A local controller and on–off switch are located near the motor. The motor is designed to run at 1250 rpm and can be adjusted depending upon the need. The fan has a low velocity as it draws the outside air across the fill; however, its exit velocity is much higher than a forced draft cooling tower. The hot **plume** is discharged high enough above the cooling tower so that it does not draw in or recirculate the hot moist air.

Plenum

The **plenum** in CTW-302 is the open area directly under the fan. In an induced draft, cross-flow cooling tower, the velocity of air creates a partial vacuum under the fan as it expels the vapor-enriched air.

Water Distribution System

CTW-302 consists of a 12-in. deep pan, 4′ × 20′, with holes equipped with stationary spiral nozzles that distribute water across the fill. A pan with holes and nozzles in it is located on the east and west sides of the cooling tower. The ends are solid and have an access door for inspection. Some systems utilize a pipe and spray nozzle design. The hot water header has the ability to divert flow to the cooling tower or **water basin,** depending on the temperature of the process flow exiting heat exchanger 204.

Water Basin

The water basin is a concrete reinforced structure designed to store water and provide a foundation upon which the rest of the cooling tower can be supported. The basin is designed to collect water as it flows across the horizontal fill and downward. The water basin provides suction for the water pump and must be able to resist chemical attack. Technicians working on the cooling tower basin must pay careful attention to water pH, temperature, parts per million (ppm), and biological problems.

Louvers

The louvers on CTW-302 are fixed and cannot be adjusted. Each set of louvers is evenly spaced and designed to direct airflow across the downward flow of water. During the winter, ice formation is carefully monitored and the cooling tower cell is kept operational. The lowest the water temperature is allowed to go is 40°F before a low-pressure steam system will warm up water in the basin.

Fill

The fill provides a vastly expanded surface area interface that enhances air–liquid contact. When evaporation occurs, air becomes saturated with water and is carried out of the cooling tower system. The fill is comprised

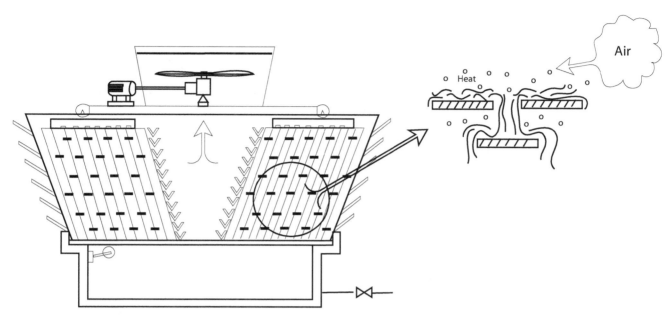

Figure 11-3 *Fill or Splashboards*

of pressure-treated materials; wooden slats run through a plastic support structure. The fill is designed to provide plenty of surface area so that good air-liquid contact is ensured. Evaporation takes place inside the fill area, transferring heat energy to the moist area, increasing the relative humidity in this area to 100%, and moving this heated plume into the atmosphere. The plume or fog is virtually harmless except that it can reduce visibility when it touches the ground. Evaporation accounts for 80–90% of the heat transfer in a cooling tower. This convective process takes place as the hot water cascades down the fill or splash boards. When the hot water spreads out across the fill, air flows over the area. Figure 11-3 illustrates how the rising hot water vapor or heat energy is carried out of the cooling tower and the cooler water drops down into the water basin.

Drift Eliminators

Drift eliminators are specifically located in the tower to block or prevent water loss from the system. As hot, moisture-rich vapor flows across the fill, which is lighter than the outside air, it accelerates as it enters the plenum. This warm air increases in velocity as it is propelled high above the cooling tower by the fan.

Blowdown

Draw-off or blowdown is primarily used to control the buildup or concentration of minerals in the recirculation water. The blowdown system is designed to control the level of suspended solids in the water basin. High levels of suspended solids will cause fouling. Blowdown is closely related to the term "concentration cycles." Blowdown automatically removes 7–10% of water in the water basin and replaces it with fresh water. In most cases, the water makeup system runs continuously due to evaporative losses, **windage or drift** losses, and draw-off.

Concentration Cycles

Concentration cycles in CTW-302 typically can range from 3 to 7. The cooling tower is typically set at six concentration cycles before it blows down the system. A cycle describes one pass from the cooling tower to the process and back to the cooling tower. Cycles of concentration display the accumulation of dissolved minerals in the recirculation system.

Induced Draft, Cross-Flow Cooling Tower

Cooling towers are classified by how they produce airflow and how they produce airflow in relation to the downward flow of water.

Cell

The **cell** is the smallest subdivision of a cooling tower that can function as an independent unit. Some cooling tower systems have multiple cells. CTW-302 has a single cell operation.

pH Control

The pH of the cold-water basin has a tendency to increase in alkalinity. The control set point is 7.8 pH. An automatic control system continuously analyzes the pH in the basin and adds small amounts of acid to maintain operational requirements. The cooling tower technicians maintain a specification range of 7.6–8.4 pH on water in the recirculation system.

Water Treatment System

The composition of the chemicals used in the water treatment system is designed to control **scale,** algae, corrosion, and wood decay, and to help suspended solids to precipitate out in the basin. A control system monitors conditions in the basin and maintains unit specifications by adding liquid chemical treatment.

Basin Temperature Control

The temperature in the basin is never allowed to drop below 40°F during winter conditions. A low-pressure steam system is designed to admit steam into the basin so that ice will not form on the exposed parts of the cooling tower. A control system monitors conditions in the basin.

Basin Level Control

Water in the basin is controlled at 75% as a level element and transmitter sends a signal to a remotely located controller. During blowdown operation, the automatic water makeup valve is in the wide-open position.

Air and Water Temperature

One of the most important variables controlled on the cooling tower is a term called "the **approach to the tower.**" This is defined as the temperature difference between water leaving the cooling tower and the WBT of air entering the tower.

Instrument Systems

To the untrained eye, a cooling tower appears to be very simple; however, a number of automated control loops are required to run this system. Instrument systems used in the cooling tower system include the following.

Temperature

TIC-301 Temperature indicating controller—located on the hot water return line. TIC-301 has the ability to divert flow into the cold-water basin or into the water distribution system of the cooling tower.

TIC-302 Temperature indicating controller—located on the cold-water basin. This control system is designed to keep the water temperature on the basin between 40°F and 85°F.

Ti-300A Temperature indicator—located near the **air intake louvers** on the cooling tower.

Ti-300B Temperature indicator—located on the water discharge line from cooling tower. The combination of Ti-300A and Ti-300B is used in calculating the approach to the tower.

Ti-300C Temperature indicator—located on the shell outlet of Ex-204. Ti-300C allows process technicians to observe how close to specification the cooling tower/heat exchanger system is operating.

Pressure

Pi-300A Pressure indicator—located on the tube inlet to Ex-204.

Pi-300B Pressure indicator—located on the tube outlet from Ex-204. The delta pressure or ΔP (difference between the inlet and outlet pressure) is used to calculate fouling inside the exchanger.

Flow

FIC-300 Flow indicating controller—controls the flow of water through Ex-204. FIC-300 is located on the discharge side of P-302. The flow rate on P-302 is carefully controlled at 525 gpm with FCV-300. Changes in the flow will cause other variables in the system to go into alarm. An orifice plate is located in the line, which allows a transmitter to measure and send a signal to the controller. The controller compares the signal to the set point and makes adjustments to the control valve. FIC-300 can be operated manually or automatically.

Fi-300 Flow indicator—located on the discharge side of Ex-204. This indicator allows the outside operator to verify the flow rate with FIC-300.

Level

LIC-300 Level indicating controller—designed to maintain a constant level (75%) in the cooling water basin. The cooling water basin is in constant movement as the heat rejection (evaporation) process occurs continuously. In most cases, the water makeup system runs continuously due to evaporative losses, windage or **drift losses,** and draw-off.

HLA-300A High-level alarm—the purpose of this alarm is to alert the technician to a problem in the water basin.

LLA-300B Low-level alarm—alerts the operator when the level in the cooling tower drops below the set point.

Analytical

AIC-300 Analytical indicating controller—the purpose of this controller is to control the pH inside the cooling water basin. The pH in the cold-water basin has a tendency to increase in alkalinity. The control set point is 7.8 pH. AIC-300 continuously analyzes the pH in the basin and adds small amounts of acid to maintain operational requirements. The cooling tower specification range is 7.6–8.4 pH on water in the recirculation system.

AIC-301 Analytical indicating controller—the purpose of this controller is to monitor and control the levels of suspended solids. A process called draw-off, or blowdown, is primarily used to control the buildup or concentration of minerals in the recirculation water. The blowdown system is designed to control the level of suspended solids in the water basin. High levels of suspended solids will cause fouling. Blowdown is closely related to the term "concentration cycles." Blowdown automatically removes 7–10% of water in the water basin and replaces it with fresh water. In most cases, the water makeup system runs continuously due to evaporative losses, windage or drift losses, and draw-off.

The concentration cycle range in CTW-302 is set at six concentration cycles before it blows down the system. A cycle describes one pass from the cooling tower to the process and back to the cooling tower. Cycles of concentration display the accumulation of dissolved minerals in the recirculation system.

AIC-302 Analytical indicating controller—the purpose of this controller is to control scale, algae, corrosion, and wood decay, and to help suspended solids to precipitate out in the basin. AIC-302 monitors conditions in the basin and maintains unit specifications by adding liquid chemical treatment. Liquid treatment is set at 4.5.

SIC-300 Speed indicating controller—the speed on the fan is set at 1250 rpm and can be adjusted by the technician. An on–off switch is also located near the motor.

The Cooling Tower System

Before operating the cooling tower system, a technician needs to be familiar with the scientific principles associated with heat transfer, evaporation, fluid flow, equipment relationships with heat exchangers, instrument systems, safety, and the basic components of the cooling tower system.

The safety aspects of the CTW-302 system include the following areas:
- chemical additives on AIC-302 (see chemical list and MSDS)
- rotating equipment (fan-300)
- hazards of hot water
- equipment failures (tube leak Ex-204)
- working at heights
- hazards of working with acid (see MSDS)
- working safely on top of the cooling tower
- confined space entry (water basin empty)
- hazardous energy.

An equipment checklist and sampling procedure are assigned to each new technician. The oil level on the fan gearbox must be checked when the fan is off. Hands and fingers must stay clear of the moving fan blades even though a protective housing surrounds the rotating equipment. Samples are taken at each shift and taken to the lab. These samples include feed water, basin water, return, and water.

Operational Specifications (SOP, SPEC Sheet, and Checklist)

The following procedures were developed using the principles of instructional system design (ISD). A group of technicians including several senior level technicians, two reasonably new technicians, and a design engineer were involved in the development team. A profile or matrix was developed during the analysis phase identifying the key competency categories and supporting objectives. The materials are designed to move from simple to complex and reflect the operational procedures for the CTW-302 system.

Action	Notes
1. Establish level on the water basin and set LIC-300 to AUTO.	
2. Sample water in basin and send to lab.	
3. Lineup pump 302.	
4. Lineup Ex-204.	
5. Start pump 302 and monitor Pi-300A (50 psig) and Pi-300B (45 psig).	
6. Set FIC-300 to 525 rpm and put in AUTO.	
7. Turn fan-300 on and set SIC-300 to 1250 rpm.	
8. Set AIC-300 to 7.8 pH and put in AUTO.	
9. Set AIC-301 to 30 ppm and put in AUTO.	
10. Set AIC-302 to 4.5 GPH and put in AUTO.	
11. Set TIC-301 to 125°F and put in AUTO.	
12. Set TIC-302 to 60°F and put in AUTO.	
13. Record Ti-300A (WBT).	
14. Record Ti-300B.	
15. Calculate the approach to the tower.	
16. Carefully monitor all conditions.	
17. Collect samples for makeup water.	
18. Collect samples for water basin.	
19. Collect sample for pump 302 discharge.	
20. Cross-check process variables with SPEC sheet.	

Specification Sheet and Checklist

Level

1.	LIC-300	75%	AUTO	Basin level
2.	LA-1	85%	High	Basin level high
3.	LA-2	65%	Low	Basin level low

Flow

4.	FIC-302	525 gpm	AUTO	Water flow
5.	Fi-300	525 gpm	—	Ex-204 tube outlet water flow

Analytical

6.	AIC-300	7.8 pH	AUTO	pH-acid
7.	AIC-301	30 ppm	AUTO	Blowdown
8.	AIC-302	4.5 GPH	AUTO	Chemical additive
9.	SIC-300	1250 rpm	On/Off	Fan speed

Pressure

10.	Pi-300A	50 psig	Water flow	—
11.	Pi-300B	45 psig	Water flow	—

Temperature

12.	TIC-301	125°F	AUTO	Hot water return
13.	TIC-302	60°F	AUTO	Cold-water basin (low)
14.	Ti-300A	—	—	Wet-bulb air temperature
15.	Ti-300B	85°F	—	P-302 discharge
16.	TR-300	125°F	—	Hot water return
17.	Ti-300C	—	—	Ex-204 shell outlet temperature

Common Cooling Tower Problems and Solutions

CTW-302 can be shut down easily; however, unless a serious problem occurs or equipment repair and turnaround are scheduled, the tower is kept in continuous operation. There are some common problems and concerns for which the cooling tower system must be monitored.

Cooling Tower Efficiency

Wet-bulb temperature (WBT) and humidity cooling must be monitored. Since wet cooling towers respond to evaporation, relative humidity has an effect on the efficiency of cooling tower system. WBT is taken using a specially designed thermometer and wick arrangement. In multicell operations, additional fans are typically turned on; however, since CTW-302 has a single fan, the RPMs can be increased to improve airflow through the cooling tower. Higher temperatures will also affect the ability of the cooling tower to expel heat.

pH Problems

Process technicians use the term pH as an expression of alkalinity or acidity. The pH scale has numerical values from 0 to 7 and then from 7 to 14. A pH reading of 7 is classified as neutral on the pH scale. As the indicator needle begins to decrease below 7, it moves into the acidic range. Readings above 7 are classified as alkaline or caustic. Examples of acids include lemon juice, sulfuric acid, and hydrochloric acid. Examples of alkaline substances include ammonia, soap, bleach, and drain cleaner. Figure 11-4 shows a typical pH scale.

Scale Formation

Scale is composed of dissolved minerals such as calcium and magnesium. Scale buildup is a process in which concentrated minerals form a solid coating on the inside of piping and tubes, reducing fluid flow and heat transfer.

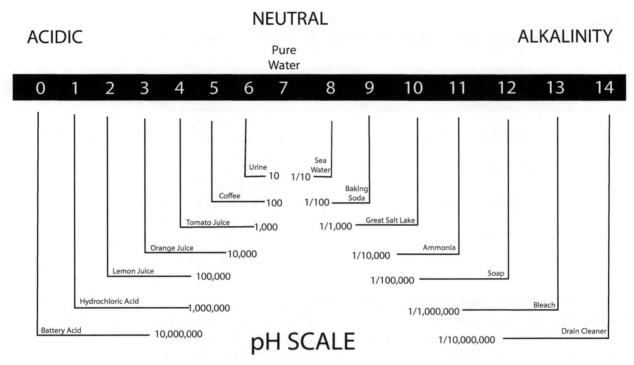

Figure 11-4 *pH Scale*

Cooling tower water that has a high pH tends to enhance scale formation. Sulfuric acid is used to treat the catch basin to lower a high pH.

Total Dissolved Solids (TDS)

Water naturally contains magnesium and calcium in the form of dissolved solids. When water **evaporates** out of the cooling tower, dissolved minerals are left behind and over time will concentrate and form scale. Dissolved solids break down the components in wood fibers that provide support for the cooling tower internal structure. Sulfuric acid can minimize problems caused by dissolved solids. Blowdown on CTW-302 is timed so that chemical injection has time to blend into the recirculation system. Chemical addition is performed in a section of the cooling tower where turbulent flow enhances mixing and where it is not drawn into pump 302's suction before adequate blending has occurred.

Dissolved Gases

Cooling tower water may have dissolved gases that chemically react with iron, such as oxygen, hydrogen sulfide, and carbon dioxide. These gases will enhance corrosion, destroy metal surfaces, damage valuable equipment, and deteriorate metal pipes, brackets, and bolts used to secure wood products together. High quantities of dissolved gases will make the catch basin water acidic.

Ruptured Tubes

If the tube ruptures on Ex-204, the butane–pentane–catalyst mix will pour into the recirculation water since it is at 100 psig and the recirculation water is at 45/50 psig. The pH of the cooling tower water would also become very alkaline and set off the high pH alarm.

Microorganisms: Chlorine or Bromine Concentrations

Cooling towers provide the perfect breeding ground for microorganisms since they are exposed to warm water and sunlight. Microorganisms form slime that can build up and restrict water flow and conductive heat transfer. Biocides such as chlorine and bromine can control the growth of these organisms. Unfortunately, chlorine is an extremely hazardous material and must be handled with caution. Bromine is a little safer to handle and use. CTW-302 utilizes a special blend of biocides to keep the cooling tower operating smoothly. Other chemical inhibitors are added to this blend and need to be thoroughly mixed with basin water.

Fan-300 Failure

CTW-302 operates the fan when temperatures and relative humidity reach predetermined levels. During the winter months, the fan is rarely operated and the hot water bypass system is frequently used. A spare motor, gearbox, and fan blade arrangement is located in the maintenance shop; however, several days are typically required for large projects. It is possible to operate the cooling tower in a natural draft condition.

Pump 302 Failure

If pump 302 fails, the distillation system will immediately go into alarm condition. The pump back-up system will need to be placed online immediately. P-302 is a vertically mounted centrifugal pump with a series of small screens installed on the suction side of the pump. These screens keep the liquid entering the pump clean and prevent internal damage. Net positive suction head (NPSH) and net positive discharge head (NPDH) must be kept within specific limits in order for the recirculation system to operate properly.

Instrument Problems

Instruments on the cooling tower include three analytical control loops: one level control loop, two temperature control loops, and one speed control loop on the fan. A variety of simple instruments are mounted on the cooling tower system.

Suspended Solids, Ex-204 Tube Fouling

As air circulates through the cooling tower, it carries solids with it in the form of small dust particles. These solids are captured by the downward flow of water and collected in the catch basin, where they form sludge. Sludge is removed through the blowdown system. Suspended solids can build up inside the tubes of a heat exchanger and cause fouling. Fouling is a term used to describe restriction or plugging.

Blowdown and Cycles of Concentration Problems

Concentration cycles in cooling towers range from 3 to 7 and are dependent upon the quality of the makeup water. Circulating water in CTW-302 is filtered and treated with **biocides and algaecides** to prevent microorganism growth and a control system for pH adjustment.

Broken or Collapsed Fill

High winds, ice buildup, corrosion, microorganisms, and other variables can cause structural damage to the cooling tower. Although a cooling tower is a simple design, it needs all of its parts to operate efficiently. Fill or splashboards can be replaced and put back in place. Bolts and brackets that have been corroded need to be repaired.

Water Distribution System Problems

The water distribution system has a variety of components including a series of valves, pans, water distributors, and pan covers that can be damaged. Pipe leaks are not uncommon and can quickly lead to larger problems. If the water distributors become dislodged from the pan, uneven amounts of water can flow over the fill before it has time to contact air flowing through the tower.

Troubleshooting Scenario 1

Cooling towers are complex devices that are subject to a variety of problems. In the first troubleshooting scenario, fan-300 fails and Ex-204 ruptures a tube. When these things occur, the system reacts. A series of questions are asked to determine how these problems affect other variables in the system. Figure 11-5 displays the cooling tower system and presents a series of what-if scenarios.

Troubleshooting Scenario 2

In Figure 11-6, the level control valve 300 fails in the closed position and Ex-204 partially plugs up or fouls. As in the first scenario, the primary cause creates a cascading effect, initiating secondary problems.

Troubleshooting Scenario 3

In Figure 11-7, the cooling towers blowdown feature is activated, creating a number of secondary responses. In the last problem, the flow control valve fails in the closed position. Carefully review each of the problems and determine whether it goes up, down, or stays the same.

Summary

CTW-302 is classified as an induced draft (or draw through), cross-flow, single cell device that is primarily designed to control the temperature on

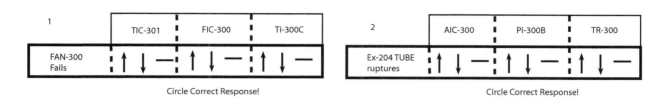

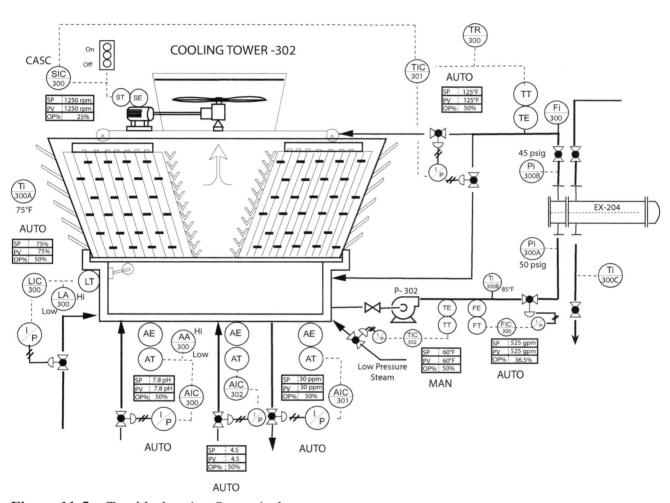

Figure 11-5 *Troubleshooting Scenario 1*

condenser Ex-204. A cooling tower is often referred to as a heat rejection device designed to extract excess heat from the returning water and expel it into the atmosphere. This type of heat transfer relies on the principle of evaporation. When this process occurs, the heat from water is absorbed by the air stream, which raises the relative humidity to near 100%. These heated currents are quickly dissipated by the wind. CTW-302 is an evaporative heat rejection device that can significantly reduce water temperatures.

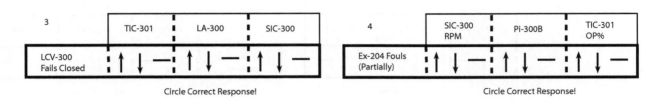

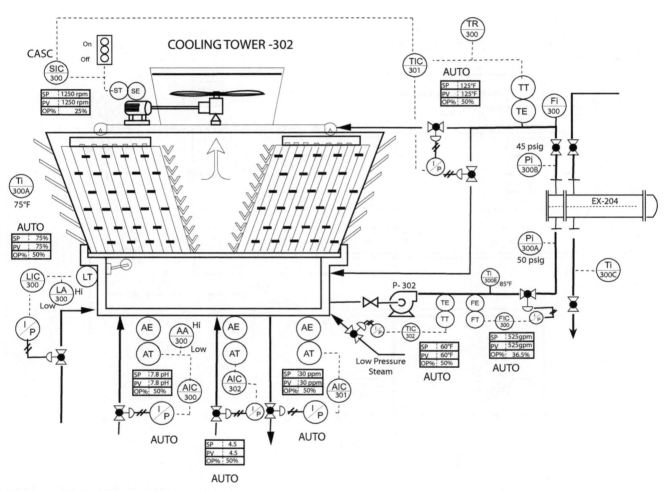

Figure 11-6 *Troubleshooting Scenario 2*

CTW-302 is an enclosed structure with a system of louvers designed to direct airflow across the fill. As warm water enters the top of the cooling tower, the water distribution system carefully sprays or directs fluid flow over a labyrinth-like honeycomb, splashboards, or fill. The purpose of the fill is to allow the hot water to spread out over the surface of the boards. The fill provides a vastly expanded surface area interface that enhances air– liquid contact. As evaporation occurs, air becomes saturated with water and is carried out of the cooling tower system. CTW-302 uses a fan to draw air into

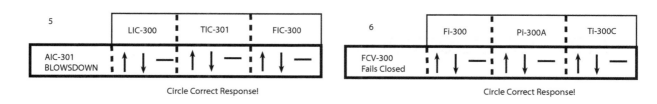

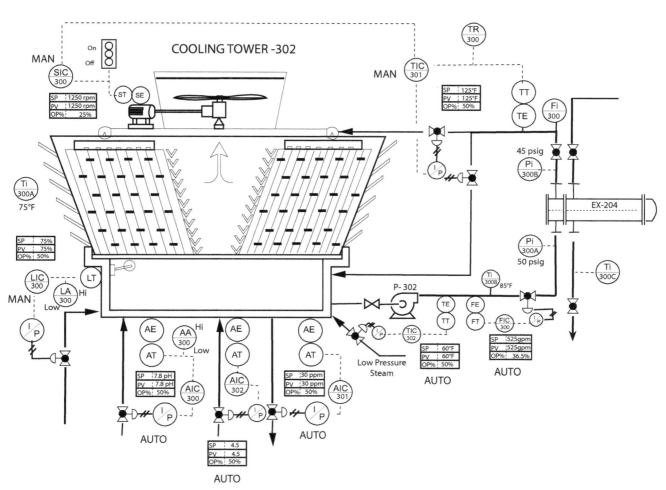

Figure 11-7 *Troubleshooting Scenario 3*

the fill and across the fill. The fan is located on the top of the cooling tower and slowly draws air into the system and rapidly discharges it at the exit point. The cooled water continues to drop through and over the fill until it enters the basin. A typical cooling tower is a heat transfer device designed to cool water so it can be reused in industrial applications.

CTW-302 is equipped with a number of automated systems designed to control pH, water level, water flow, water temperature, water treatment, fan speed, and system blowdown (TDS).

Review Questions

1. Describe the purpose of LIC-300 and how it works.

2. Describe the purpose of AIC-300 pH control and how it works.

3. Describe the purpose of AIC-301 blowdown system.

4. Describe the purpose and operation of AIC-302 water treatment control system.

5. Describe the purpose and operation of TIC-301.

6. Explain how SIC-300 regulates the temperature in the cooling tower cell.

7. Explain what the approach to the tower is and how it is calculated.

8. Evaporative cooling accounts for what percentage of the cooling effect in CTW-302.

9. Define the term "plenum."

10. Define the term "**psychrometry.**"

11. List the various ways that CTW-302 is classified.

12. Describe how the temperature in the water basin is controlled so that ice formation is reduced.

13. List the various components of CTW-302.

14. Describe the term "concentration cycles" and how it applies to CTW-302.

Boiler Model

LEARNING OBJECTIVES

After studying this chapter, the student will be able to:

- Describe the basics of boiler operation.
- Describe the scientific principles associated with steam generation.
- Draw a simple block flow diagram of the boiler system.
- Describe the main components of a water tube boiler.
- Operate the boiler system.
- Identify common problems encountered when operating a boiler.
- Solve various troubleshooting scenarios associated with boiler system operations.
- List the safety aspects associated with operating the boiler system.
- List the operational specifications of the pump system: pressures, flow rates, temperatures, etc.

Key Terms

Boiler load—plant demand for steam.

Natural gas burner—a device designed to evenly distribute air and fuel vapors over an ignition source and into a boiler firebox.

Damper—a device used to regulate airflow.

Deaerator—a device designed to remove entrained oxygen in treated boiler feed water. The deaerator is a water feed tank equipped with a boiler feed water pump, a safety valve, a vent, and a level-control system.

Desuperheating—a process applied to remove heat from superheated steam.

Downcomers—the hot water inlet tubes that connect the steam-generating drum to the mud drum. Fluid direction is from the steam-generating drum to the mud drum due to density differences between the hotter steam-generating tubes and riser tubes.

Economizer—the cooler section of the boiler located between the stack and the furnace. Feed water is preheated in this part of the boiler.

Mud drum—the lower drum of a water tube boiler. The mud drum is always run liquid-full and provides feed for the riser and steam-generating tubes.

Risers—the return tubes that discharge from the mud drum, back into the steam-generating drum with steam and liquid. The riser allows the steam-generating drum to circulate to the mud drum and back into the upper drum.

Steam-generating drum—a large upper drum partially filled with treated and preheated boiler feed water. The vapor disengaging cavity in the drum allows operating pressure to be enriched. The steam drum is connected to a liquid-filled lower drum or the "mud drum" that has two discharge ports: the riser tubes and the steam-generating tubes. Both of these tubes discharge steam into the upper steam drum.

Superheated steam—steam that is heated to a higher temperature.

Water hammer—a condition in a boiler in which slugs of condensate (water) entrained with high-velocity steam can damage process equipment.

Water tube boiler—a type of boiler that passes water-filled tubes through a heated firebox.

Introduction to Boiler Operation (Complete System)

Steam-generation systems are often referred to as boilers. A boiler is a device used by the chemical processing industry to produce steam. At first glance, it appears that a boiler is a simple device used to boil water; however, this process is very complex and an integral part of many industrial applications. Steam is produced at high, medium, and low pressure. Steam is used to protect equipment during cold periods, extrusion, stripping, preheater and heat exchanger systems, flare systems, laminating, firefighting, steam turbines, and a large variety of chemical applications.

Steam generators use a combination of radiant, conductive, and convective heat transfer methods to produce steam. A boiler is composed of a furnace, **economizer** section, stack, **deaerator,** and a steam return system. A series of tubes pass through the economizer section and the furnace. Inside the furnace is a large upper **steam-generating drum,** a lower **mud drum,** downcomer tubes, riser tubes, steam-generating tubes, and a series of burners.

Steam generators come in two basic designs: water tube and fire tube. These two designs are also referred to as direct fired and indirect fired. **Water tube boilers** are used in large industrial operations, while fire tube steam generators are used in smaller applications such as hospitals and colleges. The most complex steam-generation system is the large water tube boilers, which are part of a very elaborate matrix of equipment and systems. A water tube boiler consists of an upper and lower drum connected by a structured network of tubes. The lower drum and water tubes are filled completely with water, whereas the upper drum is only partially filled. This arrangement allows steam to pass through mechanical separators in the upper drum, flow to a superheater section, and exit the boiler. As heat is applied to the boiler firebox, water flows from the upper drum through **downcomers** into the lower drum. Tubes, called **risers,** cause water and steam to flow into the upper drum because of density differences.

Boiler water circulation operates under the principle of differential density. When fluid is heated, it expands and becomes less dense. Cooler water flows from the upper or steam-generating drum, through the downcomers, to the mud drum (the lower drum), and then rises as steam is generated. Circulation continues, and makeup water is continually added to the upper drum to replace the steam that is generated.

Water circulation continues in a water tube boiler because steam bubbles in the lower drum move up the riser tubes and cause the density of water to decrease. The cooler water in the downcomer flows into the mud drum. The riser and steam-generating tubes are physically located near the burners. Steam moves up the riser and steam-generating tubes and into the upper steam-generating drum. Steam generation causes pressure to rise. The upper steam drum has a large vapor disengaging cavity that is specifically designed to build pressure since vapors are much more compressible than liquid. A pressure-control valve is located on the steam outlet line and is designed to control system pressures. When the desired set point is achieved, the steam generator is "placed on the line." Pressure is maintained in the boiler by adding makeup water and continuously applying heat. Figure 12-1 is an illustration of the equipment and systems associated with the steam-generation system.

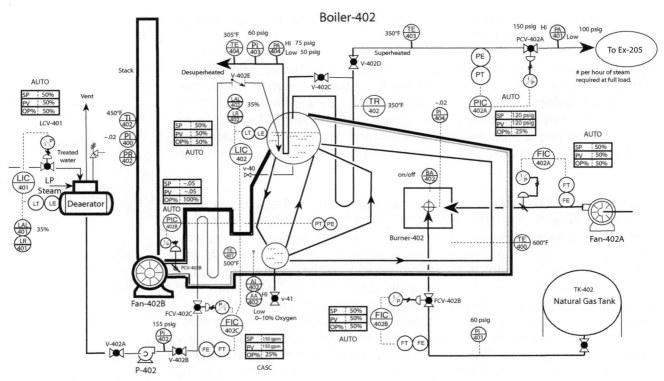

Figure 12-1 *Steam Generator*

Equipment Descriptions (Block Flow and Safety)

B-402, Furnace, Firebox, and Refractory

The water tube boiler firebox or furnace is designed to reduce the loss of heat and enhance the heat energy being applied to the boiler's internal components. Boiler furnaces have a shell, refractory lining, burners, convection-type section, radiant section, fans, and oxygen-control system. The boiler firebox is lined with a refractory brick that has a reflective and insulative property designed to hold heat in and reflect it back into the furnace. Insulation is installed between the shell and the brick. The bricks are typically stacked one on top of the other without mortar. The head joints and the bed joints are very tight and the brick is typically held in place by a stainless steel rod designed to receive a half moon-shaped notch on the brick. A type of insulative cement called castable is typically poured over the shell flooring. Four special high-temperature firebrick blocks are placed around the burners to prevent damage to the floor.

Tubes

The steam generator contains several types of tubes: a downcomer tube, a riser tube, and a steam-generating tube connecting the lower and upper drums. Water goes through the firebox and back up to the upper steam-generating drum. Downcomer tubes are classified as warm water tubes connecting the upper and lower drums. The riser tubes are classified as hot water tubes between the upper and lower drums. A water makeup line maintains the level in the upper steam-generating drum. A large vapor disengaging cavity exists above the boiling water. This vapor cavity allows pressure to build and expand as water boils at 212°F; however, the vapor disengaging cavity is exposed to the higher temperatures of the firebox. This process allows the steam to increase in temperature to temperatures as high as 1050°F. A correlation exists between pressure and temperature. The higher the temperature of the steam, the higher is the pressure.

Steam is removed from the upper steam-generating drum and heated to the desired temperature in superheater tubes. **Superheated steam** temperature can be increased as it reenters the furnace. Some processes cannot handle high temperatures, so the superheated steam is cooled off. This process is referred to as **desuperheating.** The tubes found in a boiler include:

- water makeup tubes and distributor
- riser tubes
- steam-generating drum
- superheated high-pressure steam
- desuperheated lower-pressure steam
- spuds—fuel-filled tubes.

Steam-Generating Drum and Mud Drum

The drums inside a boiler furnace are pressure cylinders connected by a complex network of tubes. The drums are classified as the upper steam-generating drum and the lower liquid-filled mud drum. The steam-generating drum contains a water-steam interface. The upper drum contains the feed water inlet distributor, a blowdown header, and water separation equipment. Steam has a number of hazards associated with it. Under different operational conditions, steam expands in a variety of ways. Figure 12-2 illustrates how steam responds to variations in pressure.

Boiler Burners

Modern steam-generation systems utilize natural gas and air to provide heat through the combustion process. Burners mix air and fuel through a distribution system designed to mix them into the correct concentrations so that combustion can occur easily. The primary components of the combustion apparatus include the following:

- **dampers** that regulate air into the burner
- air ducts with fixed blades designed to create a swirling effect as air and fuel mix

Figure 12-2
Hazards of Steam

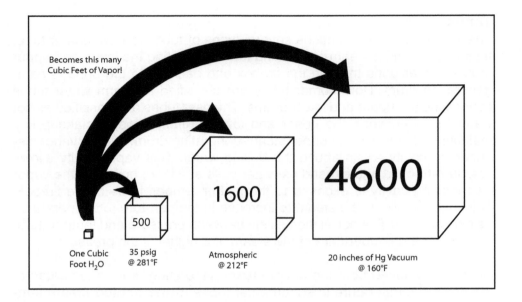

Becomes this many
Cubic Feet of Vapor!

4600

1600

500

One Cubic
Foot H₂O

35 psig
@ 281°F

Atmospheric
@ 212°F

20 inches of Hg Vacuum
@ 160°F

- spuds, which are gas-filled tubes directing flow to the mixing section of the burner
- igniter, a device equipped with a spark plug that provides the ignition source of the fuel–air mixture
- flame detection equipment equipped with an automatic fuel shutoff in the event the flame goes out
- factory mutual valves, called FM valves, used to shut off fuel gas in the event of a low drum level or flame failure
- forced draft fans to supply combustion air.

Fuel System

Tk-402 is a sphere-shaped tank designed to hold natural gas for use in the boiler's burner. Natural gas burns much cleaner than the fuel oil systems commonly used 25 years ago. The fuel pressure is maintained at 60 psig. A flow control loop, FIC-402B, controls gas flow to the burner. A burner alarm (BA-1) is ever-vigilant in monitoring whether the burner is on or off.

Economizer Section

The economizer section is used to increase boiler efficiency by preheating water as it enters the system. This section is a series of headers and tubes located between the firebox and the stack. Temperatures are typically lower in the economizer section than in the rest of the system, but the hot flue gases moving out of the firebox and into the stack still have enough heat to offset energy costs. The economizer section in a boiler is very similar to the convection section in a fired heater systems. Both operate under the energy-saving concept of using the hot flue gases before they are lost out the stack.

Stack Damper and Fan

Airflow through the steam-generating system is carefully controlled since excess oxygen will burn more fuel, expending company resources, whereas too little airflow will not provide complete combustion and temperatures will be decreased. Airflow through the furnace can be classified as (1) natural draft, (2) forced draft, (3) induced draft, and (4) balanced draft. Natural draft furnaces do not use a mechanical device to produce airflow. Forced draft fans accelerate airflow to the burner, while induced draft systems gently pull air into the furnace. Induced draft fans are typically located in the stack, with the motor located outside the stack wall. Balanced draft systems are the most complex and require control systems to help the fans work together.

The stack damper on boiler-402 is a pressure-control valve located in the inlet of the fan suction. Since the furnace is under a negative pressure, sensors are located in the firebox so that correct pressure readings can be determined.

Deaerator and Feed Water Pumps

A deaerator is a device designed to remove entrained oxygen in treated boiler feed water. The deaerator is the water feed tank equipped with a boiler feed water pump, a safety valve, a vent, and a level-control system.

Tank-402 Natural Gas Storage

Tank-402 is a cylindrical tank designed to safely store and transfer natural gas. The tank is equipped with a safety relief system that discharges to the flare in the event of an emergency. System pressure is maintained at 60 psig for transfer purposes. Pressure indicator-401 measures the pressure as the natural gas flows to the flow control valve-402B.

Instrument Systems

Steam-generating systems are very large and very complex. Modern control instrumentation makes the operation and control of this type of system much easier. There are a number of hazards associated with the boiling water and producing steam. High-pressure steam directed in a narrow beam can cut a broom stick in half. High-pressure steam can also provide rotational energy to a steam turbine. Instrument systems are only as useful as the technicians who work with them. Alarms that are ignored or bypassed, control loops that are left in manual mode, or process problems that are ignored can lead to serious consequences.

Temperature

> BA-1 Burner alarm—located near the burner and designed to alert the operator in the event the burner goes out.
> Ti-402 Temperature indicator—located in the stack.
> TR-402 Temperature recorder—located on the discharge of the steam-generating drum. Set at 350°F.

Pressure

PIC-402A Pressure indicating controller—controls the pressure on the steam-generating drum. Set at 120 psig.

PIC-402A Pressure indicating controller—controls the pressure inside the boiler. Set at −0.05 psig.

Pi-400 Pressure indicator—Located in the stack. Set at −0.02 psig.

Pi-401 Pressure indicator—Located on the natural gas supply line. Set at 60 psig.

Pi-402 Pressure indicator—Located on the discharge side of P-402. Set at 155 psig.

Pi-404 Pressure indicator—Located at the burner. Slight negative pressure, −0.2 inch of water.

PA-404 Pressure alarm Hi/Lo—Located on the discharge line of desuperheated steam.

PA-401 Pressure alarm Hi/Lo—Located on the steam discharge line.

PR-402 Pressure recorder—Located on the stack.

Flow

FIC-402A Flow indicator controller—Located on the discharge of fan-402A. Set at 50%.

FIC-402B Flow indicator controller—Located on the natural gas supply line to the boiler burner. Set at 50%.

FIC-402C Flow indicator controller—Located on the discharge of pump-402 and classified as a slave controller to LIC-402. Set at 150 gpm and cascaded to LIC-402.

Level

LIC-402 Level indicating controller—Designed to maintain a constant level in the steam-generating drum. Set at 50%.

LA-1 Low-level alarm—Designed to alert the process technician in the event of a low-level condition in the deaerator drum. Set at 35%.

LA-2 Low-level alarm—Designed to alert the process technician in the event of a low-level condition in the steam-generating drum. Set at 35%.

LR-1 Level recorder—Located on the deaerator.

LR-2 Level recorder—Located on the steam drum.

Analytical

Ai-402 Analytical indicator—The purpose of this control system is to monitor gas flow through the boiler and is located at the throat of the economizer section. 0–10% oxygen.

AA-402 Analytical analyzer—Located at the throat of the economizer section. Measures O_2 levels in boiler. Desired target: 0–10%.

The Boiler System

The primary purpose of B-402 steam-generation system is to provide steam 120-psig steam to Ex-205. Ex-205 is a kettle reboiler used to maintain energy balance on the debutanizer column. This medium pressure steam is also used in a variety of other applications. When B-402 is initially started up, a series of steps are followed. One of the most important safety concerns is to establish water flow and drum levels prior to lighting off the burner. When the burners are lit, hot combustion gases begin to flow over the generating tubes, riser tubes, downcomer tubes, and drums. Radiant, conductive, and convective heat transfer begins to take place. Hot combustion gases flow out of the firebox, into the economizer section, and out the stack. Fans provide airflow through the furnace, creating a slight draft or negative pressure. Since the furnace is hotter than the outside air, significant density differences occur. Water temperature increases at programmed rates. Pressure begins to increase inside the large vapor disengaging cavity in the upper drum. As the temperature of water inside the generating and riser tubes increases, the density of water decreases and initial circulation is established. Bubbles begin to form and rise in water, increasing circulation and pressure. Each time water passes through the tubes, it picks up more heat energy. When the pressure increases to slightly above the system pressure set point, steam will flow to the header.

Inside the upper steam-generating drum of B-402, steam and water come into physical contact, saturating the steam. This saturated condition means that for every temperature of water, a corresponding pressure of steam exists. The pressure on the water sets the temperature as long as the steam and water are in contact. Basic boiler design removes the steam from the upper steam water drum and heats it up at essentially the same pressure. This process is referred to as superheating. B-402 is designed to operate at 120 psig. Some operating facilities require low-pressure steam. This process is referred to as desuperheating. During the desuperheating process, part of the superheated steam is routed through the boiling liquid in the steam drum, cooling it down to a lower pressure. The boiling water is cooler than the 120 psig steam and reduces the pressure to around 60 psig.

A number of hazards are associated with the operation of a boiler system. Some of these hazards include:
- hazards associated with high-temperature steam, "burns"
- hazards associated with using natural gas
- hazards associated with leaks
- instrument failures
- confined space entry permit
- opening blinding permits
- isolation of hazardous energy permit, "lock-out, tag-out"
- routine work
- hazards associated with lighting burners

- exceeding boiler temperatures or pressures
- hazards associated with using water treatment chemicals
- error with valve lineup, resulting in explosion or fire.

Because a large list of potential hazards exist beyond the above list, careful training is required for all new technicians assigned to utilities.

Operational Specifications (SOP, SPEC Sheet, and Checklist)

While a number of variations are possible when looking at an operational procedure, a single, simple procedure must be developed. Following standard operating procedures is one of the most important things a technician can do. Learning the steps in starting up and operating a boiler system or any other system is an invaluable skill that makes a technician an effective member of a work team. In order to start up a steam-generation system a technician should:

Action	Notes
1. Lineup P-402	
2. Set FIC-402C to 150 gpm and place controller in CASC	
3. Set LIC-402 to 50% and place controller in AUTO	
4. Set PIC-402A to 120 psig and place in AUTO	
5. Ensure level in deaerator is at 50% and water makeup is lined up	
6. Start pump 402 and monitor system pressures. Establish proper level in upper and lower drum	
7. Open PCV-402B to 100% leave in MAN	
8. Open FCV-402A to 100% leave in MAN	
9. Ensure boiler has been purged with steam for 10 min	
10. Monitor process pressures and level in TK-402	
11. Ensure pilot light on burner 1 is active	
12. Set FIC-402A to 50% and place controller in AUTO	
13. Set PIC-402B to −0.05 and place controller in AUTO	
14. Set FIC-402B to 50% and place in AUTO	
15. Start fans-402A and 402B.	
16. Light burner and allow boiler to come up to pressure and temperature	
17. Monitor oxygen level and ensure it stays between 0–10%	
18. Monitor TR-402 video trends	
19. Monitor all process variables on the stack	
20. Cross-check process variables with SPEC sheet	

Specification Sheet and Checklist

Level

1. LIC-402	50%	AUTO	Upper steam-generating drum level
2. LIC-401	50%	AUTO	Deaerator level
3. LA-402	35%	Low	Upper steam-generating drum level
4. LR-402	—	Trend	Upper steam-generating drum level
5. LA-401	35%	Low	Deaerator level
6. LR-401	—	Trend	Deaerator level

Flow

7. FIC-402A	50%	AUTO	Air to furnace
8. FIC-402B	50%	AUTO	Natural gas feed
9. FIC-402C	150 gpm	CASC	Makeup water

Analytical

10. Ai-402	—	AUTO	Stack discharge
11. AA-402	0–10%	hi/lo	Combustion gases
12. BA-402	On/Off	—	Burner

Pressure

13. PIC-402A	120 psig	AUTO	Steam header
14. PIC-402B	−0.05 in water	AUTO	Damper
15. Pi-402	155 psig	gauge	P-402 discharge
16. Pi-400	−0.02 in water	gauge	Stack temperature
17. Pi-401	60 psig	gauge	Natural gas supply pressure
18. Pi-403	60 psig	gauge	Desuperheated steam
19. Pi-404	−0.02 in water	gauge	Fire box
20. PA-404	75/50 psig	hi/lo	Desuperheated steam pressure
21. PA-401	150/100 psig	hi/lo	Steam header
22. PR-402	—	Trend	Stack

Temperature

23. TR-402	350°F	Trend	Steam header
24. Ti-402	450°F	Gauge	Upper stack temperature
25. TE-400	600°F	—	Radiant section
26. TE-401	500°F	—	Economizer section
27. TE-403	350°F	—	Steam header
28. TE-404	305°F	—	Desuperheated steam

Common Boiler Problems and Solutions

Modern steam generation and advanced process control have eliminated many of the hazards associated with boiler operation. Steam boats running up and down the Mississippi in the nineteenth century recorded a number of events where boilers exploded, destroying the entire boat and injuring a large number of passengers and crew. In the twentieth century, industrial manufacturers experienced a number of incidents involving boiler accidents. Modern technicians are unaware of the hazards technicians faced in the past. Typical boiler problems include loss of water flow, scale, impurities in steam or water, improper water level, flame failure, tube rupture, soot build-up in superheater and economizer tubes, and flame impingement.

It is usually the technician's responsibility to control water, natural gas, steam flow rates and temperatures, and water level in the drums. The technician also checks for smoke and burner flame pattern and O_2 levels in hot combustion gases. Technicians maintain unit housekeeping and computer logs, monitor process variables, and make adjustments. Process technicians also monitor and control firebox and drum pressure; economizer, superheater and desuperheater temperatures; and fan operation.

Troubleshooting Scenario 1

Figure 12-3 displays the steam-generation system and presents a series of what-if scenarios. Refer to the figure and select the best answers.

Troubleshooting Scenario 2

Figure 12-4 displays the steam-generation system and presents a series of what-if scenarios. Refer to the figure and select the best answers.

Troubleshooting Scenario 3

Figure 12-5 displays the steam-generation system and presents a series of what-if scenarios. Refer to the figure and select the best answers.

Summary

Steam-generation systems are often referred to as boilers, a device used by the chemical processing industry to produce steam. At first glance, it appears that a boiler is a simple device used to boil water; however, this process is very complex and is an integral part of many industrial applications. Steam is used to protect equipment during cold periods, with extrusion, stripping, preheater and heat exchanger systems, flare systems, laminating, firefighting, steam turbines, and a large variety of chemical applications.

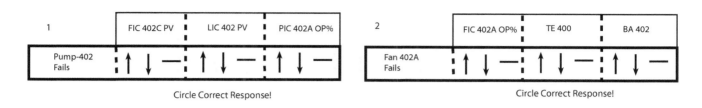

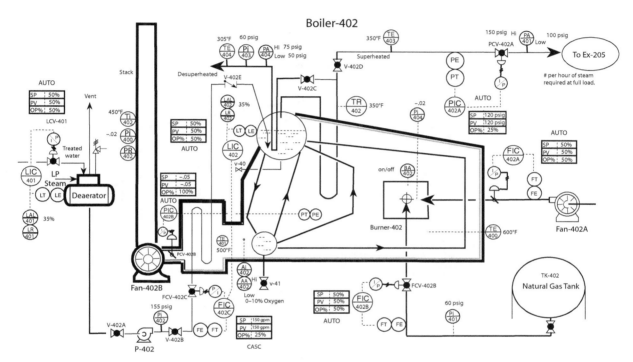

Figure 12-3 *Troubleshooting Scenario 1*

Steam generators use a combination of radiant, conductive, and convective heat transfer methods to produce steam. A commercial water tube boiler consists of an upper and a lower drum connected by a structured network of tubes. The lower drum and water tubes are filled completely with water, whereas the upper drum is only partially filled. This arrangement allows steam to pass through mechanical separators in the upper drum, flow to a superheater section, and exit the boiler. As heat is applied to the boiler firebox, water flows from the upper drum through downcomers into the lower drum. Tubes, called risers, cause water and steam to flow into the upper drum because of density differences.

Boiler water circulation operates under the principle of differential density. When fluid is heated, it expands and becomes less dense. Cooler water flows from the upper or steam-generating drum, through the downcomers, to the mud drum (the lower drum), and then rises as steam is generated.

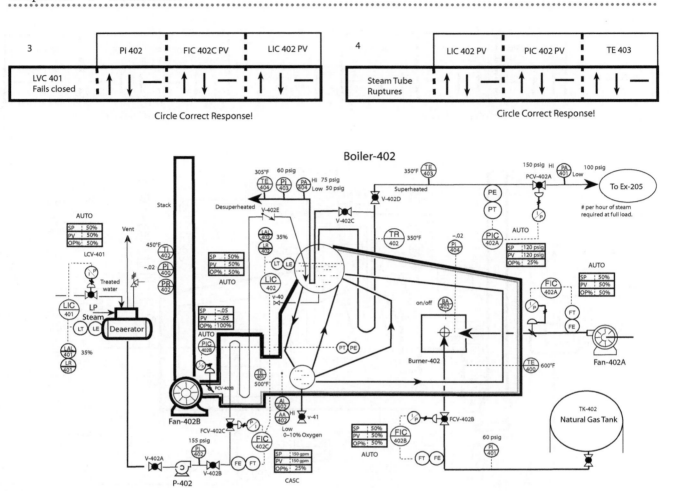

Figure 12-4 *Troubleshooting Scenario 2*

Circulation continues, and makeup water is continually added to the upper drum to replace the steam that is generated. Water circulation continues in a water tube boiler because steam bubbles in the lower drum move up the riser tubes and cause the density of water to decrease. The cooler water in the downcomer flows into the mud drum. The riser and steam-generating tubes are physically located near the burners. Steam moves up the riser and steam-generating tubes, and into the upper steam-generating drum. Steam generation causes pressure to rise. The upper steam drum has a large vapor disengaging cavity that is specifically designed to build pressure since vapors are much more compressible than liquid. A pressure-control valve is located on the steam outlet line and is designed to control system pressures. When the desired set point is achieved, the steam generator is "placed on the line." Pressure is maintained in the boiler by adding makeup water and continuously applying heat.

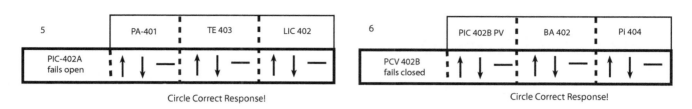

5		PA-401	TE 403	LIC 402
PIC-402A fails open		↑ ↓ —	↑ ↓ —	↑ ↓ —

Circle Correct Response!

6		PIC 402B PV	BA 402	Pi 404
PCV 402B fails closed		↑ ↓ —	↑ ↓ —	↑ ↓ —

Circle Correct Response!

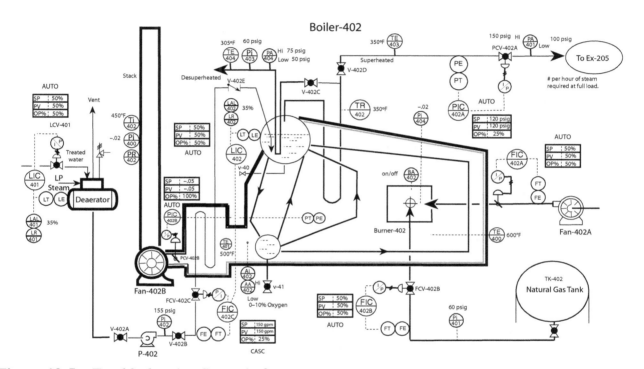

Figure 12-5 *Troubleshooting Scenario 3*

Review Questions

1. Explain the purpose of FIC-402A.
2. Explain the purpose of FIC-402B.
3. Explain the purpose of FIC-402C.
4. Describe the purpose and operation of the deaerator.
5. Draw a process flow drawing for the primary parts of a boiler system.
6. List the various components of a steam-generation system.
7. Explain how water circulation occurs between the upper and lower drum.
8. Describe how pressure builds inside the system.
9. Compare and contrast superheated steam and desuperheated steam.
10. List three operating problems associated with operating a boiler.
11. Compare and contrast the various tubes described in boiler operation.
12. Identify what the function of the economizer section is.
13. Draw and explain the purpose of LIC-402 and any associated equipment.
14. What are the purpose and operation of PIC-402B?
15. List the important aspects associated with Ai and AA-402.
16. List the three variables monitored in the stack.
17. Describe the purpose of BA-402.
18. Identify the temperature and pressure of the primary steam header.
19. Identify the temperature and pressure of the desuperheated steam header.
20. Write a paragraph or two about how to start up the boiler system.

chapter 13

Furnace Model

LEARNING OBJECTIVES

After studying this chapter, the student will be able to:

- Describe the various components of the furnace system.
- Describe the principles of heat transfer in the furnace system.
- Describe the hot oil system.
- Describe the furnace system.
- Draw a simple block flow diagram of the furnace model.
- List the safety aspects associated with the furnace system.
- Describe the furnace instrumentation system.
- List the operational specifications for the furnace system.
- Solve various troubleshooting scenarios.
- Identify common problems associated with furnace operation.
- Operate a hot oil and furnace system.

Key Terms

Air registers—devices located at the burner of a furnace used to adjust secondary airflow. Air is typically heated in the convection section before being introduced at the burner.

Bridge wall—sloping section that transitions between the radiant section and the convection section.

Convection section—the upper area of a furnace, where heat transfer is primarily through convection. Feed is typically supplied into the furnace through these tubes and exits out the radiant tubes. Tubes in this area are referred to as convection tubes and can be accessed through the header box doors at the terminal penetrations, where the return bends or rolled headers are located. Rolled headers typically have removable plugs for maintenance and tube inspection.

Draft—negative pressure of air and gas at different elevations in a furnace.

Spalled refractory—refractory that has broken loose from the sides of the furnace and fallen to the furnace floor. Caused by old refractory that has cracked or deteriorated over time, refractory that has not cured or dried properly, or broken refractory anchors.

Flame impingement—direct flame impingement occurs when the visible flame hits the tubes. Flame impingement can be classified as periodic or sustained.

Header box doors and gaskets—provide access to the terminal penetrations or bends on the convection tubes. The gaskets provide a positive seal between the inside and outside of the furnace.

Terminal penetrations—provide 180° turns or pipe bends in the convection section as the pipes scroll from one side of the furnace to the other.

Soot blowers—designed to remove soot from tubes in the convection section. Soot blowers are hollow metal rods that are inserted into the convection section. A series of timers admit nitrogen in quick bursts.

Radiant tubes—tubes located in the radiant section of a furnace and firebox that receive heat primarily through radiant heat transfer. They are also called as radiant coils.

Process heaters—typically defined as combustion devices designed to transfer convective and radiant heat energy to chemicals or chemical mixtures. Process tubes pass through the convection and radiant sections as energy is transferred to them. This transferred energy allows the liquid to be utilized in a variety of chemical processes that require higher temperatures.

Convection tubes—tubes located above the radiant section of a furnace, where heat transfer is primarily through convection. The first pass of tubes directly above the radiant section is referred to as the shock bank.

Dampers and draft control—Furnace draft can be natural, forced, induced, or balanced. Dampers can be manually or automatically operated. The damper system regulates pressure and airflow through the furnace, which is typically operated under a slight negative pressure.

Oxygen analyzer—an instrument specifically designed to detect the concentration of oxygen in an air sample. Oxygen flow rates are carefully controlled through a furnace.

Shock bank—the first pass of tubes directly above the radiant section. The shock bank is part of the convection section.

Low NOx burners—a type of gas burner, invented by John Joyce, that significantly reduces the formation of oxides of nitrogen. Low NOx burners are 100% efficient, as all heat energy released from the flame is converted to useful heat.

Preheated air—a compressed air system typically pushes air through tubes located in the upper section of the furnace. This preheated air takes full advantage of energy flow passing out of the furnace stack.

Furnace flow control—a critical feature in furnace operation, temperature control, and pressure control. A flow control loop regulates fluid feed rates in and out of the process furnace.

Feed composition—must remain uniform, otherwise furnace operational variables will be affected.

Fuel pressure control—a pressure control loop is located on the natural gas fuel line to the furnace. It is designed to maintain constant pressure in the furnace burners.

Furnace temperature control—as flow exits the process furnace, a temperature control loop carefully monitors process conditions. The natural gas flow controller (slave) is cascaded to the (master) temperature controller. The temperature controller adjusts fuel flow to the burners.

Furnace pressure control—furnace pressure is monitored in the bottom, middle, and top of the furnace. A pressure control loop is connected to the stack damper. The middle pressure reading on the furnace is compared to a set point, and adjustments are made at the damper if necessary.

Furnace Hi/Lo alarms—high and low alarms are used to keep the process flow within specifications and to prevent equipment damage or harm to the environment or human life.

Hot oil system—used to provide heat for industrial applications. Hot oil systems have a number of advantages over traditional steam systems.

Stack—outlet on the top of a furnace through which hot combustion gases escape from the furnace.

Firebox—the area in a furnace that contains the burners and open flames; the area of radiant heat transfer.

Flameout—extinguishing of a burner flame during furnace operation.

Burner alarms—designed to immediately notify technicians when a burner goes out.

Flashback—intermittent ignition of gas vapors, which then burn back in the burner; can be caused by fuel composition change.

Peepholes—holes on the side of a furnace that enable technicians to inspect visually the inner part of the furnace.

Ruptured tubes—flames coming from openings in tubes. May cause excess oxygen levels to drop and bridge wall temperatures to increase.

Vibrating tubes—tend to jump or move back and forth. Typically occurs in tubes outside the furnace. Vibrating tubes are often caused by two-phase slug-type flow inside the tubes. May be stopped by changing flow rates.

Plugged burner tips—flame pattern that appears erratic, shoots out toward a tube instead of shooting up the firebox.

Broken burner tiles—located directly around the burner and are designed to protect the burner from damage. The furnace rarely needs to be shut down to replace a broken tile unless it is affects the flame pattern.

Broken supports and guides—tend to fall to the furnace floor. Missing supports or guides will result in sagging or bowing of the tubes.

Hazy firebox or smoking stack—often occurs when not enough excess air is going into the firebox or the fuel–air mixing ratio is incorrect.

Sagging or bulged tubes—occur when guides or supports break, when the inside or outside of the tube fouls, and when there is flame impingement, reduced flow rate, or over-firing furnace. *Note*: Diameter of the tube does not change when it sags; however, it does when it bulges.

Hot tubes—glow different colors, which occurs when the inside or outside of the tubes fouls and when there is flame impingement, reduced flow rate, or over-firing furnace.

Color chart steel tubes—shows ten tube color variations associated with temperature.

Low burner turndown—may result in hazy firebox.

Introduction to Furnace Operation

A furnace, or fired heater, is a device used to heat up chemicals or chemical mixtures. Furnaces consist essentially of a battery of fluid-filled tubes that pass through a heated oven. The various parts of a process heater include a radiant section and burners, a **bridge wall** section, a **convection section** and **shock bank,** and a **stack** with damper control. These devices provide a critical function in the daily operation of the chemical processing industry. **Process heaters** are defined as combustion devices designed to transfer convective and radiant heat energy to chemicals or chemical mixtures. These process heaters are typically associated with reactors or distillation systems. Typical process heaters are available in a wide variety of

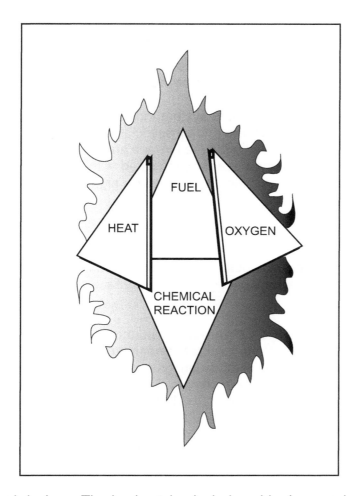

Figure 13-1
Fire Tetrahedron

shapes and designs. The basic styles include cabin, box, and cylindrical. Modern control instrumentation is used to maintain these rather large and elaborate systems.

The furnace in this chapter is called F-202 and is used to heat up hot oil for use in industrial heating applications. Process heater 202 uses natural gas, which is primarily methane (CH_4), as fuel for the low nitrogen oxide (NOx) burner. In this process, CH_4 reacts with oxygen to form carbon dioxide (CO_2) and water (H_2O): $CH_4 + 2O_2 \rightarrow CO_2 + 2H_2O$. Steam is also used to better disperse the natural gas for complete combustion.

This section focuses on the actual operation of a small process furnace that is associated with distillation unit 202. The furnace described in this chapter is used to heat up hot oil for use in various applications in the plant. Figure 13-1 shows the elements of the "fire tetrahedron."

The primary means of heat transfer in a fired heater are: radiant, conductive, and convection. Radiant heat transfer accounts for 60–70% of the total heat energy picked up by the charge in the furnace. Convective heat transfer accounts for about 30–40% of the total heat energy picked up in the furnace. Conductive heat transfer processes occur in each of these areas; however, it is easier to measure temperature differences in the actual charge than to calculate the conductive heat transfer coefficients.

The two basic elements of this system include the furnace and a hot oil reservoir. The **hot oil system** and all associated piping are well insulated. Process technicians assigned to the furnace and hot oil section need to be familiar with each of the hazards associated with operating the system. These hazards include working with extremely high temperatures and flammable material.

Controlling excess oxygen in the furnace is the single most important variable affecting efficiency. For heat transfer in the **firebox** or radiant section, the greatest efficiency is obtained when maximum furnace temperatures are achieved. Decreasing excess air in the furnace maximizes radiant heat transfer.

Excess airflow will decrease furnace temperatures around the burners and force the automatic controls to increase natural gas flow rates to the burner, wasting money. As hot combustion gases rise, cooler air is entrained, causing the temperature to decrease. Excess air enhances this process. When excess air is increased to the burner through the primary and secondary **air registers,** a temperature shift occurs as heat is moved away from the burners. Higher temperatures are found in the upper section of the firebox due to the reduced heat transfer in the lower section of the firebox. Temperatures in the convection section and stack will also rise significantly. This will reduce the amount of heat available for heating the hot oil, and more fuel will be burned in order to maintain process specifications. To be on the safe side, more air than is theoretically required for combustion is used. When this occurs, it is referred to as utilizing "excess air." The percentage of excess oxygen by volume in the flue gas can be measured using a graph. Each fuel has its own plotted curve graph. Suppose, for example, that the **oxygen analyzer** digitally indicates an O_2 reading of 3% by volume in the stack. The curve in Figure 13-2 shows that this is equal to 10% excess air for natural gas. Air can enter the furnace through:

- open **peepholes**
- leaks in furnace casing or joints
- damaged header box gaskets
- burners that have gone out.

It is important to recognize the position of the measurement, either near the burner or in the stack. Large leaks in the furnace can indicate high levels of oxygen in the system. Figure 13-2 shows the "air-to-fuel ratio" chart.

Unit specifications require the console technician to decrease the excess air so that temperatures in the process heater will increase. As the excess

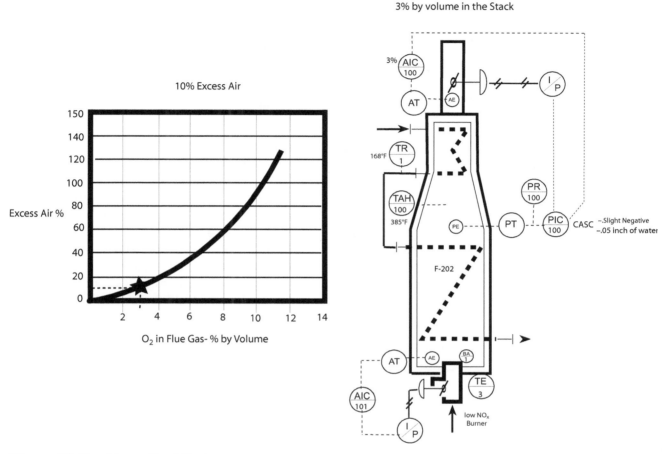

Figure 13-2 *Air-to-Fuel Ratio*

air is decreased, the heat transfer rate near the burners is increased. As the net heat released from the natural gas burner has not changed, the amount of heat contained in the hot combustion gases slows down and has time to soak into the exposed tubes. This process reduces the amount of heat transferred in the convection section and lowers the temperature in both the upper convection section and stack. This process provides a more efficient way to prevent heat energy from flowing out the stack. By decreasing the excess air flowing through the process heater, a technician can save money and more easily achieve product specification.

Some process heaters utilize advanced control instrumentation that maintains a preset ratio of air-to-fuel. For example, a ratio of 11 means that for each weight unit (kilograms or pounds) of fuel, there are eleven similar units of oxygen being supplied. Higher ratios indicate that there is more excess air, whereas a lower ratio translates to less excess air. Theoretically, air can be indicated in terms of the air-to-fuel ratio. When air is specified in terms of the air-to-fuel ratio, the amount of combustion air is calculated by adding 1 to the ratio and multiplying the results times the fuel rate. If the ratio is 11, add 1, which gives a ratio of 12. If the fuel rate is 4 lb/min, there will be 4 times 12 pounds of flue gas produced by combustion.

Fuel rate: 4 lb/min
Air rate: $11 + 1 = 12$
$4 \times 12 = 48$ lb of flue gas is produced by combustion.

Equipment Descriptions

Firebox and Refractory

The section in a furnace that contains the burners and open flames is called the firebox. The firebox is lined with a refractory layer, a brick lining that is designed to reflect heat back into the furnace. The refractory brick is classified as firebrick or insulating brick, both of which are specially designed to withstand and reflect heat. Firebrick has a density range of 131–191 lb/ft^3, and maximum temperature ranges from 2500 to 3300°F. Insulating firebrick has much lower densities, 27.3–78.7 lb/ft^3, and maximum temperature ranges from 1600 to 3250°F.

The refractory bricks are attached to stainless steel rods that are attached to 3–6 in. ceramic fiber insulation bat. The insulation bat and the metal shell of the furnace touch each other. The insulation barrier between the furnace shell and the brick prevents heat loss. In many furnace designs, the refractory runs up the vertical wall and then can be adapted to cover the sloping bridge wall section that directs flow into the convection section. Figure 13-3 is an illustration of the basic components found in a simple process heater. The upper convection section and the arch section (the narrowing neck between the convection section and the stack) are usually insulated with heavy or light high-temperature cement (castable) or firebrick. Castable peep blocks contain peepholes that allow for visual inspection. Castables have a temperature range of 1600 to 3300°F.

Typical heat loss from a furnace is between 2% and 3% of the total heat release. Since the insulation is porous, a protective coating may be applied inside the steel shell to protect it from corrosive materials such as sulfur oxides.

Temperatures inside the firebox range from 1600 to 2000°F (871–1093°C). Furnace pressures usually run below atmospheric pressure, in the range of −0.04 to −0.06 in. H$_2$O **draft** (negative pressure) at the bottom of the furnace. When flow enters the firebox, it receives radiant heat directly from the burners. Heated process flow exits out the bottom of the furnace into an insulated header.

Radiant and Convection Tube

The tubes located along the walls of the firebox are called the **radiant tubes** or coils. Radiant tubes receive direct heat from the burners. Radiant

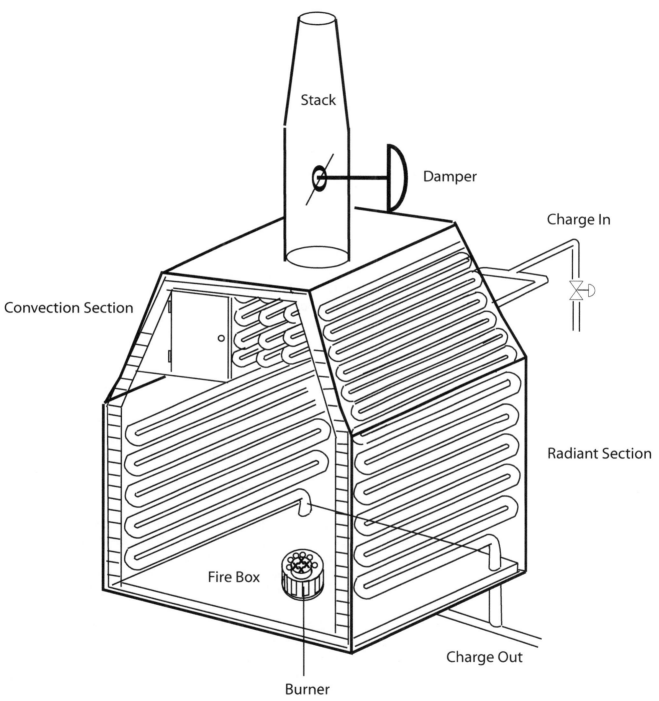

Figure 13-3 *Simple Process Heater*

heat transfer occurs best in a vacuum. These tubes operate at high temperatures and are constructed of high-alloy steels. Radiant tubes are mounted parallel to the furnace walls. Radiant heat transfer accounts for 60–70% of the total heat energy picked up by the charge in the furnace.

Convection tubes are located in the roof of the furnace and are not in direct contact with the burner flames. Only the first series of tubes, called the shock bank, are in the direct path of the hot combustion gases rising up the process heater. Hot gases transfer heat through the metal tubes and into the charge. Convection tubes are mounted horizontally and are equipped with fins to increase efficiency. Convective heat transfer to the process charge accounts for about 30–40% of the total heat energy picked up in the furnace.

Conductive heat transfer occurs as heat energy is transferred from the hot combustion gases and radiant heat source through the high-alloy steel tubes. Most technical manuals do not refer to this process inside the furnace, and it is rarely given credit for a specific percentage of heat transfer; however, this heat-transfer method occurs in both the convection and the radiant tubes.

As hot combustion gases leave the firebox, a series of tubes called the shock bank is encountered. The shock bank in a small cabin furnace receives the initial blast of hot combustion gases. The tubes in the convection section are designed to take advantage of the heat energy exiting the furnace. This process allows the feed to gradually increase the temperature as it moves through the system. This furnace has several advantages, including:
- 90–95% efficiency and
- the ability to completely drain the tubes.

Furnace F-202 utilizes a forced draft convection system by which blower 100 produces circulation through the process heater. Convection is the primary means used for heat transfer in devices such as air heaters, economizers, condensers, and any other type of heat exchanger where heat is removed from or added to air, hot combustion gases, water, brines, or various liquids or gases. In the convection section heat transfer, the resistance of the heat flow depends on the physical arrangement, the rate of circulation, the nature of the flowing material inside the tubes (hot oil), the pipe material, and other factors. There are so many contributing factors that it is difficult to calculate all the variables associated with heat transfer between a cold and a hot substance.

Soot blower
Soot blowers are devices found in the convection section of process heaters. Soot blowing is required when the efficiency of the convection section decreases. This can be calculated by looking at the temperature change from the crossover piping and at the convection section discharge.

Soot blowers utilize a transfer media such as nitrogen, water, air, or steam to remove deposits from the tubes. Air movement in the convection section is slower because of the finned tubes and close proximity of each pass. The initial blast of hot combustion gases tends to accumulate deposits in the convection system, specifically along the shock bank.

There are several different types of soot blowers, including wall blowers and finned tube blowers. Furnace wall blowers have a very short lance with a nozzle at the tip. The lance has holes drilled into it at intervals so that when it is turned on, it rotates and cleans the deposits from the wall in a circular pattern. Soot blowing continues until a preset timer goes off.

Stack Damper

Combustion gases leave the furnace through the stack and are dispersed at a height in the atmosphere where they are inert. As the hot combustion gases rise in the process heater, excess air is entrained, causing a cooling effect. Because of density variations between air inside and air outside, hot combustion gases rise through the stack. This natural draft creates a lower pressure inside the furnace, which is typically measured in inches of water. Draft is defined as the difference between atmospheric pressure and the lower pressure inside the fired heater. One inch of H_2O is equivalent to 0.036 psi.

A damper in the stack permits adjustments of stack drafts. The stack damper is typically set to give pressures from 0.05 to 0.15 in. H_2O (vacuum) draft. At 0.05 in. H_2O, approximately 350,000 lb/h of gas flow can be obtained.

The damper system utilizes a pneumatically operated butterfly valve that is controlled by a pressure-indicating controller. Butterfly valves are quarter-turn valves, which can be used to throttle airflow through the furnace. The different drafts or pressures found in a furnace are illustrated in Figure 13-4.

Burner

The single burner in this system is a low NOx system located on the floor of the furnace. **Low NOx burners** are designed to be operated with lower amounts of excess air than typical burners. The use of tertiary air registers reduces nitrogen oxides in the flue gas stream. The burner uses a small amount of steam to better disperse the fuel and oxygen.

Air shutters on the burners control primary airflow into the furnace. Air registers near the burner control secondary airflow. These registers normally are closed when excess oxygen is detected in the furnace.

Perfect mixing of air and fuel is impossible, and no practical way has been found to determine when the combustion process is complete. Incomplete combustion means that unburned vapors will be present in the hot

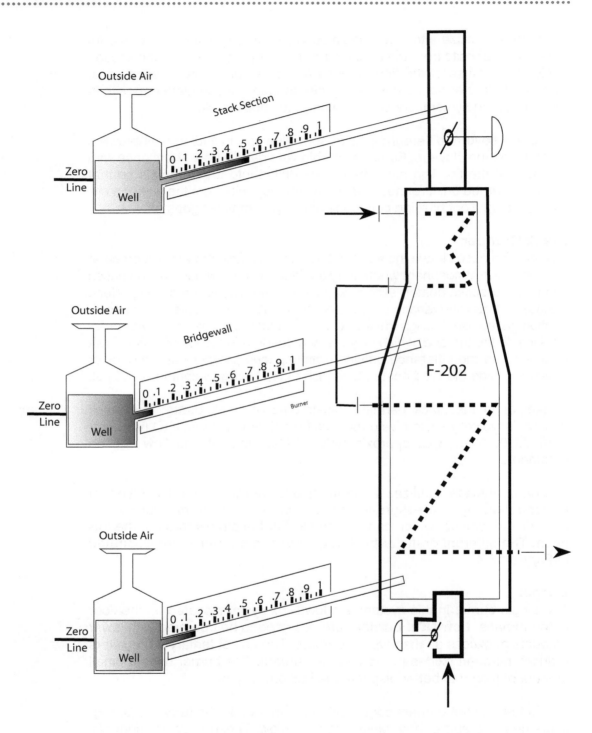

Normal Pressures in a Furnace
(One Inch of Water is Equivalent to .036 psi)

Figure 13-4 *The Inclined Furnace Tube Gauge*

combustion gases. To be on the safe side, most facilities use excess air to ensure that all of the fuel has been burned. The burners are designed to avoid direct contact of the flames with the tubes in the firebox. A space of 1.5–2 ft is considered to be a safe distance between the open flames and the radiant tubes. The flame pattern of the burners should be 60% less than the height of the firebox.

The radiant section is engineered to distribute the radiant heat energy evenly. Modern burner design consumes 100% of the fuel, with a nominal excess of 10–15% oxygen. Excess oxygen in the furnace is carefully controlled as it enters the secondary and primary registers. This control takes place as the fuel and primary air mix at the burner and is enhanced by adjustments on the secondary air registers mounted outside the burner. An oxygen monitor carefully tracks the composition of the hot combustion gases. Adjustments to the airflow rate are made at the burners and the stack damper.

The floor of the process heater has a 6-in. layer of heat-resistant castable capped with high-temperature firebrick. Four ceramic high-temperature refractory blocks are positioned around the burner. The refractory system can withstand a wide range of high-temperature conditions. The refractory layer can be over a foot thick. The convection tubes in the upper section of the furnace have a variety of return bend designs, as illustrated in Figure 13-5.

Bridge Wall Section

The bridge wall section is the sloping section of the upper furnace that connects the radiant section to the convection section. It is designed to accelerate the flow of the hot combustion gases out of the firebox and into the convection section and stack. The heat-reflective materials in this area are designed to withstand temperatures between 1600 and 3300°F. F-202 is designed to run at temperatures far below this specification range.

Fuel System

A complex network of lines that provide natural gas and air to the burners is located under the furnace. Natural gas is composed primarily of methane (CH_4). Most injuries encountered in the furnace operation occur during start-up of the fuel burning system.

Forced Draft Process Heater

Forced draft furnaces utilize a centrifugal blower to push **preheated air** to the burner for combustion. The preheated air is run through the tube coils located above the convection section and directed to the suction of blower 100, which discharges under automatic control to the burner.

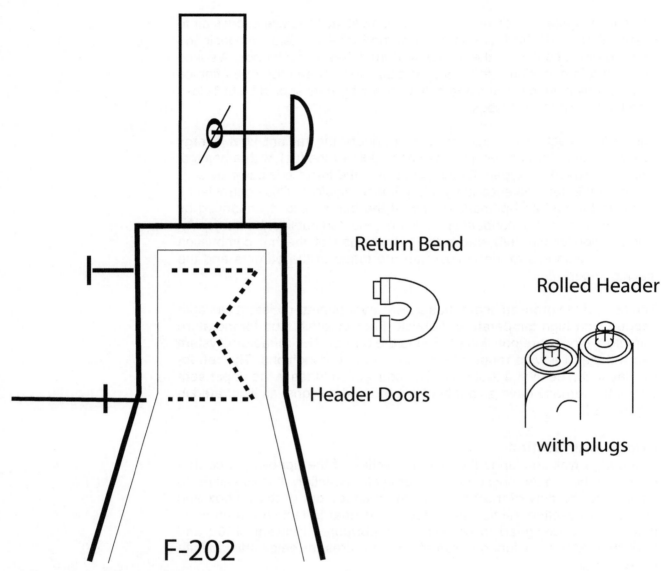

Figure 13-5 *Burner, Return Bends, and Rolled Headers*

Hot Oil Tank 100

The hot oil system is designed to provide heat to various operations that require a stable heat source. The primary use of the hot oil system is to provide a heat source to a series of heat exchangers being used to pre-heat feed in a number of distillation systems. The tank operating capacity is 20,000 gal. The tank is constructed of stainless steel and insulated to conserve and retain heat energy. The system is run under atmospheric pressure; however, the tank is sealed and caution signs are posted on the vessel. A level and pressure-relief system is attached to the tank. Process technicians are responsible for monitoring and controlling all of these variables.

Instrumentation Systems

Automatic control is the foundation for efficient continuous flow processes. F-202 utilizes a number of automated systems that allow a smaller number of process technicians to operate much larger processes. Instrument systems used in the furnace system include:

- TIC-100 has an operating temperature of 350°F. At start-up temperature, profiles are completely different from what has been achieved after final operating specifications. The convection section has a 35% heat transfer rate from 70 to 168°F after it exits the convection system. The radiant system takes temperatures from 168 to 350°F. This is an increase of 65% in energy absorption. TIC-100 is cascaded to FIC-101, which controls the flow of natural gas to the burner.

Temperature

TIC-100—temperature-indicating controller—located on the product discharge line.

TA—high (365°F) temperature alarm—located on the product discharge line.

TA—low (335°F) temperature alarm low—located on the product discharge line.

TAH-100—temperature alarm high—located in the bridge wall section. The high alarm is set at 365°F.

TE-1 3—temperature element—located in the convection section of the furnace (375°F).

TE-2 3—temperature element—located in the radiant section of the furnace (395°F).

TE-3—temperature element—located near the burner (425°F).

BA-1—burner alarm—located near the burner and is designed to alert the operator in the event the burner goes out.

Ti-1—temperature indicator—located on the insulated discharge line going into tank 100 (350°F).

Ti-2—temperature indicator—located on the discharge of hot oil pump 100. Start-up temperature is 70°F.

Pressure

Pi-1—pressure indicator—located on the suction side of P-100 (10 psig).

Pi-2—pressure indicator—located on the discharge side of P-100 (55 psig).

Pi-3—pressure indicator—located at the top stack. Slight negative pressure (0.5 in. H_2O).

PIC-100—pressure-indicating controller—measures pressure at the bridge wall section and is cascaded to the oxygen controller AIC-100 in the stack (0.05 in. H_2O).

Pi-4—pressure indicator—discharges pressure on charge outlet (55 psig).

Pi-5—pressure indicator—radiant section pressure (0.2 in. H_2O, slightly negative).

PIC-101—pressure-indicating controller—natural gas supply pressure (15 psig).

Flow

FIC-100—flow-indicating controller—hot oil feed rate (800 gpm).

FR-100—flow recorder—hot oil feed rate.

FIC-101—flow-indicating controller—natural gas supply (12,500 mbh).

Fi-1—flow indicator—discharge flow rate from F-202 (800 gpm).

FIC-102—flow-indicating controller—low-pressure steam to furnace. Used for purging the furnace prior to start-up and help with burner operation (35 psig).

Level

LIC-1—level-indicating controller—designed to maintain a constant level in the hot oil system (75%).

LA-1—level alarm—designed to alert the process technician in the event of a low-level condition in TK-100. Hot oil is often lost during process leaks and by adhering to the walls of the tubes of the system.

Analytical

AIC-100—analytical-indicating controller—the purpose of this control system is to monitor and control excess airflow through the system directly below the damper control valve. It is cascaded to the pressure control system (3% oxygen).

AIC-101—analytical-indicating controller—designed to monitor and control airflow to the burner. An auxiliary compressor system is connected to the primary and secondary air makeup systems that take suction from tubes. Natural draft provides negative pressures inside the furnace, while a forced draft air system puts positive pressure inside the furnace.

The Furnace and Hot Oil System

Operating a small process furnace requires an understanding of the scientific principles associated with heat transfer, fluid flow, pressure control, and the complex relationships associated with airflow. Review process heater 202 in Figure 13-7 and prepare to answer a variety of questions and discuss situations associated with the equipment, systems, operations, instrumentation, and a variety of troubleshooting scenarios. Figure 13-6 is an illustration of the natural gas burner located at the bottom of the furnace.

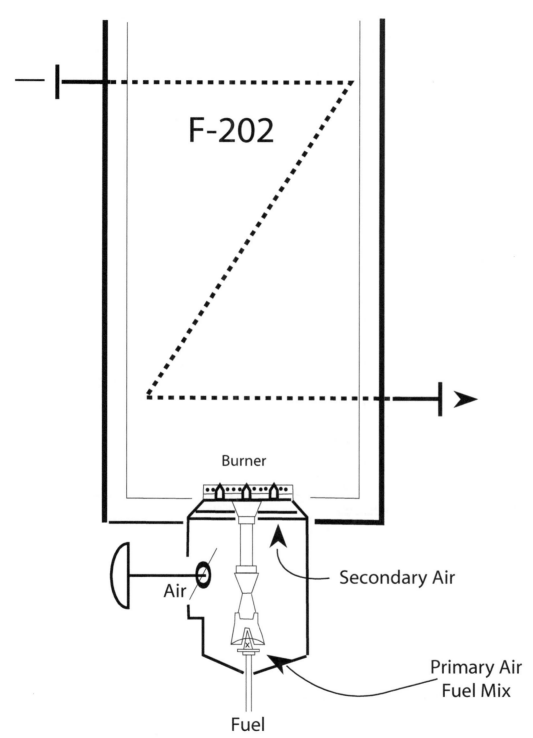

Figure 13-6 *Natural Gas Burner*

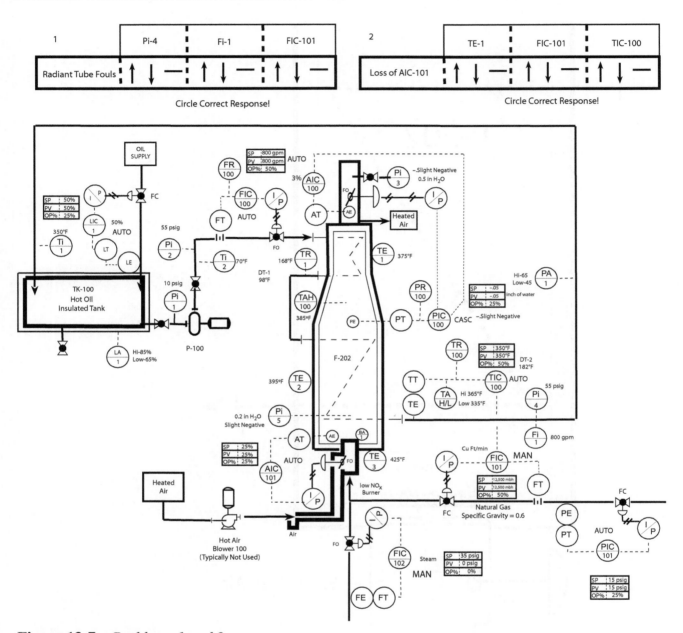

Figure 13-7 *Problems 1 and 2*

Operational Specifications

The following procedures were developed using the principles of instructional system design. A profile or matrix was developed during the analysis phase identifying the key competency categories and supporting objectives. The materials are designed to move from simple to complex and reflect the operational procedures for the F-202 system.

Action	Notes
1. Open PCV-100 to 100%.	
2. Purge furnace for 15 min with steam.	
3. Lineup P-100 to F-202 and into TK-100.	
4. Set FIC-100 to 200 gpm.	
5. Start P-100.	
6. Monitor process pressures.	
7. Increase FIC-100 to 400 gpm.	
8. Monitor process pressures and levels in TK-100.	
9. Set AIC-100 to 3% and place in AUTO.	
10. Set PIC-100 to 0.05 and CASC to AIC-100.	
11. Set FIC-101 to 3500 ft^3/h and place in MANUAL.	
12. Set AIC-101 to 21% and place in AUTO.	
13. Light burner and allow F-202 to come up to temperature.	
14. Increase FIC-100 to 600 gpm.	
15. Set FIC-101 to 8500 ft^3/h.	
16. Increase FIC-100 to 800 gpm and put controller in AUTO.	
17. Set TIC-100 to 350°F and put controller in AUTO.	
18. Increase FIC-101 to 12,500 mbh and CASC to TIC-100.	
19. Set LIC-1 to 75% and put in AUTO.	
20. Cross-check process variables with SPEC sheet.	

Specification Sheet and Checklist

Level

1.	LIC-1	75%	AUTO	TK-100 level
2.	LA-1	85%	High	TK-100 high
3.	LA-2	65%	Low	TK-100 low

Flow

4.	FIC-100	800 gpm	AUTO	Feed to furnace
5.	FIC-101	12,500 mbh	CASC	Natural gas feed
6.	FIC-102	35 psig	AUTO	Steam
7.	Fi-1	800 gpm	—	Furnace discharge

Analytical

8.	AIC-100	3%	AUTO	Stack discharge
9.	AIC-101	21%	AUTO	Air to registers
10.	BA-1	On/Off	—	Burner

Pressure

11.	PIC-100	0.05 in. H_2O	CASC	Bridge wall draft
12.	PIC-101	15 psig	AUTO	Natural gas feed
13.	Pi-3	0.5 in. H_2O	—	Top-stack
14.	Pi-5	0.2 in. H_2O	—	Radiant section
15.	Pi-1	10 psig	—	P-100 suction
16.	Pi-2	55 psig	—	P-100 discharge
17.	Pi-4	55 psig	—	Furnace discharge pressure
18.	PR-100	0.05 in. H_2O	—	Bridge wall draft pressure
19.	PA-1	Hi-65, Lo-45	—	Hot oil discharge

Temperature

20.	TR-1	168°F	—	Convection section exit temperature
21.	DT-1	98°F	Δ-temp	Delta inlet/outlet convection
22.	DT-2	182°F	Δ-temp	Delta inlet/outlet radiant
23.	TIC-100	350°F	AUTO	Furnace exit temp
24.	TAH-100	385°F	—	Bridge wall high temperature
25.	TE-1	375°F	—	Convection section
26.	TE-2	395°F	—	Radiant section
27.	TE-3	425°F	—	Burner
28.	Ti-1	350°F	—	At start-up
29.	Ti-2	70°F	—	At start-up
30.	TA-100 high	365°F	—	Hot oil discharge
31.	TA-100 low	335°F	—	Hot oil discharge

Common Furnace Problems and Solutions

Small commercial furnaces are not frequently shut down unless a serious problem occurs or equipment repair and turnaround are scheduled. There are some common problems and concerns for which equipment must be monitored.

Process Heater (Furnace) Efficiency

Running process heaters efficiently is a major operating concern because significant funds are expended on natural gas. Process heaters are also closely monitored by the Environmental Protection Agency for consistent clean operation and emissions. An advantage to using natural gas is its clean burning operation. F-202 is a clean operating system that keeps excess airflow between 8% and 10%. Ensuring proper air–fuel concentrations is critical to efficient furnace operation because the fuel burns cleaner and hotter. Incomplete combustion reduces heat output, produces waste gases, and creates a potentially hazardous condition as unburned fuel collects in the firebox. Efficient, safe process heater operation requires a technician to carefully observe and control the combustion process.

Flame Impingement

As flames reach up into the furnace, they occasionally touch the refractory wall or tube. Frequent or sustained contact is called **flame impingement.** Flame impingement can be classified as periodic or sustained. This can weaken the metal tube and cause deposits to form inside the tube. Flame impingement can sometimes be corrected by pinching back on the fuel supply to the affected burner.

Hot Spot

A hot spot can be identified inside a furnace as a glowing red spot on the metal or refractory. Sometimes hot spots can be corrected by redistributing the process flow so that additional flow goes through and cools the affected tubes.

Spalled Refractory

Refractory that has broken loose from the sides of the furnace and fallen to the furnace floor is referred to as **spalled refractory.** This is caused by old refractory that has cracked or deteriorated over time, refractory that has not cured or dried properly, or broken refractory anchors.

Ruptured Tubes

Flames come from openings in the tube. **Ruptured tubes** may cause excess oxygen levels to drop and bridge wall temperatures to increase.

Vibrating Tubes

Vibrating tubes tend to jump or move back and forth. This typically occurs in tubes outside the furnace. Vibrating tubes are often caused by two-phase slug-type flow inside the tubes. Harmonic vibrations may be stopped by changing flow rates.

Plugged Burner Tips

Plugged burner tips tend to make the flame pattern appear erratic as it shoots up and out toward a tube instead of shooting up the firebox.

Broken Burner Tiles

Broken burner tiles are located directly around the burner and are designed to protect the burner from damage. The furnace rarely needs to be shut down to replace a broken tile unless it affects the flame pattern.

Broken Supports and Guides

Broken supports tend to fall to the furnace floor. Missing supports or guides will result in sagging or bowing of the tubes. The falling supports or guides may hit a burner and damage it.

Hazy Firebox or Smoking Stack

Hazy fireboxes often occur when insufficient excess air is going into the firebox or the fuel air mixing ratio is incorrect.

Sagging or Bulged Tubes

Sagging tubes occur when guides or supports break, when the inside or outside of tube fouls, or when there is flame impingement, reduced flow rate, or an over-firing furnace. *Note*: The diameter of a tube does not change when it sags; however, it does when it bulges.

Hot tubes glow different colors, which occur when the inside or outside of the tubes foul or when there is flame impingement, reduced flow rate, or an over-firing furnace.

Fuel Composition Changes

Fuels have different heat release rates that can be identified on standard charts. Fuel compositions change occasionally, and as a result the process heater's efficiency changes. Fuel composition changes cause **flashback** (intermittent ignition of vapors, which burn back in the burner). Process technicians must be prepared for these changes and make the correct adjustments. These adjustments include fuel flow rate, damper, charge flow, and burner alignment.

Flameout

Flameout occurs when the flame on a burner goes out with fuel still being pumped to the furnace. This situation puts unburned fuel into the process heater and creates a potentially hazardous situation. F-202 has a burner alarm designed to notify the technician in the event of flameout. Flameout usually occurs when there is too much fuel being sent to the burner or when there is a loss of draft or oxygen. Occasionally soot blowers will dislodge material from the convection section that will drop down on the burner, extinguishing the flame. When this situation occurs, the process heater needs to be shut down. F-202 utilizes a steam system to purge unburned hydrocarbons out of the firebox.

Valve Failure

Furnace feed control valves (FCV-100) are designed to fail in the open position. Valve failure cannot be detected until an adjustment fails to respond. In any case, the technician will be unable to control the flow. Fortunately, bypass loops exist that allow the control valve to be isolated and repaired without shutting down the equipment. Furnace fuel gas valves are designed to fail in the closed position.

Troubleshooting Scenario 1

In the first troubleshooting scenario, Figure 13-7, the radiant tubes begin to foul and we have a loss of AIC-101. Please indicate whether you believe the selected variables will:

- go up
- go down
- stay the same, or
- other.

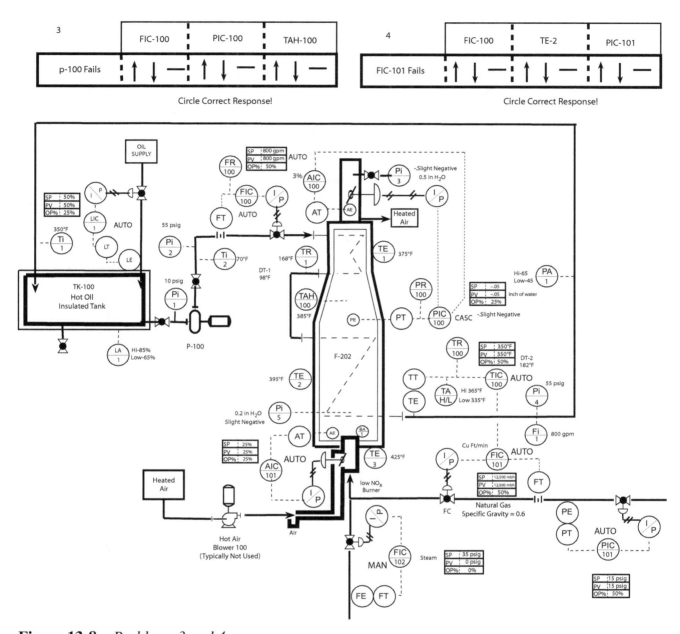

Figure 13-8 *Problems 3 and 4*

Refer to the figure and select the best answers.

Troubleshooting Scenario 2

Figure 13-8 displays the furnace and hot oil system and presents a series
of what-if scenarios. Refer to the figure and select the best answers.

Troubleshooting Scenario 3

Figure 13-9 displays the furnace and hot oil system and presents a series of what-if scenarios. Refer to the figure and select the best answers.

Troubleshooting Scenario 4

Figure 13-10 displays the furnace and hot oil system and presents a series of what-if scenarios. Refer to the figure and select the best answers.

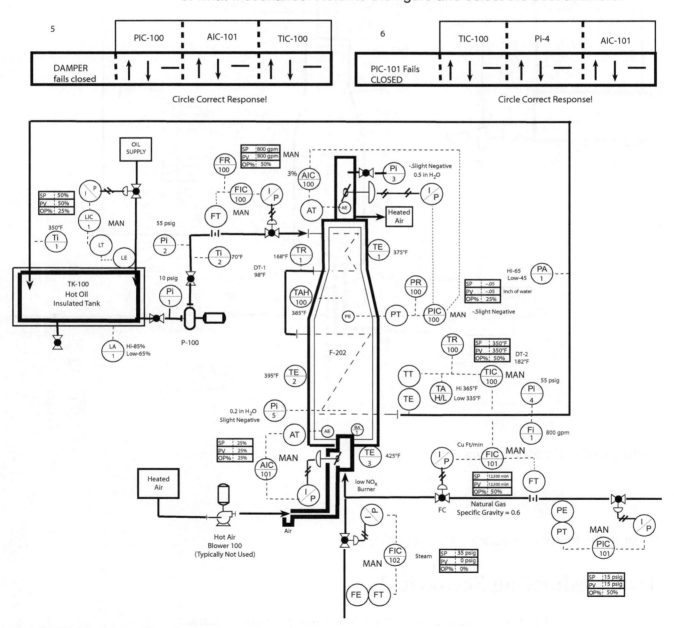

Figure 13-9   *Problems 5 and 6*

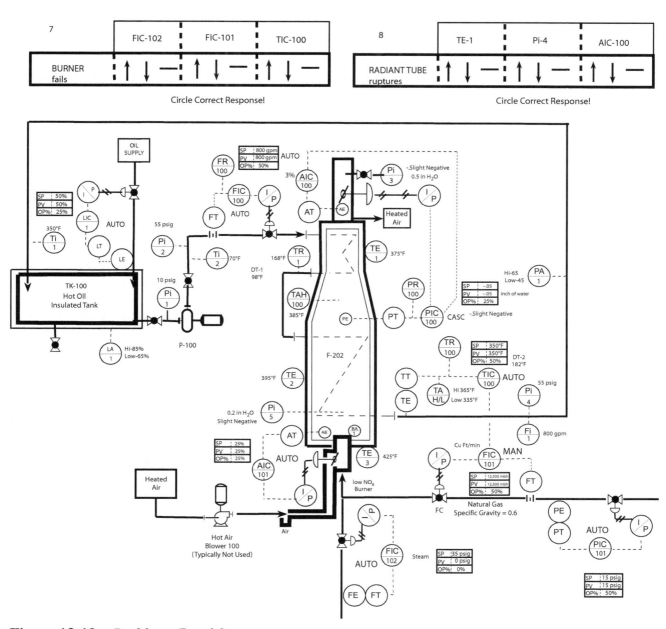

Figure 13-10 *Problems 7 and 8*

Summary

The various components of a process heater include a variety of equipment designed to function together to heat up large quantities of hydrocarbons, oil, chemicals, or chemical mixtures. A process heater is often referred to as a furnace. The basic components of a furnace include a stack, convection section, bridge wall, and radiant section. The furnace has a strong carbon steel shell that is lined with refractory and insulation. The radiant section has a firebox and burners. Secondary and primary airflows are controlled at the burner. Draft is produced by the density differences in the furnace versus the outside air. Draft is enhanced at the sloping bridge wall section. The convection section has a set of tubes called the shock bank and normal convection tubes that can be accessed through a set of doors. Return bends or rolled headers are found transitioning between one end of the convection section and the other. The tubes in this section are typically finned for greater heat transfer. The stack has a device called a damper that is used to control airflow through the furnace. A variety of instrument systems are included in the operation of the process heater system. These include flow control on the hot oil system, flow control on the natural gas system, pressure control on the gas header, pressure control on the furnace, and a flow control system on the steam.

Heat transfer in the furnace includes the control of excess air. The principles of heat transfer include conduction, radiation, and convection. Two sections in the furnace are named radiant and convection.

The hot oil system is stored in a stainless steel insulated tank. Hot oil is used as a utility to heat up other processes in the plant. Hot oil transfers heat uniformly and provides a very stable temperature control platform.

Burning natural gas, exhausting its hot combustion gases, and heating up large quantities of flammable hot oil is laced with hazards. Fire, burns, explosions, and hazardous chemical handling and storage are just a few of the hazards associated with the operation of a process heater.

Furnace operation requires a process technician to follow process guidelines and specifications. Operational procedures are associated with this device. Furnace technicians are trained for months with a qualified technician prior to being allowed to run the unit alone. General tasks typically start with very simple items such as catching samples and taking them to the lab for analysis or filling out checklists.

Troubleshooting process heater problems requires technicians to follow a prescribed methodology during the troubleshooting process. These methods will change depending on the situation; however, effective troubleshooting requires a good understanding of the furnace system, normal operational conditions, and safe work practices.

Common problems associated with furnace operation include excess airflow, spalled refractory, flame impingement, leaking header box gaskets, instrumentation problems, flameout, flashback, ruptured tubes, vibrating tubes, plugged burner tips, broken tiles, broken support grids, hazy firebox, and hot tubes.

Operating a process heater system requires more than simply following a step-by-step outline. It requires a clear understanding of the complete flow process, all of the major equipment, operating conditions, and relationships.

Review Questions

1. Describe how you would respond to a plugged burner tip.

2. Explain how you would repair a **hazy firebox or smoking stack.**

3. Describe **sagging or bulged tubes.**

4. Do broken furnace supports require immediate shutdown of the furnace?

5. Describe what you would do if the burner flame went out.

6. Explain how to start up F-202. List each step.

7. Explain how pressure is controlled in a furnace. Draw each control system.

8. Explain what causes a tube to vibrate and describe how to stop it.

9. List the pressures found in the furnace.

10. Explain the purpose of PIC-101.

11. Explain the purpose of FIC-101. Describe how it operates.

12. Describe how FIC-100 operates in manual.

13. Describe what happens when excess air enters the furnace.

14. Explain the purpose of FIC-102.

15. Explain how AIC-101 operates and its relationship to AIC-100.

chapter 14

Distillation Model

LEARNING OBJECTIVES

After studying this chapter, the student will be able to:

- Apply the principles of distillation.
- Analyze the relationship between the boiling point of a hydrocarbon and pressure, temperature, flow, and level.
- Analyze the various concepts associated with pressure in a distillation system: vapor pressure, partial pressure, relative volatility, compressibility, liquid pressure, and vacuum.
- Identify the different equipment systems used to make up a distillation system.
- Analyze how the methods of heat transfer apply to the distillation process.
- Explain the basic components of a plate column.
- Operate and troubleshoot the equipment systems associated with a distillation feed system.
- Operate and troubleshoot the equipment systems associated with a distillation preheat system.
- Describe and operate the equipment systems associated with the bottom section of a distillation column.
- Operate and troubleshoot the equipment systems at the top of a distillation column.
- Compare and contrast preheaters and condensers.
- Describe the basic principles associated with column design.
- Troubleshoot various scenarios on the distillation system.

Key Terms

Bottom product—the heavier components of the distillation process that flow into the bottom section and on to the kettle reboiler by gravity.

Distillation—the separation of components in a mixture by their boiling points.

Downcomer—downspout that allows liquid to drop to lower trays in a column.

Downcomer flooding—occurs when the liquid flow rate in the tower is so high that liquid backs up in the downcomer and overflows to the upper tray. Liquid accumulates in the column, differential pressure increases, and product separation is reduced.

Feed tray—point of entry of process fluid in a distillation column under the feed line.

Final boiling point—the temperature at which the heaviest component boils.

Heat balance—principle that heat in equals heat out.

Jet flooding—occurs when the vapor velocity is so high that liquid downflow in the column is restricted. Liquid accumulates in the tower, differential pressure increases, and product separation is reduced.

Local flooding—excessive liquid flowing down a column blocks vapor flow up the column in one section.

Material balance—principle that the sum of the products leaving equals the feed entering the distillation column.

Overall flooding—local flooding expands to entire column.

Overhead product—the lighter components in a distillation column, which rise through the column and go out the overhead line, where they are condensed.

Overlap—incomplete separation of a mixture.

Overloading—operating a column at maximum conditions.

Partial pressure—the amount of pressure per volume exerted by the various fractions in a mixture of gases.

Puking—occurs when the volume of the vapor is so high that it forces liquid up the column or out the overhead line.

Rectification—the separation of different substances from a solution by use of a distillation column.

Rectifying section—a section of the distillation column above the feed line in which a higher concentration of lighter molecules exists and the separation and segregation of molecules occurs.

Reflux—condensed vapor from the lightest component coming off the top of the column, which is cooled and condensed into a liquid so that it can be pumped back to the top tray in the column to control product purity and temperature.

Relative volatility—the characteristic associated with a liquid's tendency to change its state or vaporize inside a distillation system.

Stripping section—a section of the distillation column below the feed line in which heavier components are located.

Temperature gradient—the progressively rising temperatures from the bottom of a distillation column to the top.

Ternary mixture—three components in a mixture.

Tray column—a device located on a tray in a column that allows vapors to come into contact with condensed liquids; three basic designs are bubble-cap, sieve, and valve tray.

Vapor pressure—the outward force exerted by the molecules suspended in vapor state above a liquid at a given temperature, when the rate of liquefaction is equal to the rate of vaporization (equilibrium).

Weeping—occurs when the vapor velocity is too low to prevent liquid from flowing through the holes in the tray instead of across the tray. Differential pressure is reduced and product separation is reduced.

Preheat system—located on the front end of the distillation column, the preheat system is composed of a feed tank, a pump, two heat exchangers, a hot oil system, and a series of control loops to control flow and temperature.

Bottom system—includes the bottom of the distillation column, kettle reboiler, and pump. The bottom section is designed to control level, maintain energy balance, help separate lighter components from heavier ones, and provide heat to the tube side of Ex-202. A series of control loops are designed to control level, temperature, and flow.

Overhead system—includes the top of the distillation column, overhead condenser, drum, and pump. A series of control loops control pressure, level, and flow.

Dalton's Law—Dalton's law, $P_{total} = P_1 + P_2 + P_3 \ldots$, states that the total pressure of a gas mixture is the sum of the pressures of the individual gases.

Boyle's Law—Robert Boyle was an Irish scientist who developed this law, which describes how the volume of a gas responds to pressure changes. The basic principles are:

(a) Pressure decreases volume and moves gas molecules closer together.
(b) The higher the pressure the smaller the volume.
(c) Gas volume decreases by half when pressure doubles.

Introduction to Distillation Operation (Complete System)

Distillation is a process that separates a substance from a mixture by its boiling point. During the distillation process, a mixture is heated until it vaporizes, and then it is recondensed on the trays or at various stages of the column, where it is drawn off and collected in a variety of overhead, side stream, and bottom receivers. The condensed liquid or **overhead product** is referred to as the distillate, while the liquid that does not vaporize in a column is called the residue or **bottom product.** The composition of the feedstock is very important. Variations in the feedstock will cause significant changes to the operation of the distillation system. A variety of chemicals are typically introduced to a distillation column and separated. Two-component mixtures are referred to as binary, while three-component mixtures are called ternary.

Distillation columns are available in two designs: packed or plate. C-202 is a plate column with six trays. The basic components of a distillation column include a cylindrical shell, feed line, bottom line, side stream, and overhead line. The tray is described as a valve tray. Each valve has three riser legs and a capped dome, which is designed to enhance vapor—liquid contact. On the end of each tray is a weir or plate designed to hold a minimum liquid level on the tray. The weir allows liquid to flow over it and into the **downcomer.** The downcomer is a tube that extends into the liquid level on the next lower tray, providing a liquid seal and a passageway for the heavier liquid to flow. Rising hot vapors come into contact with the valve and lift it up to the full range of the riser leg. This action creates a pulsing action inside each sealed compartment as pressure builds and then is released into the next tray. Figure 14-1 illustrates the basic components of the C-202 **feed tray.**

Figure 14-1
C-202 Tray

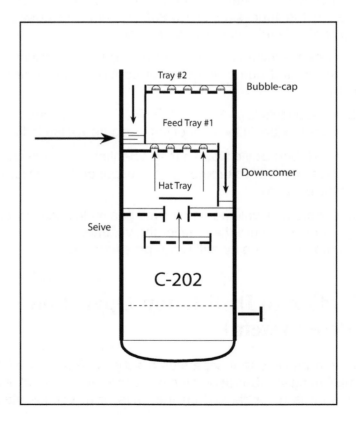

It is important to point out that a distillation unit is composed of a number of smaller systems, such as (1) preheat and feed system, (2) **bottom system,** (3) **overhead system,** and (4) tank farm system. This chapter will break down this process into three groups to simplify the training process. The **pre-heat system** has been covered in the heat exchanger chapter (Chapter 10); however, it has been included in this section to demonstrate how one system can be plugged into another. This approach is also true for the feed system. The tank farm system is designed to have prime and off-specification product tanks. This will include a feed tank equipped with an agitator, a prime bottom tank, a prime overhead tank, and an off-spec or slop tank.

Distillation column-202 is a cylindrical tower with six trays that provide contact points for the vapor and liquid. The first true distillation column was invented in 1917. As vaporization occurs, the lighter components of the mixture move up the tower and are distributed on the six trays. The lightest component goes out the top of the tower in a vapor state and flows over the cooling tubes of a shell and tube heat exchanger (condenser), Ex-204. As the hot vapor comes into contact with the tubes, it condenses and flows into a drum called an overhead accumulator. Part of this product is sent to storage, while the rest is returned to the tower as **reflux.**

Kettle reboiler-205 maintains the heat or energy balance at the bottom of the distillation column. Kettle reboiler Ex-205 takes suction off the bottom of the tower. The liquid in the column is gravity fed through the reboiler. Vaporization occurs in the reboiler, and these vapors rise up through the column. This process is called boil-up. Reboilers take suction off the bottom of the tower. The heaviest component of the column, pentane, flows into the reboiler, in which liquid catalyst and butane are stripped of smaller molecules. A kettle reboiler is equipped with a special vapor disengaging cavity that allows the lighter components to be stripped off the heavier ones and fed back to the column under the lowest tray. The stripped vapors are returned to the column and allowed to separate in the tower.

In a distillation system, the following variables have a direct relationship with each other:
- pressure and boiling point
- temperature and pressure
- composition changes in feed to column
- level in accumulator, reboiler, tanks, and tray flooding
- flow rate, and
- time.

Pressure and Boiling Point

Lighter components are more volatile than heavier components. Each tray in the column has a different molecular structure and a corresponding set of **vapor pressures.** The pressure at the bottom of the column is different from the pressure at the top; however, the pressure is controlled at the top of the column or at the overhead accumulator. Vacuum enhances the boiling point, while increased pressure pushes molecules closer together and back into the liquid state. Distillation columns feed off the vapor–liquid contact phenomenon. In a plate column, product will pulse up the column as pressure builds in the individual still and lifts or enters the upper still. When pressure increases across the column, the boiling point shifts and moves upward, causing the liquid to stop boiling.

Temperature and Pressure

It is a well-known scientific principle that higher temperatures increase or enhance molecular movement. Cooler temperatures have the opposite effect. If steam flow to the reboiler sticks in the open position, it will affect variables up and down the column, confusing a new technician. If the heat to the preheater is lost, the separation process will slow down or stop.

Composition Changes

Since it is the behavior of the mixture that is important to the distillation process, changes in the mixture will affect a wide assortment of variables. For example, if a larger percentage of pentane enters the column, bottom flow rates will increase while upper and side stream flow rates will decrease. This decrease will allow the condenser to cool the butane to such a low level that it will literally quench the separation process when it flows back into the column as reflux. Composition changes will also cause internal pressures to be different. This will also indicate molecular changes across the column and shifts in product purity.

Level Changes

When the level increases in D-204, the vapor cavity shrinks, resulting in a less stable pressure control bubble. Since vapors and gases can be compressed and liquids, as a general rule, cannot be compressed, high liquid levels result in poor column pressure control. High liquid levels in the kettle reboiler will result in unsteady energy balance as liquid is entrained and carried back into the column. High liquid levels in the reboiler will also result in product contamination of pentane and cooler temperatures at the bottom of the column. This is due to the relationship between steam flow in the reboiler tubes and the quantity of liquid pentane. Low levels will result in higher or erratic bottom temperatures and loss of flow to Ex-202. Level changes on the trays can be described as flooding or dry trays. Each of these results in serious problems for the technician. Flooded trays create serious pressure differentials above and below the tray. Product purity is also compromised since vapor–liquid contact is out of balance.

Flow Rate Changes

Flow rates are carefully calculated to C-202 at the feed tray, Ex-202, Ex-203, Ex-204, to the reactor, and to product storage. If a high steam flow rate encounters a low product flow rate in an energy transfer device, the result will be problems. Low feed flow rates to a hot column will cause the entire quantity to vaporize and carry over the top of the column. If the cooling water flow rate to Ex-204 is low, almost every alarm on the unit will go off as pressure increases; temperature increases; overhead accumulator levels drop; AT-2, AT-3, and AT-4 go into alarm; and virtually all product is off specification.

Time

Time is an important analytical variable that should not be manipulated in starting a distillation system up. Many new technicians want to take short-cuts in order to get the column on line quicker. Bypassing certain steps and not allowing the column the proper time to line out can cause trouble. For example, it takes time to get the butane analyzer on the overhead product line to read 98.5% pure butane. Some quality steps cannot and should not be bypassed in order to manipulate the system to get a sample result that is temporarily in specification. By following procedure and allowing the distillation system adequate time to clean up, a quality product can be produced.

Kettle reboilers are shell and tube heat exchangers designed to produce a two-phase vapor–liquid mixture that can be returned to a distillation column. Kettle reboilers have a removable tube bundle that uses steam or a high-temperature process medium to boil the fluid. A large vapor cavity above the heated process medium allows vapors to concentrate. Liquid that does not vaporize flows over a weir and into the liquid outlet. Hot vapors are sent back to the distillation column through the reboiler's vapor outlet ports. This process controls the level in the bottom of the distillation column, maintains product purity, strips smaller hydrocarbons from larger ones, and helps maintain the critical energy balance on the column. Kettle reboilers operate with liquid levels from 2 in. above to 2 in. below the upper tubes. Engineering designs typically allow 10–12 in. of vapor space above the tube bundle. Vapor velocity exiting the kettle reboiler must be low enough to prevent liquid entrainment. The bottom product spills over the weir that fixes the liquid level on the tube bundle.

An important concept with a distillation column is energy or **heat balance.** Reboilers are used to restore this balance by adding heat for the separation processes. Bottom products typically contain the heavier components from the tower. Reboilers suction off the bottom products and pump them through their system. Column temperatures are controlled at established set points. Product flow enters the bottom shell side of a reboiler. As flow enters the reboiler, it comes into contact with the tube bundle. The tubes have steam or hot oil flowing through them. As the bottom product comes into contact with the tubes, a portion of the liquid is flashed off (vaporized) and captured in the dome-shaped vapor cavity at the top of the reboiler shell. This vapor is sent back to the tower for further separation. A weir contains the unflashed portion of the liquid in a kettle reboiler. Excess flow goes over the weir and is recirculated through the system. Kettle reboilers are easy to control because circulation and two-phase flow rates are not considerations.

During tower operation, raw materials are pumped to a feed tank and mixed thoroughly. Mixing is usually accomplished with a pump-around loop or a mixer. This mixture is pumped to a feed preheater or furnace in which the temperature of the fluid mixture is brought up to operating conditions. Preheaters are usually shell and tube heat exchangers or fired furnaces. The fluid enters the feed tray or feed section in the distillation column. Part of the mixture vaporizes as it enters the column, while the rest flows downward, flashing and condensing over and over again until, based upon temperature and pressure, the liquid stops on a tray in which the molecular structures of the components are similar.

A distillation tower is a series of stills placed one on top of the other. As vaporization occurs, the lighter components of the mixture move up the tower and are distributed on the various trays. The lightest component goes out the top of the tower in a vapor state and is passed over the cooling coils of a shell and tube condenser. As the hot vapor comes in contact with the coils, it condenses and is collected in the overhead accumulator. Part of this product is sent to storage, while the rest is returned to the tower as reflux.

Condensed overhead product, butane, is sent to storage in either tank-204A, prime butane, or off-specification tank-204B. Flow rates are controlled at 62 gpm. Side stream flow is drawn off tray three in liquid form. The typical composition is 38% liquid catalyst, 61% butane, and 1% pentane. Flow rates are controlled at 36.5 gpm to the reactor. The bottom flow rates are controlled at 126.5 gpm, 124.6 gpm pentane, and 1.9% butane. Product flow is TK-205A, prime pentane, or TK-205B, off-specification pentane. Levels in each tank are controlled at 50%.

Equipment Descriptions (Block Flow and Safety)

Preheat and Feed System
Stirred Feed Tank-100   The hot oil system is designed to provide heat to various operations that require a stable heat source. The primary use of the hot oil system is to provide a heat source to a series of heat exchangers being used to preheat feed in a number of distillation systems. The tank operating capacity is 20,000 gal. The tank is constructed of stainless steel and insulated to conserve and retain heat energy. The system is run under atmospheric pressure; however, the tank is sealed and caution signs are posted on the vessel. A level and pressure-relief system is attached to the tank. Process technicians are responsible for monitoring and controlling all of these variables.

Pumps-202A/B   Pump-202 is a horizontally mounted centrifugal pump designed to operate at 225 gpm. Suction pressure is maintained at 40 psig, and the discharge pressure typically runs at 135 psig. The differential pressure across the pump is 75 psig. This differential is necessary for the system to

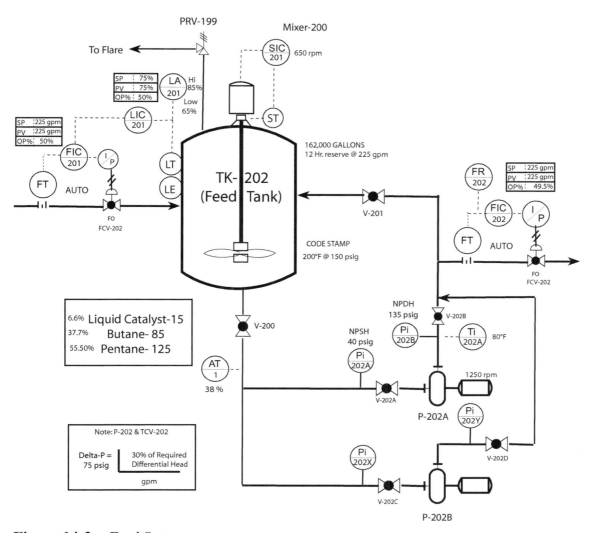

Figure 14-2 *Feed System*

operate properly. The pressure is obtained as the control valve moves through 30% of its operating range on the discharge side of the pump. Figure 14-2 shows all the elements associated with the operation of the feed system.

Flow Control Loop (FIC-202, Controls Feed Rate to the Column)
The flow control loop is designed to work with the pump and the backpressure available to the control valve. Figure 14-3 illustrates the basic layout of each component in the control loop.

The basic components of FIC-202 are:
- orifice plate
- flow transmitter
- flow-indicating controller
- transducer

Figure 14-3
FIC-202 Control Loop

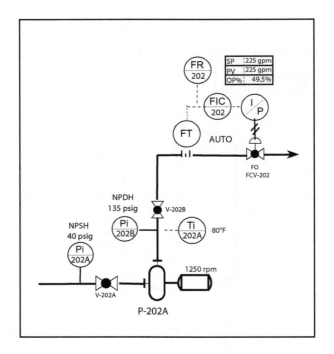

• flow control valve 202, and
• flow recorder 202.

In automatic process control, the centrifugal pump and the control valve form a unique relationship as the control valve begins to throttle flow. This throttling effect does not hurt the pump or valve, as the principle of internal slip protects the system. This is not the case when a positive displacement (PD) pump is used. PD pumps are typically not used for flow control that requires throttling, since they will damage the equipment. With a centrifugal pump, the control valve operates through a 30% value, designed to allow it to regulate flow from 0 gpm to the desired process set point.

Preheater-202 (Heats up Feed to Column) Ex-202 is a horizontally mounted two-pass system that has a shell and tube and floating head. A **ternary mixture** of butane, pentane, and liquid catalyst enters the shell side at 80°F at 225 gpm and encounters a longitudinal baffle that forces the fluid to run countercurrent to the tube side flow. A series of smaller vertical baffles allow the fluid to cascade from the longitudinal center plate to the outer shell and back until it reaches the return head and passes across the exchanger again and into the outlet. The outlet temperature of the shell side is typically 115°F, resulting in a 35°F differential. The operating pressure on the feed to the column is 135 psig. The tube side of Ex-202 receives 152 gpm at 223°F as it fills the tubes from the bottom to the top. The exiting temperature of the liquid pentane is 173°F at 130 psig. The bottom flow specifications are 92% pentane and 8% butane. The differential pressure across the tubes is 5 psi, and the temperature differential is 50°F.

Preheater 203 (Heats up Feed to Column with Hot Oil)   Ex-203 is also a horizontally mounted two-pass system that has a shell and tube and floating head. The primary difference between Ex-202 and Ex-203 is the application of a hot oil system that has been incorporated into the transfer of heat via the tube inlet and outlet of Ex-203. Like Ex-202, the shell-side parallel connection has a flow rate of 225 gpm at 115°F and 135 psig. The operating pressure on column-202 is 100 psig, so the pressure at the discharge of P-202 is monitored closely. The flow rate through Ex-202 and Ex-203 is controlled by FIC-202, which is located on the shell discharge between Ex-203 and C-202. The temperature differential across the shell side of Ex-203 is 65.5°F. This is the difference between the shell outlet (180.5°F) and shell inlet (115°F). The tube-side flow is regulated at 180°F by TIC-100 and the flow of hot oil from TK-100. TIC-100 measures the temperature of the exiting shell side flow on Ex-203 and uses this variable to compare with the set point.

Temperature Control Loop (TIC-100)   The temperature control loop utilizes a hot oil system to heat up the feed to C-202. The set point for the system is 180°F at 225 gpm at 135 psig. Figure 14-4 shows how the temperature

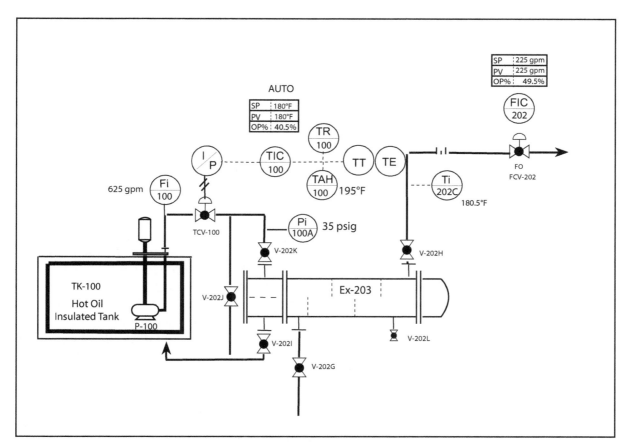

Figure 14-4   *Temperature Control Loop, TIC-100*

control loop on TIC-100 appears on a process-flow diagram. The basic components of TIC-100 include:

- temperature recorder 100
- temperature element—thermocouple in thermowell
- temperature transmitter
- temperature-indicating controller
- transducer, and
- temperature control valve.

One of the primary sections of the distillation system includes the preheat system. The preheat system illustrates how the feed is heated up before it enters the column. Figure 14-5 shows the basic components of the preheat system.

Overhead System

Condenser 204   Ex-204 closely resembles 202 and 203 except that it is used for cooling instead of heating. Cooling tower water flow through the tube side is 525 gpm at 85°F. The temperature of the exiting tube-side water is 125°F, resulting in a temperature differential of 40°F. As the hot (158°F) overhead vapors pass over the 85°F tubes, they are partially condensed and accumulate in D-204. The temperature of the cooled distillate is 128°F, resulting in a temperature differential of 30°F between vapor and liquid. Pressure on the tube inlet and the outlet are 50 psig and 45 psig, respectively.

Drum-204   The overhead accumulator is a drum designed to withstand pressures over 100 psig. Pressure on the drum is controlled at 100 psig by PIC-204. Since this is a partial condenser operation, about 10.4 ft³/min goes to the flare. A level control system, LIC-204, maintains a liquid level in the drum at 50%. Two pumps, P-204A/B, take suction off the overhead accumulator.

Pump-204A/B   P-204 is a centrifugal pump designed to operate at 228.8 gpm. The suction side of the pump is under 100 psig from D-204 plus the liquid butane level. Total suction pressure at the gauge is 105 psig. The discharge operates at 160 psig and discharges into two control valves. The upper controller, FIC-203, typically operates at 142.8 gpm unless it is in cascade and responding to changes required by TIC-203. The lower controller is LIC-204, which allows about 71.4 gpm under normal conditions. The level controller is designed to maintain the level in D-204 at 50%. The total output from P-204 is 142.8 + 60.5 = 203.3 gpm. Only 60.5 gpm of 98.5% butane is sent to the tank farm.

Flow Control Loop (FIC-203)   FIC-203 is designed to operate at 157.4 gpm and is cascaded to TIC-203. As the secondary or slave controller, FIC-203 can operate independently with its own set point or under the control of TIC-203. When engineering decides to cascade the flow controller to the

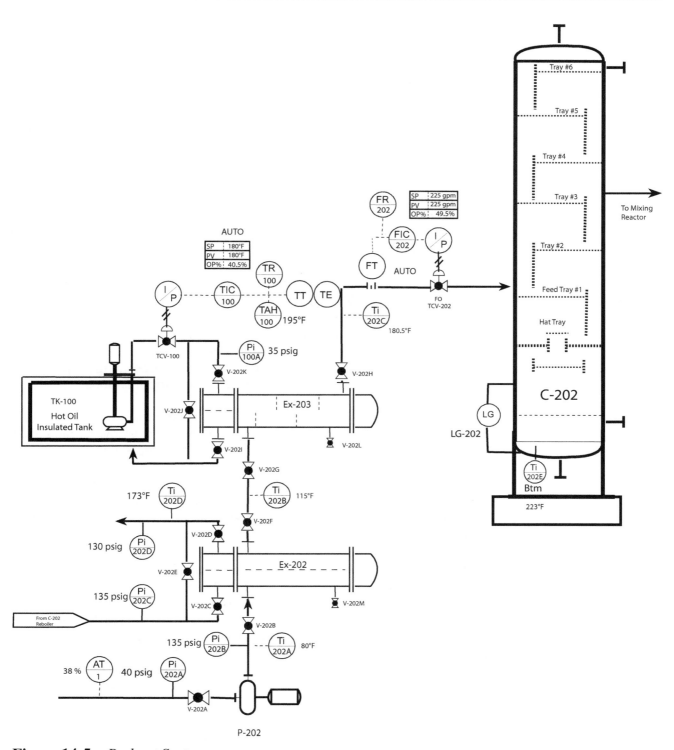

Figure 14-5 *Preheat System*

temperature controller it is because controlling the temperature is more important than controlling the flow. In this event, excess flow is sent to storage.

Temperature Control Loop (TIC-203)   The temperature control loop 203 is a master or primary controller specifically designed to control the temperature at the top of the distillation column. As mentioned earlier, TIC-203 can override the set point on the flow controller. A temperature reading is taken off the top of the column and used as a guide for establishing the **temperature gradient** of the column.

Level Control Loop (LIC-204)   Controlling the level on D-204 is critical to the efficient operation of C-202 since the reflux plays such a vital role. Reflux in theory is designed to control product purity and temperature in the top section of the column. Since pure liquid product is being pumped onto the upper tray, it will immediately come into contact with rising hot vapors and a high percentage will flow down the downcomers. This will spread the pure product across the top four trays as it vaporizes, purifies, and cools the top of the column. LIC-204 is also responsible for allowing customer product to flow to the tank farm. In normal operation, LIC-204 will not open until a 50% level has been reached in D-204. Process technicians must watch the drum level to ensure that the controller is operating properly and the P-204 is operating. This will also include the proper lineups.

Pressure Control Loop (PIC-204)   Under normal atmospheric pressure, butane, the liquid catalyst, which is a gas, and pentane will evaporate easily. To ensure that the system operates properly, it is operated under 100 psig. This is a considerable amount of pressure, and a number of safety concerns are attached to this phenomenon. The set point of PIC-204 is 100 psig; however, 10.4 ft^3/min still goes to the flare. A combination of heat and chemicals provides the pressure that allows the system to operate. This suggests a significant relationship between temperature and pressure.

Flow Control Loop (FIC-300, See Cooling Tower Chapter [Chapter 11])   FIC-300, the flow-indicating controller, controls the flow of water through Ex-204. FIC-300 is located on the discharge side of P-302. The flow rate on P-302 is carefully controlled at 525 gpm with FCV-300. Changes in the flow will cause other variables in the system to go into alarm. An orifice plate is located in the line that allows a transmitter to measure and send a signal to the controller. The controller compares the signal with the set point and makes adjustments to the control valve. FIC-300 can be operated in manual or auto mode.

Fi-300, the flow indicator, is located on the tube discharge side of Ex-204. This indicator allows the outside operator to verify the flow rate with FIC-300.

Butane Analyzer Transmitter 2   The purity of the overhead product is analyzed on the reflux line as it flows to the top of C-202. Analyzers are operated by process technicians and maintained and serviced by the instrumentation department. Samples of the overhead stream are sent to the lab for complex testing procedures and compared with the analyzer measurement. While other components are present in the overhead stream, only butane concentration is measured. The lab analyzes the other components, and these are recorded by the process technician on shift.

Bottom System   The primary purposes of the bottom system of the distillation column are to control level, maintain energy balance on the column, and strip or separate heavier components.

Kettle Reboiler (Ex-205)

Ex-205 is a heat exchanger designed to produce a two-phase vapor-liquid mixture that controls the level and energy balance column 202. A kettle reboiler uses steam or hot oil on the tube side to provide heat to the liquid in the shell. Column 202 gravity feeds liquid into the reboiler shell, in which it comes into contact with the heated tubes. The shell of a kettle reboiler has three nozzles: a liquid feed in, a heavy bottom out, and a vapor out. A special dome-shaped vapor disengaging cavity is located in the shell above the tubes, giving the kettle reboiler a unique shape. Kettle reboilers have a removable tube bundle and a channel head designed for multipass and floating head operation. Liquid that does not vaporize flows over a weir and into the suction of P-205. Hot vapors are sent back to C-202 through the reboiler's vapor nozzle outlet. These vapors enter the column under the lowest tray providing heat. Level control is located on the kettle reboiler. FIC-205 controls steam flow on the tube side of the kettle reboiler and is cascaded to a temperature control loop on the bottom of the column. A steam trap is located on the discharge side of the tubes. Condensed and low-pressure steam is routed back to the boiler.

The pressure on the shell is 102.4 psig, with a liquid temperature of 223°F. Tube flow is set at 14 mlb/h, with an inlet pressure of 120 psig and a temperature of 350°F. Level in the reboiler is controlled at 50% with the liquid level measured 2 in. above the tubes. The vapor cavity above the liquid at specifications is 8 in. Vapor velocity exiting the reboiler must be low enough to prevent liquid entrainment. Bottom product spills over the weir that fixes the liquid level on the tube bundle. A kettle reboiler is unique since it has the ability to separate lighter and heavier components in the bottom stream. Kettle reboilers are easy to control because circulation and two-phase flow rates are not considerations; however, it is possible to build up pressure in the reboiler vapor cavity and circulation can stop. This typically occurs under the following conditions:

- steam flow to tubes decreases
- high liquid level
- feed preheat problems
- large pressure variations in the column.

Centrifugal Pump(s) 205A/B   P-205 is a horizontally mounted centrifugal pump that takes suction off Ex-205 and discharges to the tube side of Ex-202 to preheat feed to the column. The suction pressure is 102.4 psig on P-205 and 135 psig on the discharge. Butane concentrations in the bottom stream are controlled around 1.5%. The bottom butane analyzer is located on the discharge side of the pump. Flow rates are controlled around 126.5 gpm.

Flow-Indicating Controller (FIC-205)   FIC-205 is used to control the flow of steam into the tube side of Ex-205. Inlet pressure is 120 psig with a steam temperature of 350°F. The control schematics for flow and temperature are linked together in a cascade arrangement with FIC-205 operating as the secondary or slave controller. Operational specifications require a set point of 14 mlb/h.

Temperature-Indicating Controller (TIC-205)   TIC-205 is a master or primary controller designed to maintain the temperature at the base of column 202 by adjusting the flow of steam on FIC-205. A cascade relationship exists between temperature controller 205 and flow controller 205. The temperature at the base of the column is controlled at 221.7°F. A thermowell and thermocouple are located directly below the lowest tray in C-202. Data collected from this point are transmitted to temperature controller 205. TIC-205 is linked to FIC-205. While the temperature control loop only uses three elements of a control loop, the flow control loop utilizes all five.

Level-Indicating Controller (LIC-205)   The primary purpose of LIC-205 is to control the level in C-202 and Ex-205. Since a gravity flow line is connected to the column and kettle reboiler, bottom liquid level has a direct correlation. A high level in the column will result in a high level in the reboiler. The control valve for the level control system is located on the discharge side of P-205 and is designed to open or close depending upon the height of the liquid on the discharge side of the weir.

Differential Pressure Transmitter (DPT-202)   The differential pressure (DP) cell is designed with a high-pressure leg and a low-pressure leg. The high pressure leg is connected to the gravity flow line and measures pressure fluctuations. The low pressure leg is connected to the top of column 202 and measures pressure across the top tray. The difference between the two measurements is controlled around 2.4 psig. Figure 14-6 shows the basic components associated with the bottom system.

Instrument Systems

The distillation system is one of the most complex operations that a process technician could be assigned to. Automatic process control makes it possible for a handful of talented technicians to operate the system. In a

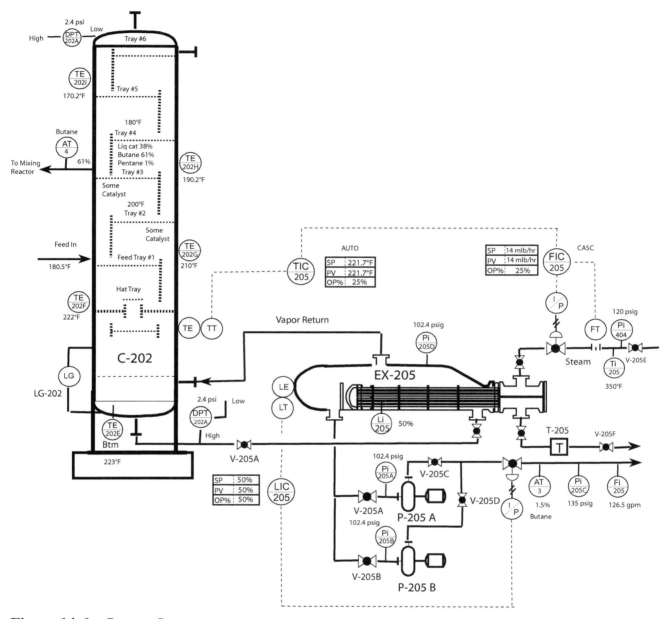

Figure 14-6 *Bottom System*

distillation process, there are four variables that have immediate or delayed effects on each other. These variables include temperature, flow, pressure, and analytical variables such as feed composition.

Temperature
Feed System

       TIC-100 temperature-indicating controller—located on the tube inlet of Ex-203.

TR-100 temperature recorder—provides a video trend of variations in the feed entering C-202.

Ti-202A temperature indicator—located on the discharge of P-202. Runs at 80°F.

Ti-202B temperature indicator—located on the shell outlet line from Ex-202. Runs at 115°F.

Ti-202C temperature indicator—located on the shell outlet line from Ex-203. Runs at 180.5°F.

Ti-202D temperature indicator—located on the tube outlet line from Ex-202. Runs at 173°F.

TAH-100 temperature alarm high—located on P-100 hot oil discharge. Alarms at 195°F.

Overhead System

TIC-203 temperature-indicating controller—takes temperature measurement at the top of the column and transmits signal to controller. TIC-203 is classified as a master or a primary controller. The temperature at the top of the column is the most important variable, and TIC-203 uses the reflux flow "slave" controller FIC-203 to regulate temperature.

TE-202H temperature element—measures the temperature on tray three. Operates at 190.2°F. Sends a signal to a remote location in the control room.

TE-202i temperature element—measures the temperature on tray five. Operates at 170.2°F. Sends a signal to a remote location in the control room.

TE-202F temperature element—measures the temperature on the hat tray. Operates at 222°F. Sends a signal to a remote location in the control room.

Ti-204 temperature indicator—located on the discharge of P-204. Runs at 128°F.

Ti-300 temperature indicator—located on the discharge of pump-300. Runs at 125°F.

Bottom System

TIC-205 temperature-indicating controller—takes a temperature measurement below the hat tray on the column and transmits signal to the controller. TIC-205 is classified as a master or a primary controller. The temperature at the bottom of the column is controlled by TIC-205 "master controller" adjusting the set point on FIC-205. This process adjusts the steam flowing through the tube bundle on the kettle reboiler.

TE-202E temperature element—located at the bottom of the column. Operates at 223°F. Sends a signal to a remote location in the control room.

TE-202F temperature element—measures the temperature on the hat tray. Operates at 222°F. Sends a signal to a remote location in the control room.

TE-202G temperature element—measures the temperature on the feed tray. Operates at 210°F. Sends a signal to a remote location in the control room.

Ti-205 temperature indicator—located on the tube inlet to Ex-205. Runs at 350°F.

Pressure
Feed System

Pi-202A pressure indicator—located on the suction side of P-202. Runs at 40 psig.

Pi-202B pressure indicator—located on the discharge side of P-202. Runs at 135 psig.

Pi-202C pressure indicator—located on the tube inlet of Ex-202. Runs at 135 psig.

Pi-202D pressure indicator—located on the tube outlet of Ex-202. Runs at 130 psig.

Pi-100A pressure indicator—located on the tube inlet of Ex-203. Pressure typically runs at 35 psig.

Overhead System

PIC-204 pressure-indicating controller—the pressure element is located on the top of D-204 and transmits a signal to the controller. A vapor cavity at the top of the drum allows pressure to be controlled at 100 psig on the top tray of the column. Pressure on each tray is slightly different due to variations in molecular structure. PIC-204 is positioned so that it can control the pressure from the base of the column to the top.

Pi-300A pressure indicator—located on the tube inlet to Ex-204. Runs at 50 psig.

Pi-300B pressure indicator—located on the tube outlet from Ex-204. Runs at 45 psig.

Pi-204A pressure indicator—located on the suction side of P-204A. Runs at 105 psig.

Pi-204B pressure indicator—located on the suction side of P-204B. Runs at 105 psig.

Pi-204C pressure indicator—located on the discharge side of P-204. Runs at 160 psig.

Bottom System

DPT-202 differential pressure transmitter—located on the column bottom line. Runs at 2.4 psig. The other leg connects to the top tray of the column.

Pi-205A/B pressure indicator—located on the suction side of P-205A/B. Typically runs at 102.4 psig.

Pi-205C pressure indicator—located on the discharge side of P-205. Operates at 135 psig.

Pi-205D pressure indicator—located on the top of Ex-205. Runs at 102.4 psig.

Pi-205A pressure indicator—located on the tube inlet to Ex-205. Steam pressure runs at 120 psig.

Flow
Feed System

FIC-202 flow-indicating controller—located on the feed line between Ex-203 and C-202. Designed to regulate the flow of feed at 225 gpm to the column.

FR-202 flow recorder—located on the FIC-202 control loop. Designed to monitor variations in flow rate.

Overhead System

FIC-300 flow-indicating controller—located on the tube inlet to Ex-204. Designed to regulate cooling tower water flow rate through the condenser at 525 gpm.

Fi-300 flow indicator—located on the tube outlet of Ex-204. Typically operates at 525 gpm.

Fi-204B flow indicator—located on the flare vapor discharge line from D-204. Runs at 10.4 ft^3/min.

Fi-204A flow indicator—located on the overhead liquid butane to storage line. Runs at 71.4 gpm.

FIC-203 flow-indicating controller—located on the discharge line or reflux line from P-204 to the top tray of C-202. FIC-203 is a secondary or slave controller to TIC-203. Runs at 142.8 gpm. The basic equipment used in Figure 14-7 includes Ex-204 condenser, D-204 overhead accumulator, P-204A/B, and a variety of modern process control systems.

Bottom System

FIC-205 flow-indicating controller—located on the tube inlet to Ex-205. FIC-205 is classified as a secondary or slave controller. During normal operation, it is cascaded to TIC-205. Runs at 14 mlb/h.

Fi-205 flow indicator—located on the discharge side of P-205. Liquid pentane flow rate to storage runs at 126.5 gpm.

Analytical
Feed System

AT-1 analytical transmitter—located on the suction side of P-202. Measures the percentage of butane in the feed. Set point is 38% and operational specifications allow 36–38%.

Overhead System

AT-2 analytical transmitter—located on the discharge side of P-204. Measures the percentage of butane in the overhead stream. Set point is set at 98.5%.

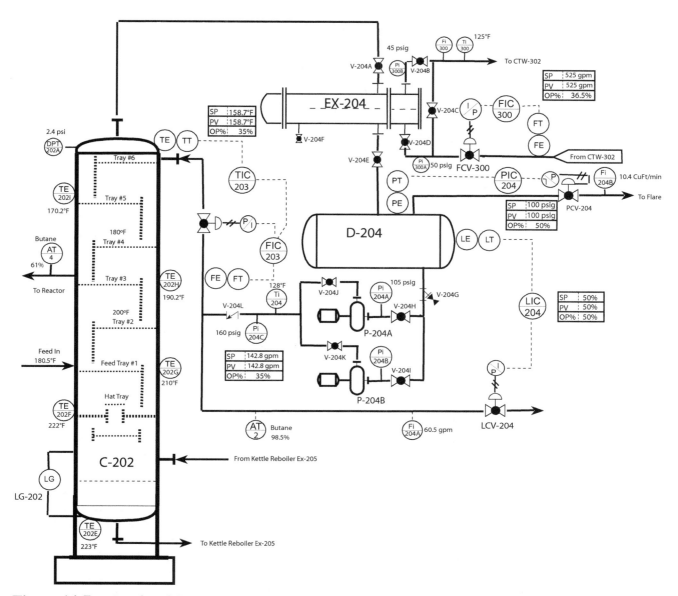

Figure 14-7 *Overhead System*

Bottom System

AT-3 analytical transmitter—located on the discharge side of P-205. Measures the percentage of butane in the feed. Set point is set at 1.5%.

To Reactor System

AT-4 analytical transmitter—located on tray three, the side stream feed line to reactor 210. Measures the percentage of butane in the reactor feed. Set point is 61%. Feed concentrations typically include liquid catalyst 38%, butane 61%, and pentane 1%.

Operating Distillation Systems

Operating a distillation system is a very complex process that combines a series of smaller systems into a large operation. The basic systems that make up the distillation process include the feed system, the preheat system, the upper section of the column, the bottom section of the column, and the product storage system.

Distillation Feed System

As feed is brought into the unit, it is sampled and blended to ensure correct composition. The primary feed tank, TK-202, is equipped with a stirrer. The ternary feed is composed of 6.6% liquid catalyst, 37.7% butane, and 55.5% pentane.

Distillation Preheat System

After the feed is blended, it is pumped out of the feed tank in a continuous operation to the preheat section, Ex-202 and Ex-203. Heat energy is transferred to the feed stock, which results in a temperature delta of 100°F. The basic equipment used in the preheat system includes Ex-202 and Ex-203, a hot oil system, and a variety of instrumentation.

Distillation System

When the heated feed enters column 202 at 225 gpm and 180°F, it partially flashes as the lighter components move up the column and the heavier components in the form of liquid move down the column, contacting the rising vapor. In a distillation system, vapors go up and liquids go down. In this system the pentane concentrates in the lower section of the column, the liquid catalyst around the third tray, and the butane vaporizes and flows out the top of the column. The basic equipment associated with the top of the column includes Ex-204, D-204, P-204A/B, and a variety of instrumentation. Figure 14-8 illustrates the basic equipment associated with the bottom pentane product to storage. The primary components include Ex-205, P-205A/B, and a variety of instrumentation. Figure 14-9 illustrates all the basic equipment and simple flow paths to the butane storage units.

Operational Specifications (SOP, SPEC Sheet, and Checklist)

Operating the three sections of the distillation system requires a technician to move from the feed system to the bottom system and then to the overhead system. To simplify the process, three short procedures have been

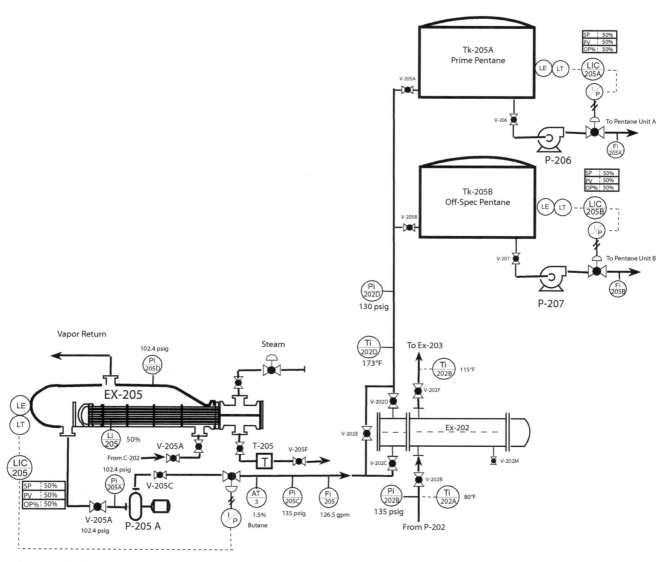

Figure 14-8 *Pentane 205A/B*

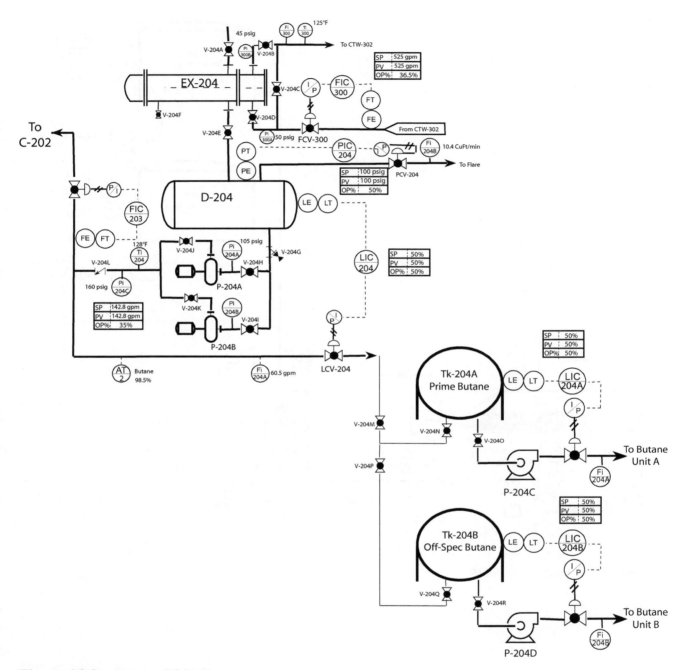

Figure 14-9 *Butane 204A/B*

set up to guide the new technician. The operating procedures associated with the distillation preheat system include:

Action	Notes
1. Verify pump recirculation to TK-202.	This includes looking at pump suction and discharge pressures and level in TK-202.
2. Line up heat exchanger system to C-202; open V-202F, V-202G, and V202H.	
3. Open V-202C and V-202D.	This will admit flow into the lower tube side of Ex-202. Flow comes from the kettle reboiler.
4. Set FIC-202 to 75 gpm and set in AUTO. Ensure C-202 has been prepared prior to this step.	
5. Close V-201 and ensure the feed tank system is operating properly and the analyzer (AT-202A) is between 36% and 38%.	
6. Set FIC-202 to 225 gpm.	
7. Inspect hot oil system and ensure it is operating properly.	
8. Open V-202J.	
9. Set TIC-100 to 180°F and put in AUTO.	
10. Open V-202K and V-202L. Close V-202J. This will admit flow into the tube side of Ex-203.	
11. Monitor and ensure that both control systems are operating properly. This includes the flow and temperature recorder.	
12. Place the system on-spec when operating temperatures approach set point.	
13. Cross-check process variables with SPEC sheet.	

The operating procedures associated with the distillation bottom system include:

Action	Notes
1. Set LIC-205 to 50% and set in AUTO.	
2. Line up P-205A to Ex-202 to TK-205B. Open V-205A, V-205C, V-202C, V-202D, and V-205B.	
3. Set LIC-205A/B to 50% and put in AUTO.	
4. Ensure level in Ex-205 is at 25% and rising.	
5. Set FIC-205 to 14 mlb/h and put in AUTO.	
6. Set TIC-205 to 221.7°F and put in AUTO.	
7. Start P-205A and ensure all pressures are correct and TK-205B is filling.	
8. Monitor temperature profile at the bottom of the column.	
9. Line up and start P-207 when level reaches 50%.	
10. Monitor AT-3 and switch to TK-205A when analyzer reads below 2.5%. Open V-205A and close V-205B.	
11. Shut down P-207 and line up P-206.	
12. Cascade FIC-205 to TIC-205.	
13. Cross-check process variables with SPEC sheet.	

The operating procedures associated with the distillation overhead system include:

Action	Notes
1. Set FIC-300 to 525 gpm and put in AUTO.	
2. Set PIC-204 to 100 psig and set in AUTO.	
3. Set LIC-204 to 50% and put in AUTO.	
4. Set FIC-203 to 142.8 gpm and set in AUTO.	
5. Set TIC-203 to 158.7°F and set in AUTO.	
6. Set LIC-204A/B to 50% and put in AUTO.	
7. Line up P-204A to reflux and to TK-204B off-spec. Open V-204H, V-204J, V-204M, V-204P, and V-204Q.	
8. Start P-204 when level reaches 50% in D-204 and ensure all pressures are correct and TK-204B is filling.	
9. Lineup and start P-204D when level reaches 50%.	
10. Ensure reflux is flowing to C-202 and each controller is operating.	
11. Monitor AT-2 and switch to TK-204A prime butane when analyzer reaches 97.5%. Open V-204N, and close V-204P and V-204Q.	
12. Shut down P-204D and lineup and start P-204C when level in tank reaches 50%.	
13. Cascade FIC-203 to TIC-203.	
14. Cross-check process variables with SPEC sheet.	

Specification Sheet and Checklist

Flow

1.	FIC-201	225 gpm	CASC	Feed to tank-202
2.	FIC-202	225 gpm	AUTO	Feed flow
3.	FR-202	225 gpm	AUTO	Feed flow to C-202
4.	FIC-203	142.8 gpm	CASC	Reflux to C-202
5.	Fi-204A	60.5 gpm	Gauge	Butane to storage
6.	FIC-300	525 gpm	AUTO	Cooling tower water flow
7.	Fi-300	525 gpm	Gauge	Cooling tower water flow
8.	FIC-205	14 mlb/h	CASC	Steam to reboiler
9.	Fi-205	126.5 gpm	Gauge	Pentane to storage
10.	Fi-204B	10.4 ft³/min	Butane/liquid catalyst to flare	
11.	Fi-205A	126.5 gpm	Gauge	P-206 discharge
12.	Fi-205B	126.5 gpm	Gauge	P-207 discharge
13.	Fi-204A	60.5 gpm	Gauge	P-204C discharge
14.	Fi-204B	60.5 gpm	Gauge	P-204D discharge

Analytical

15.	AT-1	38%	—	Butane—feed
16.	AT-2	98.5%	—	Butane—reflux
17.	AT-3	1.5%	—	Butane—bottom
18.	AT-4	61%	—	Butane—reactor
19.	SIC-201	650 rpm	AUTO	Agitator motor on TK-202

Pressure

20.	PIC-204	100 psig	AUTO	D-204 to flare
21.	Pi-202A	40 psig	Gauge	Suction P-202A
22.	Pi-202B	135 psig	Gauge	Discharge P-202A
23.	Pi-202C	40 psig	Gauge	Suction P-202B
24.	Pi-202D	130 psig	Gauge	Discharge P-202B
25.	PI-100A	35 psig	Gauge	Tube inlet Ex-203
26.	Pi-204A	105 psig	Gauge	Suction P-204A
27.	Pi-204B	105 psig	Gauge	Suction P-204B
28.	Pi-204C	160 psig	Gauge	Discharge P-204A/B
29.	Pi-300A	50 psig	Gauge	Lower tube inlet Ex-204
30.	Pi-300B	45 psig	Gauge	Upper tube outlet Ex-204
31.	DPT-202A	2.4 psig	ΔP cell	Bottom and top of the column
32.	Pi-404	120 psig	Gauge	Steam pressure
33.	Pi-205A	102.4 psig	Gauge	P-205A suction
34.	Pi-205B	102.4 psig	Gauge	P-205A suction
35.	Pi-205C	135 psig	Gauge	P-205A/B discharge
36.	Pi-205D	102.4 psig	Gauge	Ex-205 vapor cavity

Temperature

37.	TIC-100	180°F	AUTO	Tube inlet Ex-203 hot oil system
38.	Ti-202A	80°F	Gauge	Discharge P-202
39.	Ti-202B	115°F	Gauge	Discharge shell side Ex-202
40.	Ti-202C	180.5°F	Gauge	Discharge shell side Ex-203
41.	Ti-202D	173°F	Gauge	Tube outlet Ex-202
42.	TR-100	180°F	—	Temperature recorder hot oil system
43.	TAH-100	195°F	—	High-temperature alarm
44.	TIC-203	158.7°F	AUTO	Top tray of C-202
45.	TE-202i	170.2°F	—	Tray five
46.	TE-202H	190.2°F	—	Tray three
47.	TE-202G	210°F	—	Feed tray one
48.	TE-202F	222°F	—	Hat tray
49.	TE-202E	223°F	—	Bottom of C-202
50.	Ti-300	125°F	Gauge	Tube outlet Ex-204 cooling water
51.	Ti-205	350°F	Gauge	Upper tube inlet on Ex-205 steam
52.	TIC-205	221.7°F	AUTO	Below hat tray, bottom of C-202
53.	Ti-204	128°F	Gauge	P-204A/B discharge

Level

54.	LG-202	50%	—	Bottom of column
55.	LIC-204	50%	AUTO	Drum-204 level control
56.	LIC-205	50%	AUTO	Kettle reboiler level
57.	LIC-204A	50%	AUTO	TK-204A level
58.	LIC-204B	50%	AUTO	TK-204B level
59.	LIC-205A	50%	AUTO	TK-205A level
60.	LIC-205B	50%	AUTO	TK-205B level
61.	Li-205	50%	Gauge	Level over tube bundle in Ex-205
62.	LIC-201	75%	AUTO	Level control TK-202
63.	LA-201	Hi-85%/Lo-65%	—	Located on LIC-201

Troubleshooting Scenario 1

Refer to Figure 14-10 and select the best answers. A series of what-if troubleshooting scenarios have been developed to test your knowledge of distillation theory.

Troubleshooting Scenario 2

Refer to Figure 14-11 and select the best answers to the various troubleshooting scenarios listed.

Troubleshooting Scenario 3

Refer to Figure 14-12 and select the best answers to the various troubleshooting scenarios listed.

Summary

Distillation separates the basic components of a mixture by their individual boiling points. During the distillation process, a mixture is heated until it vaporizes, and then it is recondensed on the trays in the column, where it is drawn off and collected in a variety of overhead, side stream, and bottom receivers. The condensed liquid is referred to as the distillate; the liquid that does not vaporize in a column is called the residue. A distillation column can be compared to a series of stills placed one on top of the other. As vaporization occurs, the lighter components of the mixture move up the column. The lightest component goes out the top of the column in a vapor state and is passed over a condenser. As the hot vapor comes in contact

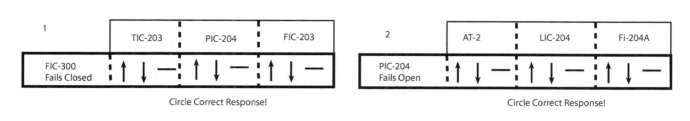

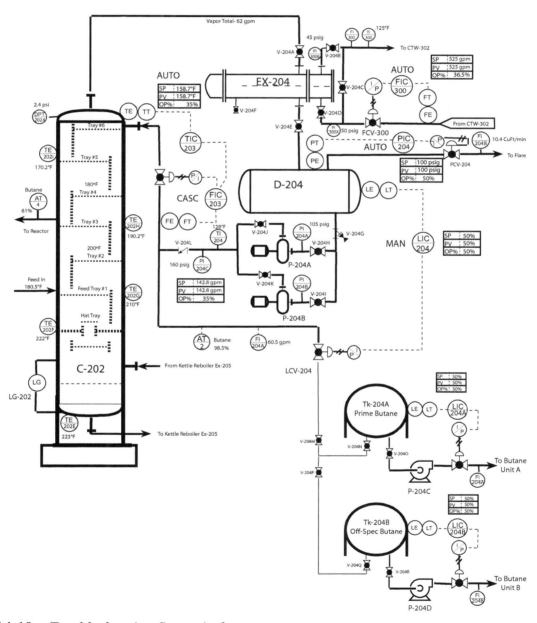

Figure 14-10   *Troubleshooting Scenario 1*

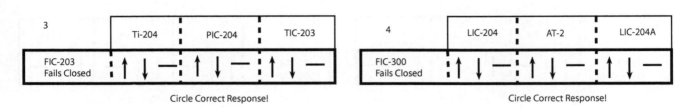

3		Ti-204	PIC-204	TIC-203
FIC-203 Fails Closed		↑ ↓ —	↑ ↓ —	↑ ↓ —

Circle Correct Response!

4		LIC-204	AT-2	LIC-204A
FIC-300 Fails Closed		↑ ↓ —	↑ ↓ —	↑ ↓ —

Circle Correct Response!

Figure 14-11 *Troubleshooting Scenario 2*

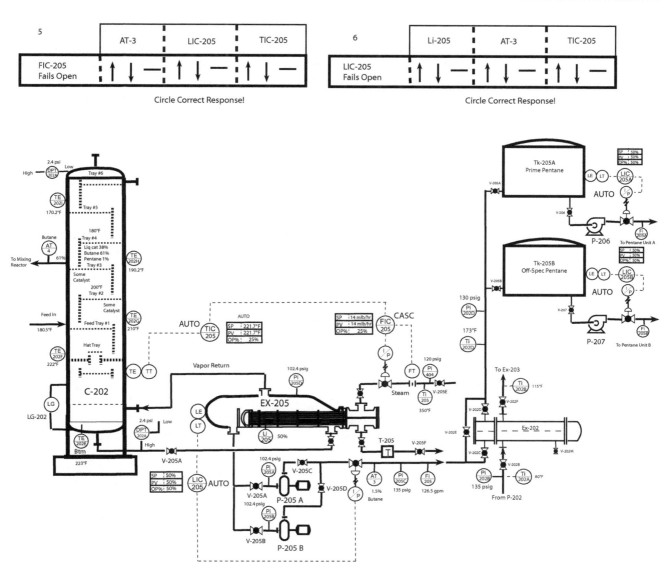

Figure 14-12 *Troubleshooting Scenario 3*

with the cool condenser, it changes back into a liquid and accumulates in D-204.

Distillation systems are composed of a number of smaller systems: (1) pre-heat and feed system, (2) bottom system, (3) overhead system, and (4) tank farm system.

Distillation is a process that separates a substance from a mixture by its boiling point. In this system, three components are used, making them a ternary mixture. During the distillation process, the mixture is heated until it vaporizes, and then it is recondensed on the trays or at various stages of the column, where it is drawn off and collected in a variety of overhead, side stream, and bottom receivers. The condensed liquid or overhead product is referred to as the distillate while the liquid that does not vaporize in a column is called the residue or bottom product. The composition of the feedstock is very important. Variations in the feedstock will cause significant changes to the operation of the distillation system. A variety of chemicals are typically introduced to a distillation column and separated. Two or more components in a mixture are referred to as binary, while three component mixtures are called ternary.

During tower operation, raw materials are pumped to tank-202 and mixed thoroughly. Mixing is usually accomplished with a pump-around loop or a mixer. This mixture is pumped to a feed preheater, Ex-202 and Ex-203, in which the temperature of the fluid mixture is brought up to operating conditions. Preheaters are usually shell and tube heat exchangers or fired furnaces. The fluid enters the feed tray or feed section in the distillation column. Part of the mixture vaporizes as it enters the column, while the rest flows downward, flashing and condensing over and over again until, based upon temperature and pressure, the liquid stops on a tray in which the molecular structures of the components are similar.

Distillation column-202 is a cylindrical tower with a series of trays that provide contact points for the vapor and the liquid. The first true distillation column was invented in 1917. As vaporization occurs, the lighter components of the mixture move up the tower and are distributed on the various trays. The lightest component goes out the top of the tower in a vapor state and flows over the cooling tubes of a shell and tube heat exchanger (condenser), Ex-204. As the hot vapor comes into contact with the tubes, it condenses and flows into a drum called an overhead accumulator. Part of this product is sent to storage, while the rest is returned to the tower as reflux. A large number of control loops are located at the top of the column. These include control systems that regulate cooling water flow to the condenser, reflux to the top of the column, temperature control at the top of the column, and level control on the overhead accumulator.

At the base of the distillation column, C-202 utilizes kettle reboiler Ex-205 to maintain the energy balance on the separation system. The kettle reboiler takes suction off the bottom of C-202. The liquid in the column is gravity fed through the reboiler. Vaporization occurs in the reboiler, and these vapors rise up through the column. This process is called boil-up. The heaviest components of the column flow into the reboiler and are stripped off the smaller molecules. A kettle reboiler is equipped with a special vapor disengaging cavity that allows the lighter components to be stripped off the heavier ones and fed back to the column under the lowest tray. The stripped vapors are returned to the column and allowed to separate in the tower.

Kettle reboilers are shell and tube heat exchangers designed to produce a two-phase vapor-liquid mixture that can be returned to a distillation column.

Kettle reboilers have a removable tube bundle that uses steam or a high-temperature process medium to boil the fluid. A large vapor cavity above the heated process medium allows vapors to concentrate. Liquid that does not vaporize flows over a weir and into the liquid outlet. Hot vapors are sent back to the distillation column through the reboiler's vapor outlet ports. This process controls the level in the bottom of the distillation column, maintains product purity, strips smaller hydrocarbons from larger ones, and helps maintain the critical energy balance on the column. Kettle reboilers operate with liquid levels from 2 in. above to 2 in. below the upper tubes. Engineering designs typically allow 10–12 in. of vapor space above the tube bundle. Vapor velocity exiting the kettle reboiler must be low enough to prevent liquid entrainment. The bottom product spills over the weir that fixes the liquid level on the tube bundle.

An important concept with a distillation column is energy or heat balance. Reboilers are used to restore this balance by adding heat for the separation processes. Bottom products typically contain the heavier components from the tower. Reboilers suction off the bottom products and pump them through their system. Column temperatures are controlled at established set points. Product flow enters the bottom shell side of a reboiler. As flow enters the reboiler, it comes into contact with the tube bundle. The tubes have steam or hot oil flowing through them. As the bottom product comes into contact with the tubes, a portion of the liquid is flashed off (vaporized) and captured in the dome-shaped vapor cavity at the top of the reboiler shell. This vapor is sent back to the tower for further separation. A weir contains the unflashed portion of the liquid in a kettle reboiler. Excess flow goes over the weir and is recirculated through the system. Kettle reboilers are easy to control because circulation and two-phase flow rates are not considerations.

Condensed overhead product, butane, is sent to storage in either tank-204A, prime butane, or off-specification tank-204B. Flow rates are controlled at 62 gpm. Side stream flow is drawn off tray three in liquid form. The typical composition is 38% liquid catalyst, 61% butane, and 1% pentane. Flow rates are controlled at 36.5 gpm to the reactor. The bottom flow rates are controlled at 126.5 gpm, 124.6 gpm pentane, and 1.9% butane. Product flow is TK-205A, prime pentane, or TK-205B, off-specification pentane. Levels in each tank are controlled at 50%.

Review Questions

1. Draw and label the feed system. Show all instruments. List start-up procedures.

2. Draw and label the preheat system. Show all instruments. List start-up procedures.

3. Draw and label the bottom section of the distillation system. Show all instruments. List start-up procedures.

4. Draw and label the upper section of the column. Show all instruments. List start-up procedures.

5. List the various components in the feed mixture.

6. List the required specification of components on tray three.

7. Draw and label the butane storage system.

8. Draw and label the pentane storage system.

9. Describe the function of each of the four butane analyzers and identify where each is located in the system.

10. Explain how the following variables relate to each other: pressure and boiling point, temperature, flow rate, level, time, and feed composition.

Reactor Model

LEARNING OBJECTIVES

After studying this chapter, the student will be able to:

- Draw the reactor system, show all flows, and instruments.
- Draw each of the control loops in the reactor.
- Explain the primary purpose of each control loop.
- Describe the basic principles of reaction as applied to Rx-202.
- Operate the reactor system.
- Identify the common problems associated with the reactor system.
- Solve various troubleshooting scenarios associated with the reactor system.

Key Terms

Balanced equation—chemical equation in which the sum of the reactants (atoms) equals the sum of the products (atoms).

Chemical reaction—a term used to describe the breaking of chemical bonds, forming of chemical bonds, or breaking and forming of chemical bonds.

Endothermic reaction—a chemical reaction that must have heat added to make the reactants combine to form the product.

Exothermic reaction—a chemical reaction that produces heat.

Material balancing—a method for calculating reactant amounts versus product target rates.

Products—the final result of a chemical reaction.

Reactants—the raw materials in a chemical reaction.

Reaction rate—the amount of time taken by a given amount of reactants to form a product or products.

Stirred reactor—a reactor designed to mix two or more components into a homogeneous mixture; also called autoclave.

Solvent—a liquid that dissolves a liquid, a solid, or a gaseous solute, resulting in a solution.

Toluene—known as phenylmethane or methylbenzene; it is an aromatic hydrocarbon, that is, a water-insoluble liquid. The molecular formula of toluene is C_7H_8 and it has a boiling point of 231°F, 110.6°C. Toluene is also used as an octane booster in gasoline.

Butane—the boiling point of butane is 31.1°F (-0.5°C), C_4H_{10}.

Pentane—used mainly as a solvent, fuel, or hydrogen source in steam reforming. It is classified as the most volatile hydrocarbon, that is, liquid at 72°F. The boiling point of pentane is 96.98°F, 36.1°C (308°K), C_5H_{12}.

Principles of Chemical Reactions (Include Q-Control, Data Collection, Org, Analyze, Reactor Chemistry)

A reactor is a vessel in which a controlled **chemical reaction** takes place. The factors that have an effect on a chemical reaction are called reaction variables. The design and operation of Rx-202 enhance molecular contact between four **reactants: pentane, butane,** liquid catalyst, and **solvent.** Feed to the reactor is controlled at 36.5 gpm. The composition of feed from the column is 38% liquid catalyst, 61% butane, and 1% pentane. Solvent feed to the reactor is controlled at 68 gpm. The materials in the reactor are chilled to 120°F at 85 psig. The reactants are designed to form a new product with an excess of pure butane. A separator is used to remove the new product and isolate butane for storage. The effect of pressure on a reaction cannot be generalized as easily as the effect of temperature. In a liquid-phase reaction, pressure can increase or decrease the **reaction rate,** depending upon the readiness of the reactants or **products** to vaporize. In a gaseous reaction, pressure forces molecules closer together, causing them to collide more frequently. Therefore, the higher the pressure in a gaseous reaction, the higher the reaction rate. In a reaction mixture, gas or liquid, the temperature determines how fast the molecules of the reactants move. At a high temperature, the molecules move rapidly through the mixture, colliding with each other frequently. The more often they collide, the more apt they are to react with one another. A good rule of thumb for chemical reactions is that the speed with which two chemicals will react doubles for each 10°C increase in temperature. This rule assumes, of course, that other variables do not change. For example, if 10% of the chemicals in a reaction form products at 50°C in 1 h, then 20% will form products at 60°C in 1 h. The danger in this type of reaction is that twice as much heat will be given off at the higher temperature. If the extra heat cannot be removed from the reactor as fast as it is generated, then the temperature will rise, causing the reaction to proceed at even higher rates. Obviously, there could be an uncontrollable reaction and, possibly, an explosion.

The concentration of reactants in the reactor has a major effect on how fast the reaction will take place, what products will be produced, and how much heat will have to be added to or taken away from the reaction. We will generalize here for the sake of simplicity and say that for most reactions the higher the concentration of reactants the faster the reaction and thus the more heat that is generated or needed. Figure 15-1 illustrates the basic components of the **stirred reactor** system.

Agitation provided by mechanical agitators or by the turbulent flow of the reactants may affect concentration. In general, good mixing or good agitation produces an efficient reaction with the desired products. Poor agitation may produce pockets that have a high concentration of one reactant and a low concentration of the other. This uneven distribution may produce undesirable as well as unpredictable products.

Many chemicals will not react when placed together at high concentrations and heated, or will produce unwanted products. Therefore, an additional substance called a catalyst is added to stimulate the reaction and to produce the more desirable product. Most catalysts do not react with the reactants, so usually they can be reused several times until they become dirty or ineffective or poisoned. Then they must be replaced or regenerated in some way. Other types of catalysts, usually liquids or gases, will react with the reactant, forming an intermediate product. The reaction will continue until the desired product is formed, releasing the catalyst to be used again. It is not unusual for some catalysts to be destroyed in the reaction or discarded because of the difficulty in recovering them.

A common type of reactor is the mixing or stirred reactor. The basic components of this device will include a mixer or agitator mounted on a tank. The mixer will have a direct or an indirect drive. The reactor shell is designed to

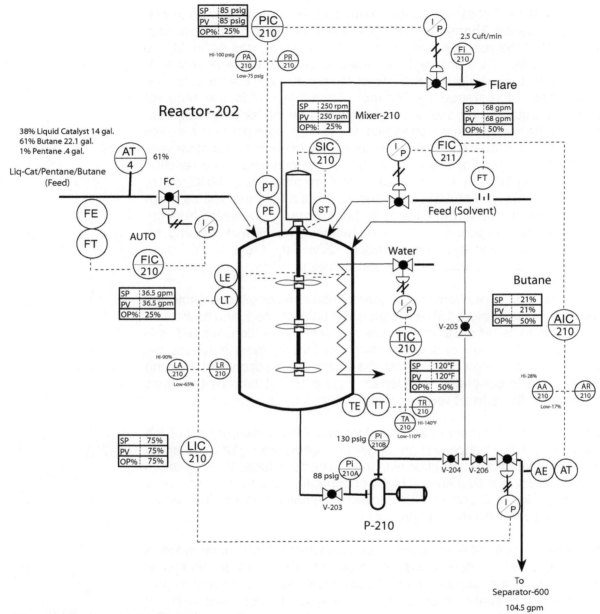

Figure 15-1 *Typical Stirred Reactor*

be heated or cooled and withstands operational pressures, temperatures, and flow rates. The design may include tubing coils wrapped around the vessel, internal or external heat transfer plates, or jacketed vessels. Hot oil, steam, or cooling water may be the heating or cooling medium. The stirred reactor is designed to mix two or more components into a homogeneous mixture. As these components blend together, chemical reactions take place, which create a new product. Blending time and exact operating conditions are critical to the efficient operation of a stirred reactor.

Stirred reactors are equipped with a number of safety features: pressure relief systems, quench systems, process variable alarms, and automatic shutdown controls. Quench systems are designed to stop the reaction process. Pressure relief systems are sized to handle and contain any release from the reactor. The relief system may be designed for liquid, vapor, or a liquid–vapor combination. Safety relief systems allow process releases to go to the plant flare system. Sometimes a chemical scrubber system is used before a product is sent to the flare system.

Process variable alarms will be activated by analytical (composition), pressure, temperature, flow, level, and time variables. Rotational speed on the agitator may be fixed or variable. A series of interlocks, permissives, and alarms will engage during operation and will provide a support network for the technician. A series of process video trends will be displayed to track each of the critical variables. Samples are taken frequently to ensure product quality. Stirred reactors are connected to off-specification (off-test) systems that allow flexibility in switching between prime and off-test operations. The automatic shutdown allows a technician to push one button and shut down the system in the event of a runaway reaction or emergency.

A stirred tank reactor is often referred to as an "autoclave" or batch reactor. This type of reactor is used in both batch processes, such as the suspension vinyl resins unit and the phenolic resins unit, and continuous feed processes, such as the solvent vinyl resins unit and the polyethylene unit. The basic features of a stirred reactor are designed to provide long reaction times, mechanical agitation, and a method of cooling or heating the reaction. Stirred tank reactors are used for liquid–liquid reactions, gas–liquid reactions, and liquid–solid reactions. Stirred tank reactors are used for reactions that require a relatively long reaction time and for reactions of slurry or thick liquid in which mechanical agitation must be used to provide a uniform reaction mixture. Such reactors are also used when the production rate is to be variable and several different reactions or products are to be made in the same reactor. The early research and development of most processes are conducted in batch stirred reactors because of their versatility.

Other considerations that must be taken into account are how much agitation is required to produce a uniform reaction mixture; what type of heat transfer equipment will be needed to prevent excessive fouling of the heat transfer surface; the minimum size of reactor that is needed to obtain the necessary reaction time; and whether a catalyst is needed to promote the reaction and how will it be added to prevent a localized reaction.

Equipment Descriptions (Block Flow and Safety)

Pump-210

One of the most important components of the stirred reactor system is the pump. Pump-210 is a horizontally mounted centrifugal pump. The internal components of the pump are designed to resist corrosion and the effects of the chemicals being introduced into the system. The suction pressure on the pump is 88 psig and the discharge pressure is controlled around 130 psig. Level control valve 210 provides backpressure on the pump's discharge. The three-phase motor and pump are specially sized to control flow rates between 100 and 110 gpm. P-210 A/B is a redundant system that is used to maintain consistent flow rates to the separator. The pump has a recirculation line designed to be used during reactor startup. Pump flow rates are controlled at 104.5 gpm to separator 600. Figure 15-2 is a simple block flow diagram used to explain the various flows in and out of the stirred reactor.

TIC-210

Temperature indicating controller-210 uses water to cool the temperature of the reactor's contents to 120°F. A thermowell is located at the base of the

Figure 15-2
Reactor System Block Flow Diagram

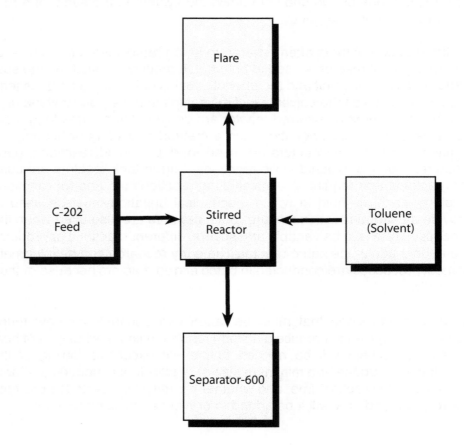

reactor, which protects and secures a thermocouple used to detect the temperature of the exiting product. A signal is transmitted to the controller that opens or closes the cooling water coils to the reactor.

FIC-211

Flow indicating controller-211 controls the solvent feed to the top of the reactor. Flow rates are carefully controlled at 68 gpm. The solvent is an important part of the reaction and is necessary in the formation of the new product. FIC-211 is classified as a slave or secondary controller to AIC-210.

Mixer-200

Speed indicating controller-210 (SIC-210) is a variable speed controller designed to regulate the rotation of the agitation blades on Mixer-200. The rotational speed of the blades is important in order for the reaction to occur. Revolutions per minute (rpm) are controlled at 250 on Reactor-202 (Rx-202). If agitation falls below 200 rpm, the reaction will not occur and the system's pressure and temperature will increase dramatically. The agitator blades have three specific sections designed for complete product mixing. Damage to the blades will have the same effect as a decrease in rpm.

AIC-210

The overall concentration of butane in the reactor system after the addition of the solvent is controlled at 21%. An alarm in the control room indicates high or low concentrations. Butane concentrations are measured at the reactor outlet downstream LCV-210.

AT-4

The composition of the feed to the reactor is carefully controlled at 38% (14 gal.) liquid catalyst, 61% (22.1 gal.) butane, and 1% (4 gal.) pentane. Variations in the feed composition will dramatically affect the operation of the reactor. If the butane composition increases, both pressure and temperature in the reactor will increase. If the butane composition drops below 55%, the reaction will fall short of the operating directive.

AA-210

Analyzer alarm-210 has a high limit of 28% butane and a low limit of 17% butane. The butane concentration is taken at the discharge line from Rx-202, as a reaction has been initiated that includes butane, pentane, liquid catalyst, and **toluene.**

AR-210

Analyzer recorder-210 provides a video trend reading of the selected concentration. This trend is monitored continuously in the control room by the reactor technician.

FIC-210

Flow indicating controller-210 controls the rate of feed to the reactor. Flow rates are carefully controlled at 36.5 gpm. The composition of the feed is determined by the operation of the distillation column. Samples of the reactor feedstock are sent to the lab for analysis every 2 hours during normal operation and more frequently during reactor startup. Flow rates from tray three in column-202 (C-202) are intentionally kept low so as not to affect the operation of the column and to allow the removal of the liquid catalyst. The liquid catalyst is designed to enhance the separation of butane and pentane in C-202. Once the liquid catalyst comes into contact with butane and pentane, its molecular structure is slightly modified.

LIC-210

Level in the reactor is controlled at 75% in order to completely cover the agitator blades. Level control valve 210 throttles flow exiting the reactor. If the level in the reactor drops below 70%, the upper blades will be exposed, reducing agitation and the cooling effect of the temperature-control system.

LA-210

Level alarm-210 has a high limit of 90% and a low limit of 65%. Level in the reactor is carefully controlled above each of the three agitation sections. If the level drops below the upper agitators, the reaction is diminished.

PIC-210

Pressure indicating controller-210 is designed to control the pressure on the reactor at 85 psig. Excess flow through this line is directed into the flare header to be disposed of safely. Pressure control on the reactor is critical. If the pressure increases, the boiling point shifts upward. If the pressure decreases, the heated mixture will vaporize and the flare header will be illuminated by the burning flame. A loss of cooling water will also cause pressure swings in the reactor and will affect the reactor.

PA-210

Pressure alarm-210 has a high limit of 100 psig and low limit of 75 psig. If the high level alarm (HLA) goes off, flow will be decreased from tray three. If the low level alarm (LLA) goes off, flow from the distillation column will increase as vapor flows from C-202.

PR-210

Pressure recorder-210 is an electronic video trend designed to provide continuous feed to the reactor technician. A pressure drop of 15 psi is experienced from the column to the reactor. The line is sized so that flow rates are artificially controlled in order to keep from draining tray three in the column.

Pi-210A/B

Pressure indicator A/B is located on the suction and discharge side of P-210. The suction pressure is typically 88 psig on Pi-210A and 130 psig on Pi-210B.

Rx-202

Reactor-202 has a shell that is composed of stainless steel and can withstand temperatures in excess of 650°F and corresponding pressures around 500 psig. The reactor has a special cooling water jacket and mixer hub that is fitted with a product mixer motor. The motor is connected to the flange on the top of the reactor, and a gasket and set of bolts hold it firmly in place. The agitator and blades are coupled to the motor. An exit port to the flare system is located at the top of the reactor. A pressure-control system is incorporated into the flare header line. A column feed line, a solvent feed line, and a recirculation line are located at the top of the reactor. A single bottom exit port is located at the lowest point in the center of the reactor.

Reactor Instrument Systems

Reactor systems are very complex and are designed to make, break, or make and break chemical bonds. Stirred reactors are specially designed to enhance molecular contact and to do this under specific operational conditions.

Temperature

TIC-210 Temperature indicating controller—located on the cooling water jacket to Rx-202. The temperature is controlled at 120°F. Temperature on the reactor is measured at the bottom of the reactor.

Pressure

PIC-210 Pressure indicating controller—located on the reactor to flare line at the top of the reactor. PCV-210 throttles flow to the flare. Pressure on the reactor is controlled at 85 psig.

Pi-210A Pressure indicator—located on the suction side of pump 210. Pressure typically runs at 88 psig.

Pi-210B Pressure indicator—located on the discharge side of pump 210. Pressure typically runs at 130 psig.

Flow

FIC-210 Flow indicating controller—located on the reactor feed inlet line. Flow rates are controlled at 36.5 gpm. Feed composition is 38% liquid catalyst, 61% butane, and 1% pentane.

FIC-211 Flow indicating controller—located on the solvent feed inlet line. Flow rates are controlled at 68 gpm. The solvent is a critical element of the reaction process. FIC-211 is classified as a slave or secondary controller.

Fi-210 Flow indicator—located on the pressure control line to the flare. Flow to the flare after start-up is 2.5 ft^3/min.

Level

LIC-210 Level indicating controller—located on the lower reactor outlet. The level element takes its measurement slightly above the upper agitator blades. Level is controlled at 75%.

Analytical

AT-4 Analyzer transmitter—located on the feed line to the reactor. AT-4 is designed to measure the butane concentration in the feed from tray three. Butane concentration at this point is 61%.

AIC-210 Analyzer indicating controller—measures the concentration of butane in the exiting product. AIC-210 is classified as a master or primary controller. The solvent flow rate is used to control the overall percentage of butane in the reactor product.

The Stirred Reactor System

Reactors utilize the basic principles of chemistry. Chemistry is described as the study of the characteristics or structure of elements and the changes that take place when they combine to form other substances. Reactor technicians play a major role in the production and manufacture of raw materials that are used to make finished products. Modern chemistry is an essential part of the process environment and for this reason a vital part is the initial training for reactor technicians. A simple explanation of the information found on the periodic table can be found in Figure 15-3.

An atom is the smallest particle of an element that still retains the characteristics of an element. Atoms are composed of positively charged particles called protons, a variable but usually equal number of neutral particles called neutrons, and negatively charged particles called electrons. Protons and neutrons make up the majority of the mass in an atom and reside in an area referred to as the nucleus. The sum of the masses in the nucleus (protons and neutrons) is called the atomic mass unit (amu). The atomic number of an element is determined by the number of protons in its nucleus. The atomic number is used to place the element in its proper place on the periodic table. Electrons are negatively charged particles that orbit the nucleus.

Figure 15-3
Periodic Table
Information Box

Periodic Table
INFORMATION BOX

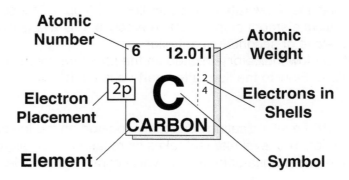

Protons and electrons are typically equally balanced in an atom. This is important because it ensures that the element is equally neutral. Valence electrons reside in the outermost shell of an atom. Valence electrons are important to chemistry because they provide the links in virtually every chemical reaction. Atoms share their valence electrons to form chemical bonds.

Chemical bonding is typically classified as ionic or covalent. Covalent bonds occur when elements react with each other by sharing electrons. This forms an electrically neutral molecule because the protons and electrons electrically balance each other.

Matter is anything that occupies space and has mass. The four physical states of matter are solid, liquid, gas, and plasma. Plasma can be found in powerful magnetic fields. Molecules and compounds are products of chemical reactions. A compound is defined as a substance formed by the chemical combination of two or more substances in definite proportions by weight. A molecule is the smallest particle that retains the properties of the compound. Solutions are a type of homogenous mixture. The term homogenous refers to the evenly mixed composition of the solution. A common example of a homogenous solution are red dye and water. When the red dye comes into contact with water, it is evenly dispersed throughout the solution. Mixtures do not have a definite composition. A mixture is composed of two or more substances that are only mixed physically. Because a mixture is not chemically combined, it can be separated through physical means, such as boiling or magnetic attraction. Crude oil is a simple example of a mixture. It is composed of hundreds of different hydrocarbons. Reactor technicians separate the different components in the crude oil by heating it to the boiling point in a distillation column.

Chemical elements found in nature are listed on the periodic table. These elements can be described as the building blocks of all substances. Each element is composed of atoms from only one kind of element. Chemists describe elements with letters of the alphabet. The letter symbol for hydrogen is H. The letter symbol for carbon is C. A list of all known chemical symbols can be found in a periodic table. A good understanding of the periodic table is required for all reactor technicians. A chemical reaction can be described by associated numbers and symbols. The chemical number identifies how many protons are in an atom, and the amu identifies how many units of an element are present. In a chemical equation, the raw materials or reactants are placed on the left side. As the reactants are mixed together, they yield predictable products. A yield sign or arrow immediately follows the reactants. The products are placed on the right side of the equation. Because atoms cannot be created or destroyed, a common rule of thumb is "what goes into a chemical equation must come out." The sum of the reactants must equal the sum of the products. Figure 15-4 includes the standard information found on a periodic table and additional information useful to process technicians.

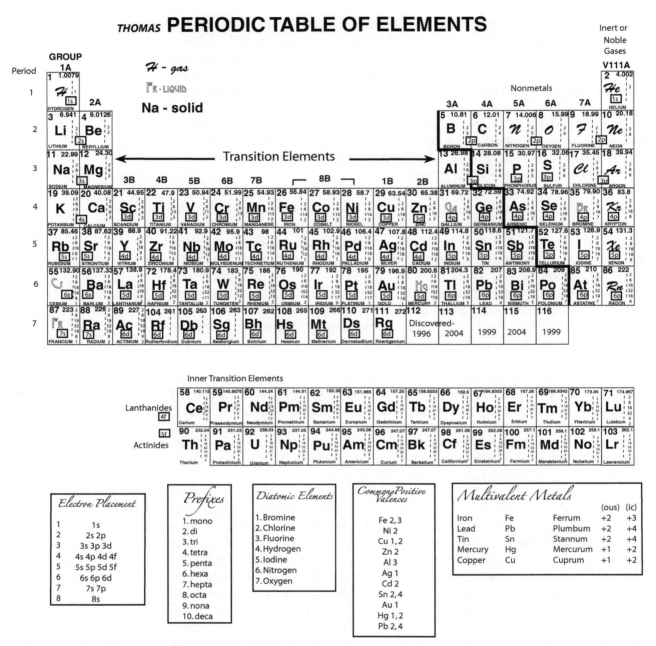

Figure 15-4 *Thomas Periodic Table*

When working out mass relationships, you need to have a good understanding of the periodic table. Certain elements combine to form chemicals that you will recognize easily.

Example:

(Butane)   (Pentane)

C_4H_{10} + C_5H_{12} +   Liquid catalyst + Solvent : New product

Butane has:	Pentane has:
4 carbon atoms = 48 amu 10 hydrogen atoms = 10 amu	5 carbon atoms = 60 amu 12 hydrogen atoms = 12 amu
58 amu	72 amu

In order for the reaction to occur in the stirred reactor, a specific amount of butane, pentane, liquid catalyst, and solvent must be mixed at 250 rpm at 120°F at 85 psig. Under these conditions a new product will be formed.

An important aspect of reactor operation is material balance. For example:

$$Cu + H_2SO_4 : CuSO_4 + H_2$$

1 Copper	1 Copper
2 Hydrogen	2 Hydrogen
1 Sulfur	1 Sulfur
4 Oxygen	4 Oxygen

A simple example of an atom can be found in Figure 15-5. This figure shows the various components of an atom.

Another important aspect of reactor operation is "mass relationships." When working out mass relationships, you need to have a good understanding of the periodic table. Certain elements combine to form chemicals that are easily recognized. Water (H_2O) or carbon dioxide (CO_2) are a good examples of elements and proportions and amus that can be found on the periodic table. A good example of working a mass relationship is H_3PO_4 + $3NaOH : Na_3PO_4 + 3H_2O$. In this reaction, phosphoric acid and sodium hydroxide react to form sodium phosphate and water. By looking at the reactants and products relationship and total weights, it is easy to see how a chemical reaction works.

$$H_3PO_4 + 3NaOH : Na_3PO_4 + 3H_2O$$

Phosphoric acid H_3PO_4 (**reactant 1**)

$$
\begin{aligned}
&\text{3 hydrogen atoms } = 3 \times 1.008 = 3.024 \text{ amu} \\
&\text{1 phosphorus atom } = 1 \times 30.98 = 30.98 \text{ amu} \\
&\text{4 oxygen atoms } \quad = 4 \times 16 \quad = 64 \text{ amu} \\
&\hphantom{\text{4 oxygen atoms } = 4 \times 16 = } \overline{98 \text{ gm, lb, or t.}}
\end{aligned}
$$

Sodium hydroxide $3NaOH$ (**reactant 2**)

$$
\begin{aligned}
&\text{3 sodium atoms } \quad = 3 \times 2 \quad\;\; = 69 \text{ amu} \\
&\text{3 oxygen atoms } \quad = 3 \times 16 \quad = 48 \text{ amu} \\
&\text{3 hydrogen atoms } = 3 \times 1.008 = 3.024 \text{ amu} \\
&\hphantom{\text{3 hydrogen atoms } = 3 \times 1.008 = } \overline{120.02 \text{ gm, lb, or t.}}
\end{aligned}
$$

98 + 120 = 218 amu. Reactant's total molecular weight.

Figure 15-5
Carbon Atom

CARBON ATOM

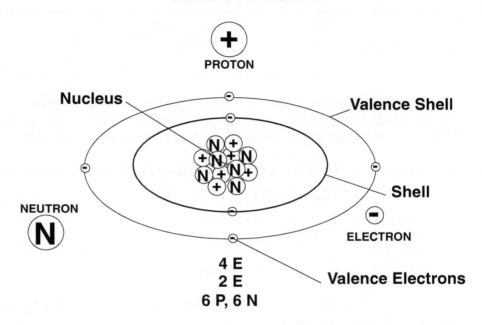

PROTON

Nucleus

Valence Shell

Shell

NEUTRON

ELECTRON

Valence Electrons

4 E
2 E
6 P, 6 N

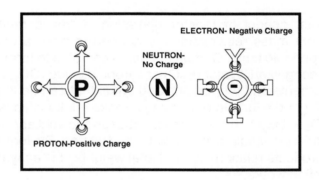

ELECTRON- Negative Charge

NEUTRON-
No Charge

PROTON-Positive Charge

Sodium phosphate Na_3PO_4 (**product 1**)

3 sodium atoms	$= 3 \times 23$	$= 69$ amu
1 phosphorus atoms	$= 1 \times 30.98$	$= 30.98$ amu
4 oxygen atoms	$= 4 \times 16$	$= 64$ amu

163.98 gm, lb, or t.

Water $3H_2O$ (**product 2**)

6 hydrogen atoms	$= 6 \times 1.008$	$= 6.048$ amu
3 oxygen atoms	$= 3 \times 16$	$= 48$ amu

54.05 gm, lb, or t.

$163.98 + 54.05 = 218.03$ amu. Product's total molecular weight.

AMUs are very flexible and can be substituted for other weights such as grams, pounds, or tons.

Reactor technicians are required to look carefully at mass relationship problems. For most of these situations, a chemist has already calculated the correct reactant amounts and specified the correct operating conditions. Technicians see mass relationship problems in actual weights—grams, pounds, or tons. An example of this is found in the following reaction: $N_2 + 3H_2$ yields $2NH_3$.

N_2	+	$3H_2$	:	$2NH_3$
120 lb		39 lb		159 lb

This is the normal operation of the unit; however, at the end of the month production needs to be increased. When this occurs the nitrogen is increased to 420 lb. This will require an adjustment in the hydrogen in order for the product to remain within operational guidelines. When this occurs, the reactor technician is required to make a small calculation.

N_2	+	$3H_2$	:	$2NH_3$
120 lb		39 lb		159 lb
420 lb		? lb		? lb

The relative weight is 420 lb. The original actual weight is 120 lb. To solve this problem, simply divide the relative weight by the actual weight. This will give you a factor of 3.5 lb. By multiplying the hydrogen feed rate of 39 lb by 3.5 lb, the new flow rate will be calculated. $420 \div 120 = 3.5$ lb.

$$3.5 \times 39 = 136.5 \text{ lb}$$
$$3.5 \times 159 = 556.5 \text{ lb}$$

This process will take production rates from 159 to 556.5 lb/h. This simple process can be used to calculate most reactor problems that deal with relationships.

Solvent
A solvent is described as a liquid that dissolves a liquid, a solid, or a gaseous solute, resulting in a solution. Rx-202 utilizes toluene, which is described as a (carbon-containing) organic chemical. Organic solvents evaporate easily and have low boiling points that can be separated by distillation or other separation methods. In most reactions solvents are inert and do not chemically react with the dissolved compounds. Solvents can be used to separate or extract soluble compounds from a mixture. Solvents are usually colorless and have a heavy aromatic odor.

Toluene (Solvent)
Toluene is known as phenylmethane or methylbenzene and is an aromatic hydrocarbon that is a water-insoluble liquid. The molecular formula of toluene is C_7H_8, and it has a boiling point of 231°F (110.6°C).

Toluene is classified as a common solvent that is able to dissolve specific chemicals and is used as a feedstock for toluene diisocyanate, which is used to manufacture TNT, phenol, and polyurethane. Toluene is also used as an octane booster in gasoline.

Butane (Reactant)

In the presence of oxygen, butane burns to form water vapor and carbon dioxide. The boiling point of butane is 31.1°F (-0.5°C). The chemical reaction for this process is:

$$2C_4H_{10} + 13O_2 : 8CO_2 + 10H_2O$$

DuPont chemical company's catalytic process makes maleic anhydride using n-butane as a feedstock: The chemical reaction looks like this:

$$CH_3CH_2CH_2CH_3 + 3.5O_2 : C_2H_2(CO)_2O + 4H_2O$$

Pentane (Reactant)

Pentane is used mainly as a solvent, fuel, or hydrogen source in steam reforming. It is classified as the most volatile hydrocarbon, that is, liquid at 72°F. The boiling point of pentane is 96.98°F, or 36.1°C (308°K).

Pentane burns to form carbon dioxide and water. The chemical reaction looks like this:

$$C_5H_{12} + 8O_2 : 5CO_2 + 6H_2O$$

Since n-butane is the conventional feedstock in DuPont's chemical synthesis, maleic anhydride, pentane is also a substrate:

$$CH_3CH_2CH_2CH_2CH_3 + 5O_2 : C_2H_2(CO)_2O + 5H_2O + CO_2$$

Rx-202 has a stainless steel shell designed to withstand temperatures in excess of 650°F at 500 psig. A dimpled water jacket is used to control the temperature at 120°F in order to maximize reaction rates. Product agitation is maintained by mixer 210 at 250 rpm. The composition of the feed to the reactor from C-202 is 1% pentane, 38% liquid catalyst, and 61% butane. A solvent (toluene, C_7H_8) is introduced to the reactor and blended with the column feed. Unlike other processes, the solvent (toluene) and liquid catalyst react with butane and pentane to form a new product that is separated in Separator-600. The liquid catalyst enhances the separation of butane and pentane in the column and is easily separated at tray three; once the liquid catalyst is exposed to butane and pentane feed, it is slightly modified at the molecular level. The operating conditions in C-202 also help with this process. The agitation process in the reactor is also a critical variable responsible for the reaction that forms a new product.

Problems associated with the operation of Rx-202 include:
* feed composition changes
* concentration increase—increases reaction factors
* agitation problems—reduce reaction
* loss of cooling water—temperature increases
* loss of level control
* instrument problems
* loss of pressure control—increase or decrease
* reaction time in reactor—reaction incomplete
* column and solvent flow rates
* temperature increase—doubles reaction rate for every 10°C increase
* loss of catalyst—reaction stops.

Operational Specifications

The operating procedures associated with the reactor system include:

Action	Notes
1. Ensure C-202 is operating properly and at operational specification.	
2. Sample product on tray three and send to lab for analysis.	
3. Set LIC-210 to 75% and put in AUTO.	
4. Set PIC-210 to 85 psig and put in AUTO.	
5. Set TIC-210 to 120°F and put controller in AUTO.	
6. Set FIC-210 to 36.5°gpm and put in AUTO.	
7. Set FIC-211 to 68°gpm and put in AUTO.	
8. Set Mixer-210 to 250°rpm and set in AUTO.	
9. Open V-203, V-204, and V-205.	
10. Leave V-206 closed.	
11. Start P-210 and circulate to Rx-202.	Monitor pressures and level.
12. Open V-206 and close V-205.	
13. Compare AT-4 to lab sample and to AIC-210.	
14. Cross-check process variables with SPEC sheet.	
15. Set FIC-211 to CASCADE.	

Specification Sheet and Checklist

Flow

FIC-210	36.5 gpm	AUTO	Feed flow to reactor
FIC-211	68 gpm	CASCADE	Solvent (toluene)

Analytical

AT-4	61%	—	Butane
AIC-210	21%	AUTO	Butane
SIC-210	250 rpm	AUTO	Mixer-210
AA-210	Hi 28% Lo 17%	—	AIC control loop
AR-210	Video trend	AUTO	AIC control loop

Pressure

Pi-210A	88 psig	—	Suction P-210
Pi-210B	130 psig	—	Discharge P-210
PIC-210	85 psig	—	Reactor-202
PA-210	Hi 100 psig/Lo 75 psig	—	Pressure control
PR-210	Video trend	AUTO	Pressure control

Temperature

TIC-210	120°F	AUTO	Cooling water to reactor shell
TA-210	Hi 140°F/Lo 110°F	—	TIC control loop
TR-210	Video trend	AUTO	TIC control loop

Level

LIC-210	75%	AUTO	Reactor level control
LA-210	Hi 90% Lo 65%	—	LIC control loop
LR-210	Video trend	AUTO	LIC control loop

Troubleshooting Scenario 1

Refer to Figure 15-6 and select the best answer to the what-if scenarios.

Troubleshooting Scenario 2

Refer to Figure 15-7 and select the best answer to the what-if scenarios.

Troubleshooting Scenario 3

Refer to Figure 15-8 and select the best answer to the what-if scenarios.

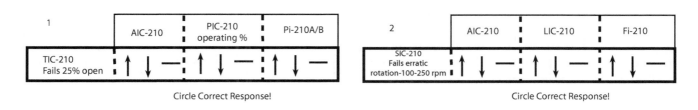

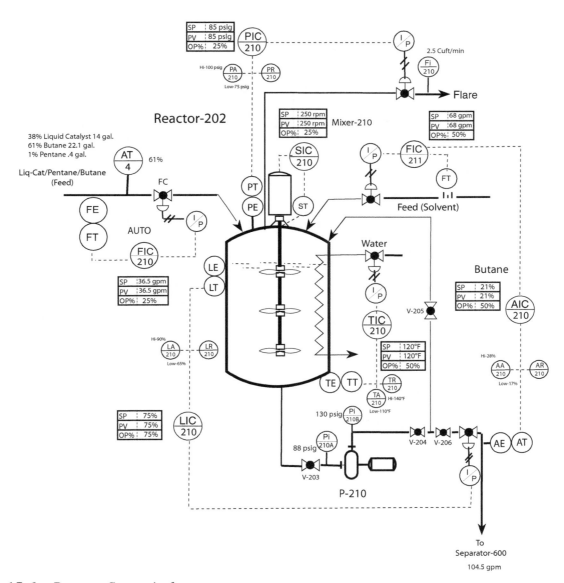

Figure 15-6   *Reactor Scenario 1*

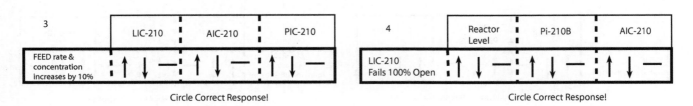

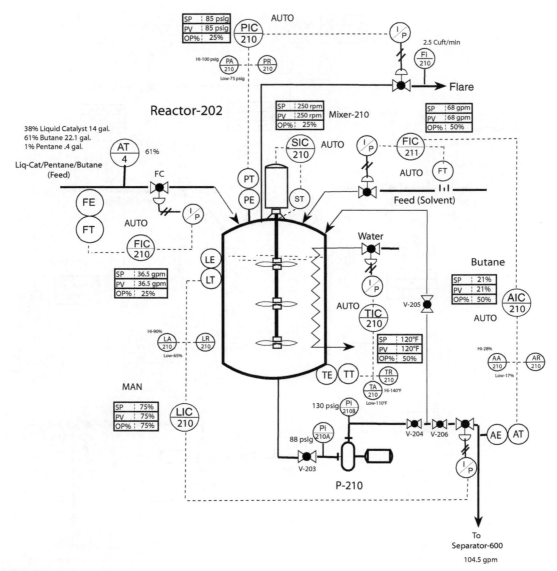

Figure 15-7 *Reactor Scenario 2*

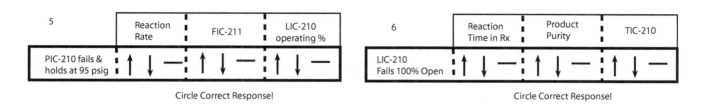

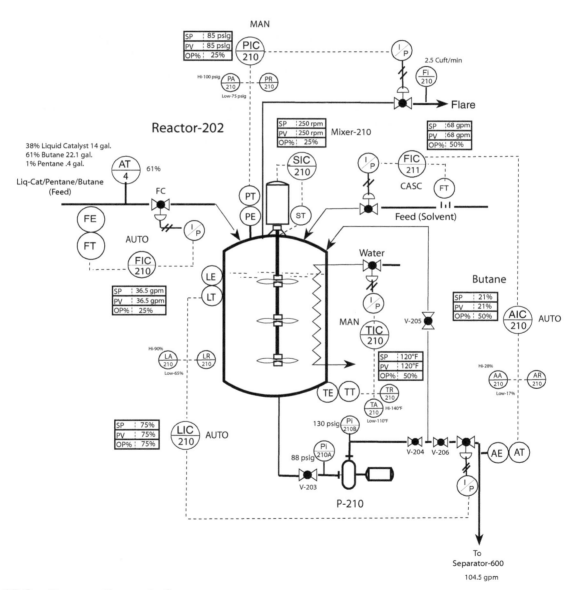

Figure 15-8   *Reactor Scenario 3*

Summary

A stirred reactor is a device used to convert raw materials into useful products through chemical reactions. Pressure, flow rates, liquid catalyst, agitation speed, surface area, heat, and reaction time can affect reaction rates. Reactor technicians are responsible for establishing correct flow or feed rates to the reactor; ensuring correct concentrations, pressures, and levels; monitoring and controlling reaction times; ensuring that specified blending or agitation is occurring; and maintaining primary and secondary equipment. Stirred reactors combine reactants with a variety of process variables. The liquid catalyst increases the separation process and enhances the formation of a new product and actually becomes part of the finished product. Typically a catalyst does not become part of the final product; however, in this reaction it does until it is separated after it exits the separator. **Exothermic reactions** generate heat, while **endothermic reactions** produce heat. The process described in this chapter is exothermic and requires a cooling medium to keep the reaction under control.

Review Questions

1. Explain the primary purpose of stirred Rx-202.

2. Draw and label the basic components and equipment associated with Rx-202.

3. Identify the instruments associated with Rx-202.

4. List the basic components in the two feed stocks.

5. List the safety features found on Rx-202.

6. Explain how the reaction is controlled in the Rx-202 system.

7. Draw each of the control loops associated with the stirred reactor system.

8. List the desired set points on each control loop on the stirred Rx-202.

9. List the primary reactor variables in effect on Rx-202.

10. Explain how to safely start up the stirred reactor system.

11. List the various operational problems reactor technicians encounter on a stirred Rx-202.

Separation Model

LEARNING OBJECTIVES

After studying this chapter, the student will be able to:

- Apply the scientific principles associated with chemical separation process.
- Describe the operation of the separator-600 system.
- List the composition of the separator feedstock.
- Draw the separator-600 system.
- Identify the simple instruments used to operate the separator system.
- Draw each of the control loops in the separator system and list their typical set points.
- Operate the separator system.
- Troubleshoot problems associated with the operation of the separator system.

Key Terms

Liquid–liquid extraction—a process for separating two materials in a mixture or solution by introducing a third material that will dissolve one of the first two materials but not the other.

Feedstock—the original solution to be separated in liquid–liquid extraction, which is composed of toluene, butane, pentane, and liquid catalyst.

Layer out—a process in which two liquids that are not soluble separate naturally from each other.

Separator—a device that is designed to separate two liquids from each other by density differences; typically, a solvent is introduced that will dissolve one of the components in the mixture, enhancing the separation process. A separator has a shell, a weir, a vapor cavity, a feed inlet, an extract pump, and a raffinate pump.

Solute—a material that is dissolved in liquid–liquid extraction.

Solvent—a chemical that will dissolve another chemical.

Separation—a method to separate the various components in a mixture. For example, the chemical reaction using a multicomponent mixture is completed in the separator and the new product, an octane booster (extract), is separated from the pure butane (raffinate) using a device called a separator.

Principles of Separation

One of the most frequently encountered problems in chemical process operations is that of separating two materials from a mixture or a solution. Distillation is one way of carrying out such a **separation** and is perhaps the most frequently used method. Another useful method is extraction. Extraction is the process of separating two materials in a mixture by introducing a third material that will dissolve one of the first two materials but not the other. In **liquid–liquid extraction,** all the four materials are liquids, and the mixture is separated by allowing them to **layer out** by weight or density.

In many cases, it is impractical to separate two chemicals by distillation because the boiling points of the materials are too close together. In such a case, it is frequently possible to find a third chemical that will dissolve one of the two chemicals. In this situation, extraction would be a better method of making the separation than distillation.

Many chemicals are sensitive to heat and will degrade or decompose if raised to a temperature high enough for distillation. In this case, extraction, which can usually be carried out at normal temperatures, will be a practical alternative. Often, one of the materials to be separated is present in very small amounts. It might be possible to recover such a material by distillation, but it is usually much easier and more economical to do so by extraction. Finally, the key requirement of any commercial process is that it should be economical. In situations where several alternative means of separating two chemicals could be used, the one that is the most economical is chosen. Many relatively inexpensive **solvents** are available, but because the equipment required for an extraction operation is relatively simple, economic considerations often favor liquid–liquid extraction.

There are basically three steps in the liquid–liquid extraction process: (1) contact the solvent with the feed solution, (2) separate the raffinate from the extract, and (3) separate the solvent from the **solute.** Step 3, that is, recovery of the solvent and solute, is left to be carried out by some other process such as distillation. In liquid–liquid extraction, the feed is the original solution. The feed solution, containing the solute (the material that will be dissolved), is fed to the lower portion of the extraction column. The solvent (the material that dissolves the solute) is added near the top of the column. Because of density differences, the lighter feed solution tends to rise to the top, while the heavier solvent sinks to the bottom. As the two streams mix, the solvent dissolves the solute. Thus, the solute, which originally rises with the feed solution, actually reverses its direction of flow and mixes with the solvent at the bottom of the column. This new solution, consisting of solvent and solute, is called the extract. The other chemical in the feed stream, now free of the solute, separates out at the top as the raffinate. The raffinate and extract streams are not soluble in each other and will layer out. The butane in the "octane booster" has been chemically reacted and does not separate; only the excess butane not needed for the reaction separates. Figure 16-1 shows the basic flow path and equipment and instruments associated with the **separator** system.

The solvent must be able to dissolve the solute, but it should not be a substance that will dissolve the raffinate or contaminate it. It also must be insoluble so that it will layer out. The density of the solvent should vary sufficiently from the density of the raffinate so that both solvent and raffinate can layer out by the effects of gravity. The solvent must be a substance that can be separated from the solute. It should be inexpensive and readily available, and it should not be hazardous or corrosive.

Equipment Descriptions (Block Flow and Safety)

Basically, the equipment for commercial operations is designed to ensure contact between the solvent and the feed and to separate the extract from the raffinate. This process takes place initially in the reactor and later in the separator. The simplest extraction apparatus is the single-stage batch unit. In such an operation, the feed and the solvent are added to a tank or some other suitable container. In this process, the solvent (toluene) is added in the stirred reactor. They are then thoroughly mixed by a mixer in the tank or by circulation in the tank. After the materials have been thoroughly mixed, mixing is stopped and the materials are allowed to layer out. The mixer in this system runs continuously as the reacted feed flows into the separator and is allowed to layer out. The extract (octane booster) and the raffinate (99.9% butane) layers can be removed.

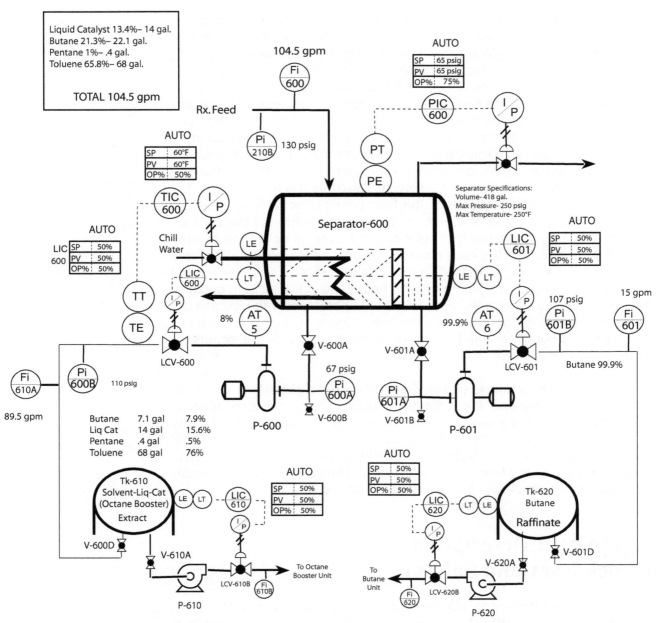

Figure 16-1 *Separator System*

Separator-600 has been used for continuous operations by continuously adding the feed and the solvent and continuously withdrawing the raffinate and extract. For such a process to be successful, mixer-210 in the reactor runs continuously, mixing or contacting the materials and allowing ample space for the raffinate and extract to layer out. Many simple yet ingenious methods have been employed for this purpose.

Most frequently, a single-stage device as described will not provide a perfect separation; however, with the stirred reactor and separator working together, the raffinate is 99.9% pure butane and can be sent to the storage tank.

Pump-600

The separator utilizes two centrifugal pumps: P-600 and P-601. Centrifugal pumps provide smooth, even flow that can be throttled or regulated. In many cases, flow can be stopped without causing serious damage to the pump. Starting torque on a centrifugal pump is low, allowing the use of inexpensive drivers such as a starting torque squirrel cage induction motor. This type of pump utilizes a simple design, few working parts, and can use a multistage approach with high differential pressure applications. P-600 takes suction off the heavier (denser) octane booster. Flow rates on the pump must be maintained such that lighter materials are not pulled into the suction of the pump. Each of these processes is carefully controlled. The components in the extract or octane booster include toluene, liquid catalyst, and pentane. These components have chemically reacted together to form a new product. While the raw materials or reactants are known, the actual formula for the octane booster is proprietary. The suction pressure on the pump is 67 psig, and the discharge pressure is 110 psig with a ΔP of 43 psi. Typical flow rates should be around 89.5 gpm.

Pump-601

P-601 is a small centrifugal pump designed to take suction off the effluent that flows over the weir and into the raffinate line. The product pumped by P-601 is 99.9% pure butane. The suction pressure is slightly more than 65 psig. The discharge pressure is 107 psig with a flow rate of 15 gpm.

TIC-600

One of the most critical components of the separator is the chill water system that is designed to take the blended **feedstock** from 120°F to 60°F. This step is a critical part of the reaction and should be allowed to finish. The cooler temperature allows the newly formed product to layer out. Temperature indicator controller-600 controls the flow of chilled water through the coils that can be found inside the separator. The temperature element is located on the discharge line of P-600 and transmits information back to the controller in the control room. As mentioned earlier, the set point on the controller is 120°F and is typically kept in auto mode.

PIC-600

Pressure in the separator must be kept within a narrow window, 63–68 psig. The desired set point is 65 psig. Pressure in the separator is slightly less than in the reactor. A pressure element is located at the top of the separator, which transmits a signal to the controller. This signal is compared to the set point and adjustments are made. The pressure control line is directed to the flare where excess vapors can be burned safely.

LIC-600

Level-indicating controller-600 is one of two level control systems on the separator. LIC-600 is designed to control the level of the two phases by ensuring that the level in the separator stays within specific guidelines. In this case, the control set point is 50%. This measurement is taken primarily at the top of the lighter component in the separator.

LIC-601

Butane flows over the internal weir in the separator pools above the suction line. A level element is located in this smaller section of the separator and is required to have a quicker response time than LIC-600. The set point on the controller is set at 50%. A control valve is located on the discharge side of P-601. The control valve is designed to operate so that the pump discharge will utilize approximately 30% of the total pressure differential that can be produced by the pump between controlled pressure locations within the system. Typically, every centrifugal pump process has the ability to operate below 30% or above 30% of the operating design.

AT-5

The separator has two analyzer systems that are designed to measure the presence of butane in the process stream. AT-5 is located on the discharge side of P-600. As the new product, octane booster, flows to the storage tank, the butane within the new product is detected. This reading is typically around 8%.

AT-6

Analyzer-6 is located at the discharge of P-601. The desired setting is 99.9% pure butane. The flow rate that this analyzer is tapped into is low, around 15 gpm.

Separator-600

The most important piece of equipment in this process is the separator. It is designed to perform a specific liquid–liquid separation. The internal design has a set of cooling coils, a weir, a vapor disengaging cavity, an upper feed line, an upper pressure release line, two analyzers, two pumps, and two level control systems.

Instrument Systems

Temperature

TIC-600 Temperature-indicating controller—located on the chill water coil to the separator. The temperature is controlled at 60°F. Temperature on the separator is measured on the discharge line from P-600.

Pressure

PIC-600 Pressure-indicating controller—located on the upper vapor outlet from the separator to the flare. PCV-600 throttles the flow to the flare. Pressure on the separator is controlled at 65 psig.

Pi-210B Pressure indicator—located on the feed inlet to the separator. Pressure typically runs at 130 psig.

Pi-600A Pressure indicator—located on the suction side of P-600. The pressure typically operates at 67 psig.

Pi-600B Pressure indicator—located on the discharge side of P-600. Pressure typically operates at 110 psig.

Pi-601A Pressure indicator—located on the suction side of P-601. Pressure typically operates slightly above 65 psig.

Pi-601B Pressure indicator—located on the discharge side of P-601. Pressure typically operates at 107 psig.

Level

LIC-210 Level-indicating controller—located on the primary feed line to the separator. LIC-210 is designed to control the flow rate to the separator based on the level in reactor-202. The level is controlled at 75%.

LIC-600 Level-indicating controller—located on the extract side of the separator, designed to control the level of the separator as the extract and raffinate layer out. The level must be controlled at 50% for the raffinate (butane) to flow over the weir.

LIC-601 Level-indicating controller—located on the raffinate (butane) side of the separator. The level is controlled at 50% of the measured span on butane that has flowed over the weir.

LIC-620 Level-indicating controller—controls the level on tank-620 at 50%.

LIC-610 Level-indicating controller—controls the level on tank-610 at 50%.

Flow

Fi-600 Flow indicator—located on primary feed line to separator. Typically operates at 104.5 gpm.

Fi-610A Flow indicator—located on the discharge side of P-600. Typically operates at 89.5 gpm.

Fi-601 Flow indicator—located on the discharge side of P-601. Typically operates at 15 gpm.

Fi-620 Flow indicator—located on the discharge side of P-620. Typically operates at 15 gpm while running.

Fi-610B Flow indicator—located on the discharge side of P-610. Typically operates at 89.5 gpm while running.

Analyzer

> AT-5 Analyzer transmitter—is a remote located device designed to measure the concentration of butane in the discharge flow from P-600. Butane concentration is typically around 8%.
>
> AT-6 Analyzer transmitter—is a remote located device designed to measure the concentration of butane in the discharge of P-601. Product purity is typically 99.9%.

The Separator System

Inside the separator, a unique set of scientific principles become operative as the product begins to cool from 120°F to 60°F. During the reaction phase, solvent (toluene), butane, liquid catalyst, and a small amount of pentane are blended together under specific conditions. These conditions are important and must be followed precisely. These operating conditions include:

- specific concentration of reacted feedstock
- flow rate of 104.5 gpm
- temperature of 60°F (cooling is critical)
- pressure of 65 psig (decreased pressure is critical)
- Correct agitation of reacted feed.

Under these conditions, a unique reaction begins to take place between these feedstocks with an excess of butane that is separated in the separation system. As the reactor feed begins to cool, it separates into two phases, with butane on the top and a new product, octane booster. The reaction looks like this:

Butane	Pentane	Toluene	Liquid cat	Octane booster + Butane
$C_4H_{12}+$	$C_5H_{12}+$	C_7H_8+	14 gal	$\rightarrow OB + C_4H_{12}$

The primary separation or layering out occurs in the separator. The lighter raffinate flows over the weir in the separator and to TK-620 at approximately 15 gpm. The heavier solvent and octane booster flow out of the separator at 89.5 gpm to the extract storage facility, TK-610. The following concentrations can be found in the octane booster product.

7.9%	7.1 gal.	Butane
15.6%	14 gal.	Liquid catalyst
0.5%	0.4 gal.	Pentane
76%	68 gal.	Toluene

Characteristics of various components:

Boiling point	Toluene 231°F	C_7H_8
Boiling point	Butane 31.1°F	C_4H_{12}
Boiling Point	Pentane 96.98°F	C_5H_{12}

Operational Specifications (SOP, SPEC Sheet, and Checklist)

The operating procedures associated with the separation system include:

Action	Notes
1. Set PIC-600 to 65 psig and set in AUTO.	
2. Set TIC-600 to 60°F and set in AUTO.	
3. Set LIC-600 to 50% and set in AUTO.	
4. Set LIC-601 to 50% and set in AUTO.	
5. Set LIC-610 to 50% and set in AUTO.	
6. Set LIC-620 to 50% and set in AUTO.	
7. Establish flow to separator-600.	
8. Open V-600D and A and start P-600.	
9. Open V-601D and A and start P-601.	
10. Open V-601D and V-620A and start P-620.	
11. Open V-600D and V-610A and start P-610.	
12. Monitor AT-5 and ensure it is reading between SPC guidelines.	
13. Monitor AT-6 and ensure it is reading between SPC guidelines.	
14. Cross-check process variables with SPEC sheet.	

Specification Sheet and Checklist

Flow

1.	Fi-600	104.5 gpm	gauge	Feed flow
2.	Fi-610A	89.5 gpm	gauge	Feed flow to TK-610
3.	Fi-601	15 gpm	gauge	Feed to TK-620
4.	Fi-610B	89.5 gpm	gauge	Product "octane booster"
5.	Fi-620	15 gpm	gauge	Product "butane"

Analytical

6.	AT-5	8%	—	Octane booster
7.	AT-6	99.9%	—	Butane

Pressure

8.	Pi-210B	130 psig	—	Feed inlet to separator
9.	PIC-600	65 psig	—	Separator pressure control
10.	Pi-600A	67 psig	—	Suction P-600
11.	Pi-601A	65 psig	—	Suction P-601
12.	PI-600B	110 psig	—	Discharge P-600
13.	Pi-601B	107 psig	—	Discharge P-601

Temperature

14.	TIC-600	60°F	AUTO	Chill water system on separator

Level

15.	LIC-600	50%	AUTO	Level on extract side of separator
16.	LIC-601	50%	AUTO	Level on raffinate side of separator
17.	LIC-610	50%	AUTO	TK-610 level control
18.	LIC-620	50%	AUTO	TK-620 level control

Common Separations Problems
- feed composition changes
- unreacted feed
- loss of cooling water
- loss of level control
- instrument problems
- loss of pressure control
- equipment failures—e.g. pump.

Troubleshooting Scenario 1

Refer to Figure 16-2 and select the best answers.

Troubleshooting Scenario 2

Refer to Figure 16-3 and select the best answers.

Troubleshooting Scenario 3

Refer to Figure16-4 and select the best answers.

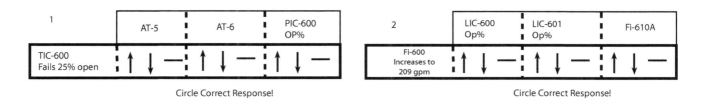

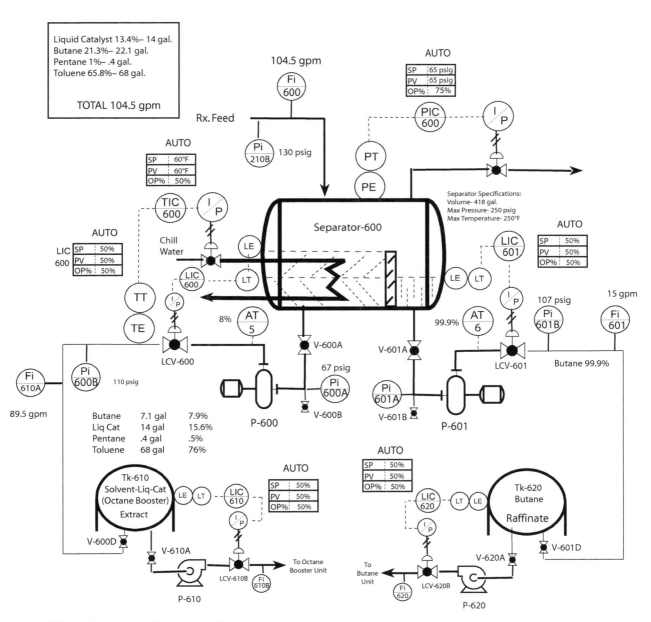

Figure 16-2   *Separator Scenario 1*

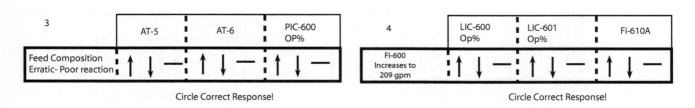

3	AT-5	AT-6	PIC-600 OP%
Feed Composition Erratic- Poor reaction	↑ ↓ —	↑ ↓ —	↑ ↓ —

Circle Correct Response!

4	LIC-600 Op%	LIC-601 Op%	Fi-610A
Fi-600 Increases to 209 gpm	↑ ↓ —	↑ ↓ —	↑ ↓ —

Circle Correct Response!

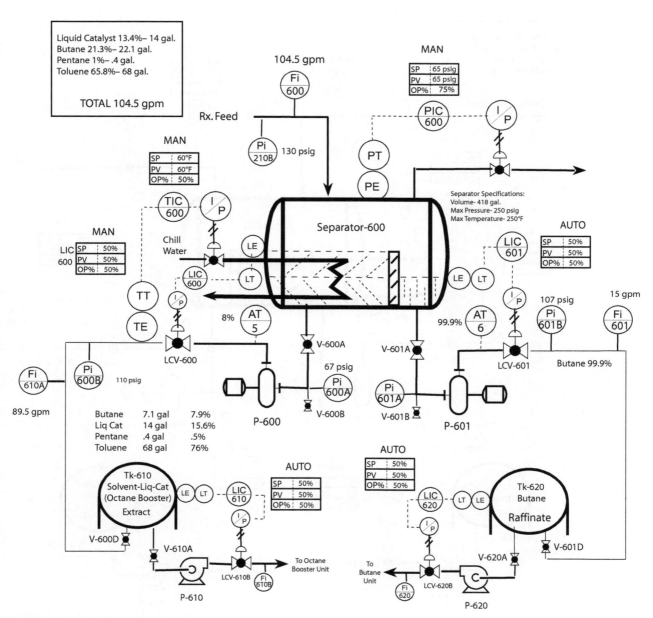

Figure 16-3 *Separator Scenario 2*

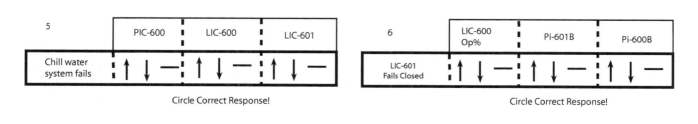

5	PIC-600	LIC-600	LIC-601
Chill water system fails	↑ ↓ —	↑ ↓ —	↑ ↓ —

Circle Correct Response!

6	LIC-600 Op%	Pi-601B	Pi-600B
LIC-601 Fails Closed	↑ ↓ —	↑ ↓ —	↑ ↓ —

Circle Correct Response!

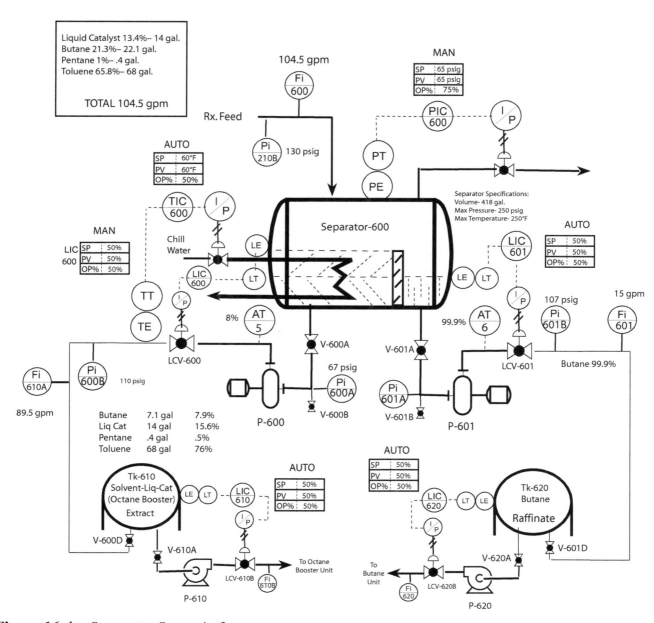

Figure 16-4 *Separator Scenario 3*

Summary

Inside separator-600, a unique set of scientific principles operates as the product begins to cool from 120°F to 60°F. During the reaction phase, solvent (toluene), butane, liquid catalyst, and a small amount of pentane are blended together under specific conditions. These conditions are important and must be followed precisely. Inside the separator, another set of critical variables exist and also must be followed exactly. These operating conditions include: (1) specific concentration of reacted feedstock, (2) flow rate of 104.5 gpm, (3) temperature control at 60°F, (4) pressure control at 65 psig, and (5) time for materials to layer out for separation.

Under these conditions, a unique reaction begins to take place as the reactor feed begins to cool and separate into two phases: butane at the top and a new product, "octane booster," at the bottom. The simulated reaction looks like this:

Butane   Pentane   Toluene   Liquid cat   Octane booster + Butane

$$C_4H_{12}+ \quad C_5H_{12}+ \quad C_7H_8+ \quad 14\text{ gal} \quad \rightarrow OB + C_4H_{12}$$

The primary separation or layering out occurs in the separator at 15 gpm as raffinate flows to the system, TK-620. The heavier solvent and octane booster flows out of the separator at 89.5 gpm to the extract storage facility, TK-610. The following concentrations can be found in the octane booster product:

7.9%	7.1 gal.	Butane
15.6%	14 gal.	Liquid catalyst
0.5%	0.4 gal.	Pentane
76%	68 gal.	Toluene

Review Questions

1. List the basic equipment associated with the separation unit.
2. List the basic instrumentation associated with the separation unit.
3. Draw a simple sketch of the separator unit.
4. Draw each control loop associated with the separation unit and explain how each one works.
5. List the basic steps associated with operating the separation unit.
6. Identify the common problems associated with separation operation.

Multivariable Plant

LEARNING OBJECTIVES

After studying this chapter, the student will be able to:

- Describe the various problems associated with operating process equipment and systems.
- Describe troubleshooting pumps and tanks.
- Describe the various troubleshooting models.
- Describe how different variables affect each other.
- Explain how problems with process equipment affect other systems.
- Analyze process problems and provide solutions.
- Troubleshoot specific operational scenarios.
- Apply various instrumentation used to troubleshoot process problems.
- Distinguish between primary and secondary problems.
- Use statistical methods and collect, organize, and analyze data.
- Respond to alarms and control systems that are outside operational guidelines.
- Compare troubleshooting methods and models.

Key Terms

Process variables—instruments used to detect process variables provide clues that can be used to complete the big picture.

Troubleshooting models—tools used to teach troubleshooting techniques. Basic models include distillation, reaction, absorption and stripping, or combination of any of these three.

Primary operational problems—the first problem that created a process upset.

Secondary operational problems—that are created or respond to a primary problem.

Troubleshooting methods—educational, instrumental, experiential, and scientific.

Basic Troubleshooting Principles

Process technology students must take up the study and application of process principles seriously in order to successfully troubleshoot process problems. This includes math, chemistry, technology, physics, safety, quality, equipment, systems, operations, instrumentation, and troubleshooting. This specifically includes how scientific methods relate to **process variables** such as pressure, temperature, fluid flow, level, and analytical variables. These methods are typically taught in lower-level classes.

In this text ten different **troubleshooting models** have been studied. These models include:
- pump and tank model
- compressor model
- heat exchanger model
- cooling tower model
- boiler model
- furnace model
- distillation model
- reactor model
- separator model, and
- combinations of models.

Pumps are susceptible to corrosion, which can damage or destroy internal components such as impellers or exposed metal parts. This can seriously change the operating characteristics of the pump. Examples of operating characteristics include fluid flow, pressure, and differential pressure drop across the pump. Occasionally the motor will overload and trip off or the coupling will break. Motors tend to trip when the bearings or seals on the rotating parts in the pump fail. When these parts fail, the rotating parts bind down and friction is generated. Sometimes these devices will generate enough heat to catch fire. In automated systems, a control valve is located on the discharge side of the pump. When the pump starts to fail, the control valve will attempt to make up or compensate for the loss or increase in flow. As the pump fails, the system will go completely out of control.

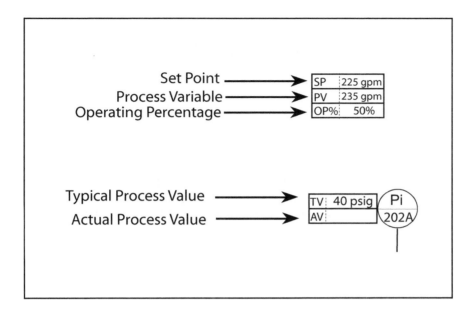

Figure 17-1
*Simple
Troubleshooting
Blocks*

The basic instruments and controls found in an automated system have specific characteristics when they fail. For example, when each part of the control loop fails, it typically does so in the following pattern:

- Gauge indicator—typically will freeze at one position or read high or low. Examples of indicators include pressure indicators, level indicators, temperature indicators, flow indicators, and analytical indicators.
- Primary element—typically will fail in a way that causes the transmitted value to drop to zero.
- Transmitter—typically will fail in a way that causes the transmitted value to decrease to zero, or increase to 100%, or freeze at one value.
- Controller—a device designed to compare a signal to a set point and transmit a signal to a final control element.
- Control valve—typically will fail open (FO), or fail closed (FC), or will be in an incorrect position.

Figure 17-1 illustrates some of the information a process technician looks at while operating and troubleshooting a system.

Troubleshooting Methods

There are a number of **troubleshooting methods** that can be used with these models. Methods vary depending on individual educational faculty, consultants, and industry. The basic approach to most methods includes the development of a good educational foundation.

Method One: Educational
- Basic knowledge of the equipment and technology
- Understanding of the math, physics, and chemistry associated with the equipment

- Study of equipment arrangements in systems
- Study of process control instrumentation
- Ability to operate equipment in complex arrangements
- Ability to troubleshoot process problems.

Troubleshooting is a process that requires a wide array of skills and techniques. Modern control instrumentation includes indicators, alarms, transmitters, controllers, control valves, transducers, analyzers, interlocks, etc. The primary goal is to control variables such as temperature, pressure, flow, level, or analytical variables. It is possible to control large, complex processes from a single room. In these types of systems, process set points (SPs) and process variables on controllers should clearly reflect each other. Process problems are quickly identified when these two do not line up. Example: If the flow rate is set at 200 gpm and the process variable is 175 gpm, a 25 gpm difference exists. This could indicate a serious problem.

Method Two: Instrumental
- Basic understanding of process control instrumentation
- Basic understanding of the unit process flow plan
- Advanced training in controller operation—programmable logic controller (PLC) and distributed control system (DCS)
- Training to troubleshoot process problems.

Method Three: Experiential
- Experience in operating specific equipment and system
- Familiarity with past problems and solutions
- Ability to think outside the box
- Ability to think critically—identify and challenge assumptions
- Ability to evaluate, monitor, measure, and test alternatives
- Ability to troubleshoot process problems.

Method Four: Scientific
- Grounded in principles of mathematics, physics, and chemistry
- Knowledge of theory-based operations
- Good understanding of equipment design and operation
- Ability to view the problem from the outside in
- Ability to utilize outside information and expertise, and reflective thinking
- Ability to generate alternatives, brainstorm, rank alternatives
- Ability to troubleshoot process problems.

Figure 17-2 shows a simple example of a steam heater. The instrument blocks have been added to enhance operations and troubleshooting.

Aside from the ten different models and four distinct methods, a few other ideas or concepts need to be discussed. Some of these include the concept of FO and FC. When a valve is installed in a unit or process, the engineers take into account whether the automated valves should fail open or fail

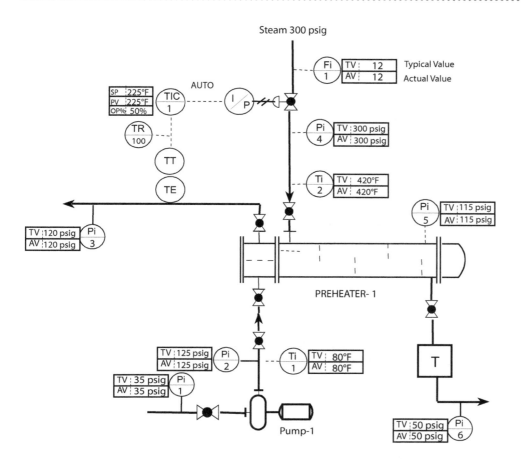

Figure 17-2
Steam Heated Shell and Tube Heat Exchanger

closed. Each of these valves operates differently; for example, when a valve is designed to fail closed, a heavy spring causes the flow control element to move to the closed position. It takes the instrument air to open the valve. For example, an FC valve position would assume the following positions:

Fail Closed (FC)
- 0%   Closed
- 25%   25% open
- 50%   50% open
- 75%   75% open
- 100%   100% open

When a valve is designed to fail in the open position, like an emergency water system, the valve will respond to the following:

Fail Open (FO)
- 0%   100% open
- 25%   75% open
- 50%   50% open
- 75%   25% open
- 100%   closed

Figure 17-3 compares these two systems and illustrates how each works.

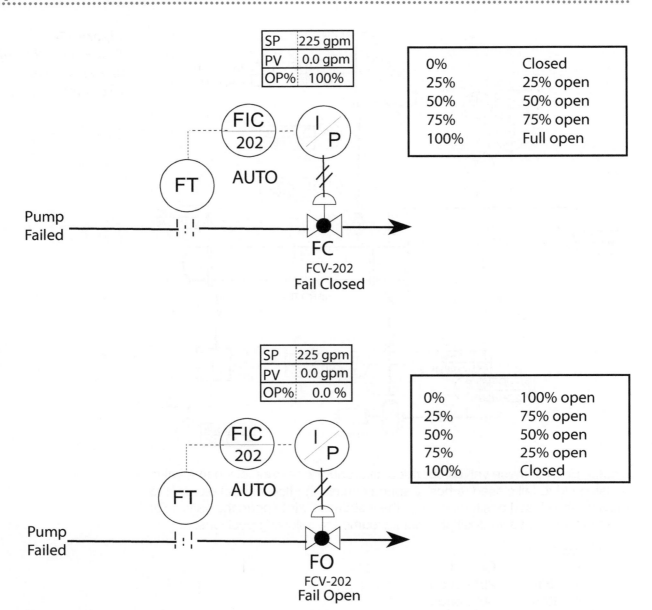

SP	225 gpm
PV	0.0 gpm
OP%	100%

0%	Closed
25%	25% open
50%	50% open
75%	75% open
100%	Full open

FIC
202

I
P

FT

AUTO

Pump
Failed

FC
FCV-202
Fail Closed

SP	225 gpm
PV	0.0 gpm
OP%	0.0 %

0%	100% open
25%	75% open
50%	50% open
75%	25% open
100%	Closed

FIC
202

I
P

FT

AUTO

Pump
Failed

FO
FCV-202
Fail Open

Figure 17-3   *Fail Open-Fail Closed*

Multivariable Plant—Systematic Approach to Troubleshooting

The multivariable unit being described in this chapter combines all nine of the troubleshooting models for the first time. It shows how complicated an operation could be if it were assigned to a new technician on his first day. Figure 17-4 illustrates how each of these parts and a variety of operational data fit together. By combining the technical drawings found in Chapters 8–16, a complex piping and instrumentation diagram (PID) could be developed. A set of operational specifications can be found following this diagram.

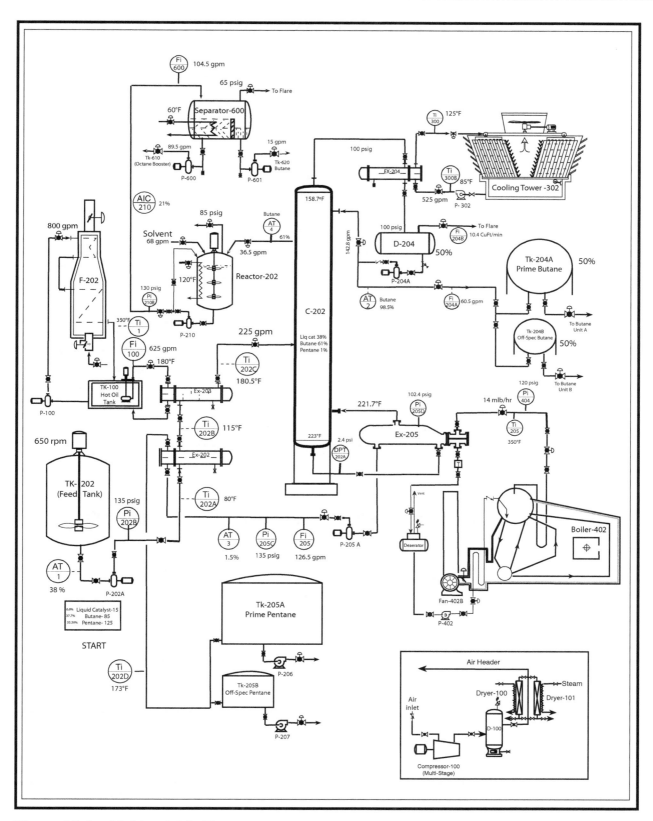

Figure 17-4 *Multivariable Plant*

Pump System Specification Sheet and Checklist

Level

1.	LIC-201	75%	AUTO	Feed tank level
2.	LA-201	85%	High	Feed tank level high
3.	LA-201	65%	Low	Feed tank level low

Flow

4.	FIC-201	225 gpm	CASC	Feed flow
5.	FIC-202	225 gpm	AUTO	Feed flow to C-202
6.	FR-202	—	—	Feed flow to C-202

Analytical

7.	AT-1	38%	—	Butane–liquid catalyst
8.	SIC-201	650 rpm	AUTO	Mixer-200

Pressure

9.	Pi-202A	40 psig	—	Suction P-202A
10.	Pi-202B	135 psig	—	Discharge P-202A
11.	Pi-202C	40 psig	—	Suction P-202B
12.	Pi-202D	135 psig	—	Discharge P-202B
13.	PRV-199	150 psig	—	Tank-202

Temperature

14.	Ti-202A	125°F	AUTO	Hot water return

Compressor System Specification Sheet and Checklist

Pressure

1.	PIC-100	100 psig	AUTO	Instrument air header
2.	Pi-103	100 psig	—	Instrument air header

Heat Exchanger Specification Sheet and Checklist

Flow

1.	FIC-202	225 gpm	AUTO	Feed flow
2.	FR-202	225 gpm	AUTO	Feed flow to C-202

Analytical

3.	AT-1	38%	—	Butane-liquid catalyst

Pressure

4.	Pi-202A	40 psig	—	Suction P-202A
5.	Pi-202B	135 psig	—	Discharge P-202A
6.	Pi-202C	40 psig	—	Suction P-202B
7.	Pi-202D	130 psig	—	Discharge P-202B
8.	PI-100A	35 psig	—	Tube inlet Ex-203

Temperature

9.	TIC-100	180°F	AUTO	Tube inlet Ex-203 hot oil system
10.	Ti-202A	80°F	Gauge	Discharge P-202
11.	Ti-202B	115°F	Gauge	Discharge shell-side Ex-202
12.	Ti-202D	173°F	Gauge	Tube outlet Ex-202
13.	TR-100	180°F	—	Temperature recorder hot oil system
14.	TAH-100	195°F	—	High temperature alarm

Cooling Tower Specification Sheet and Checklist

Level

1.	LIC-300	75%	AUTO	Basin level
2.	LA-1	85%	High	Basin level high
3.	LA-2	65%	Low	Basin level low

Flow

4.	FIC-302	525 gpm	AUTO	Water flow
5.	Fi-300	525 gpm	—	Ex-204 tube outlet water flow

Analytical

6.	AIC-300	7.8 pH	AUTO	pH—acid
7.	AIC-301	30 ppm	AUTO	Blow down
8.	AIC-302	4.5 gph	AUTO	Chemical additive
9.	SIC-300	1250 rpm	On/Off	Fan speed

Pressure

10.	Pi-300A	50 psig	—	Water flow
11.	Pi-300B	45 psig	—	Water flow

Temperature

12.	TIC-301	125°F	AUTO	Hot water return
13.	TIC-302	60°F	AUTO	Cold water basin (low)
14.	Ti-300A	—	—	Wet bulb air temp
15.	Ti-300B	85°F	—	P-302 discharge
16.	TR-300	125°F	—	Hot water return
17.	Ti-300C	—	—	Ex-204 shell outlet temperature

Boiler Specification Sheet and Checklist

Level

1.	LIC-402	50%	AUTO	Upper steam generating drum level
2.	LIC-401	50%	AUTO	Deaerator level
3.	LA-402	35%	Low	Upper steam generating drum level
4.	LR-402	—	Trend	Upper steam generating drum level
5.	LA-401	35%	Low	Deaerator level
6.	LR-401	—	Trend	Deaerator level

Flow

7.	FIC-402A	50%	AUTO	Air to furnace
8.	FIC-402B	50%	AUTO	Natural gas feed
9.	FIC-402C	150 gpm	CASC	Makeup water

Analytical

10.	Ai-402	—	AUTO	Stack discharge
11.	AA-402	0–10%	Hi/Lo	Combustion gases
12.	BA-402	On/Off	—	Burner

Pressure

13.	PIC-402A	120 psig	AUTO	Steam header
14.	PIC-402B	−.05 in water	AUTO	Damper
15.	Pi-402	155 psig	Gauge	P-402 discharge
16.	Pi-400	−.02 in water	Gauge	Stack temperature
17.	Pi-401	60 psig	Gauge	Natural gas supply pressure
18.	Pi-403	60 psig	Gauge	Desuperheated steam
19.	Pi-404	−.02 in water	Gauge	Fire box
20.	PA-404	75/50 psig	Hi/lo	Desuperheated steam pressure
21.	PA-401	150/100 psig	Hi/lo	Steam header
22.	PR-402	—	Trend	Stack

Temperature

23.	TR-402	350°F	Trend	Steam header
24.	Ti-402	450°F	Gauge	Upper stack temperature
25.	TE-400	600°F	—	Radiant section
26.	TE-401	500°F	—	Economizer section
27.	TE-403	350°F	—	Steam header
28.	TE-404	305°F	—	Desuperheated steam

Furnace Specification Sheet and Checklist

Level

1.	LIC-1	75%	AUTO	Tk-100 level
2.	LA-1	85%	High	Tk-100 high
3.	LA-2	65%	Low	Tk-100 low

Flow

4.	FIC-100	800 gpm	AUTO	Feed to furnace
5.	FIC-101	12,500 mbh	CASC	Natural gas feed
6.	FIC-102	35 psig	AUTO	Steam
7.	Fi-1	800 gpm	—	Furnace discharge

Analytical

8.	AIC-100	3%	AUTO	Stack discharge
9.	AIC-101	21%	AUTO	Air to registers
10.	BA-1	—	On/off	Burner

Pressure

11.	PIC-100	0.05 in water	CASC	Bridgewall draft
12.	PIC-101	15 psig	AUTO	Natural gas feed
13.	Pi-3	0.5 in water	—	Top-stack
14.	Pi-5	0.2 in water	—	Radiant section
15.	Pi-1	10 psig	—	P-100 suction
16.	Pi-2	55 psig	—	P-100 discharge
17.	Pi-4	55 psig	—	Furnace discharge pressure
18.	PR-100	0.05 in water	—	Bridgewall draft pressure
19.	PA-1	Hi-65/Lo-45	—	Hot oil discharge

Temperature

20.	TR-1	168°F	—	Convection section exit temp.
21.	DT-1	98°F	Δ-temp	Delta inlet/outlet convection
22.	DT-2	182°F	Δ-temp	Delta inlet/outlet radiant
23.	TIC-100	350°F	AUTO	Furnace exit temperature
24.	TAH-100	385°F	—	Bridgewall high temperature
25.	TE-1	375°F	—	Conv—sect
26.	TE-2	395°F	—	Rad—sect
27.	TE-3	425°F	—	Burner
28.	Ti-1	350°F	—	At start-up
29.	Ti-2	70°F	—	At start-up
30.	TA-100 high	365°F	—	Hot oil discharge
31.	TA-100 low	335°F	—	Hot oil discharge

Distillation Specification Sheet and Checklist

Flow

1.	FIC-201	225 gpm	CASC	Feed to tank 202
2.	FIC-202	225 gpm	AUTO	Feed flow
3.	FR-202	225 gpm	AUTO	Feed flow to C-202
4.	FIC-203	142.8 gpm	CASC	Reflux to C-202
5.	Fi-204A	60.5 gpm	Gauge	Butane to storage
6.	FIC-300	525 gpm	AUTO	Cooling tower water flow
7.	Fi-300	525 gpm	Gauge	Cooling tower water flow
8.	FIC-205	14 mlb/h	CASC	Steam to reboiler
9.	Fi-205	126.5 gpm	Gauge	Pentane to storage
10.	Fi-204B	10.4 ft^3/min	—	Butane/liquid catalyst to flare
11.	Fi-205A	126.5 gpm	Gauge	P-206 discharge
12.	Fi-205B	126.5 gpm	Gauge	P-207 discharge
13.	Fi-204A	60.5 gpm	Gauge	P-204C discharge
14.	Fi-204B	60.5 gpm	Gauge	P-204D discharge

Analytical

15.	AT-1	38%	—	Butane—feed
16.	AT-2	98.5%	—	Butane—reflux
17.	AT-3	1.5%	—	Butane—bottom
18.	AT-4	61%	—	Butane—reactor
19.	SIC-201	650 rpm	AUTO	Agitator motor on Tk-202

Pressure

20.	PIC-204	100 psig	AUTO	D-204 to flare
21.	Pi-202A	40 psig	Gauge	Suction P-202A
22.	Pi-202B	135 psig	Gauge	Discharge P-202A
23.	Pi-202C	40 psig	Gauge	Suction P-202B
24.	Pi-202D	130 psig	Gauge	Discharge P-202B
25.	PI-100A	35 psig	Gauge	Tube inlet Ex-203
26.	Pi-204A	105 psig	Gauge	Suction P-204A.
27.	Pi-204B	105 psig	Gauge	Suction P-204B
28.	Pi-204C	160 psig	Gauge	Discharge P-204A/B
29.	Pi-300A	50 psig	Gauge	Lower tube inlet Ex-204
30.	Pi-300B	45 psig	Gauge	Upper tube outlet Ex-204
31.	DPT-202A	2.4 psig	ΔP cell	Bottom and top of column
32.	Pi-404	120 psig	Gauge	Steam pressure
33.	Pi-205A	102.4 psig	Gauge	P-205A suction
34.	Pi-205B	102.4 psig	Gauge	P-205A suction
35.	Pi-205C	135 psig	Gauge	P-205A/B discharge
36.	Pi-205D	102.4 psig	Gauge	Ex-205 vapor cavity

Temperature

37.	TIC-100	180°F	AUTO	Tube inlet Ex-203 hot oil system
38.	Ti-202A	80°F	Gauge	Discharge P-202
39.	Ti-202B	115°F	Gauge	Discharge shell-side Ex-202
40.	Ti-202C	180.5°F	Gauge	Discharge shell-side Ex-203
41.	Ti-202D	173°F	Gauge	Tube outlet Ex-202
42.	TR-100	180°F	—	Temperature recorder hot oil system
43.	TAH-100	195°F	—	High-temperature alarm
44.	TIC-203	158.7°F	AUTO	Top tray of C-202
45.	TE-202i	170.2°F	—	Tray five
46.	TE-202H	190.2°F	—	Tray three
47.	TE-202G	210°F	—	Feed tray one
48.	TE-202F	222°F	—	Hat tray
49.	TE-202E	223°F	—	Bottom of C-202
50.	Ti-300	125°F	Gauge	Tube outlet Ex-204 cooling water
51.	Ti-205	350°F	Gauge	Upper tube inlet on Ex-205 steam
52.	TIC-205	221.7°F	AUTO	Below hat tray, bottom of C-202
53.	Ti-204	128°F	Gauge	P-204A/B discharge

Level

54.	LG-202	50%	—	Bottom of column
55.	LIC-204	50%	AUTO	Drum-204 level control
56.	LIC-205	50%	AUTO	Kettle reboiler level
57.	LIC-204A	50%	AUTO	Tk-204A level
58.	LIC-204B	50%	AUTO	Tk-204B level
59.	LIC-205A	50%	AUTO	Tk-205A level
60.	LIC-205B	50%	AUTO	Tk-205B level
61.	Li-205	50%	Gauge	Level over tube bundle in Ex-205
62.	LIC-201	75%	AUTO	Level control Tk-202
63.	LA-201	Hi-85%/ Lo-65%	—	Located on LIC-201

Reactor Specification Sheet and Checklist

Flow

1.	FIC-210	36.5 gpm	AUTO	Feed flow to reactor
2.	FIC-211	68 gpm	CASC	Solvent (toluene)

Analytical

3.	AT-4	61%	—	Butane
4.	AIC-210	21%	AUTO	Butane
5.	SIC-210	250 rpm	AUTO	Mixer-210
6.	AA-210	Hi-28%/Lo-17%	—	AIC control loop
7.	AR-210	Video trend	AUTO	AIC control loop

Pressure

8.	Pi-210A	88 psig	—	Suction P-210
9.	Pi-210B	130 psig	—	Discharge P-210
10.	PIC-210	85 psig	—	Reactor-202
11.	PA-210	Hi-100/Lo-75 psig	—	Pressure control
12.	PR-210	Video trend	AUTO	Pressure control

Temperature

13.	TIC-210	120°F	AUTO	Cooling water to reactor shell
14.	TA-210	Hi-140°F/Lo-110°F	—	TIC control loop
15.	TR-210	Video trend	AUTO	TIC control loop

Level

16.	LIC-210	75%	AUTO	Reactor level control
17.	LA-210	Hi-90%/Lo-65%	—	LIC control loop
18.	LR-210	Video trend	AUTO	LIC control loop

Separator Specification Sheet and Checklist

Flow

1.	Fi-600	104.5 gpm	Gauge	Feed flow
2.	Fi-610A	89.5 gpm	Gauge	Feed flow to Tk-610
3.	Fi-601	15 gpm	Gauge	Feed to Tk-620
4.	Fi-610B	89.5 gpm	Gauge	Product "octane booster"
5.	Fi-620	15 gpm	Gauge	Product "butane"

Analytical

6.	AT-5	8%	—	Octane booster
7.	AT-6	99.9%	—	Butane

Pressure

8.	Pi-210B	130 psig	—	Feed inlet to separator
9.	PIC-600	65 psig	—	Separator pressure control
10.	Pi-600A	67 psig	—	Suction P-600
11.	Pi-601A	65 psig	—	Suction P-601
12.	PI-600B	110 psig	—	Discharge P-600
13.	Pi-601B	107 psig	—	Discharge P-601

Temperature

14.	TIC-600	60°F	AUTO	Chill water system on separator

Level

15.	LIC-600	50%	AUTO	Level on extract side of separator
16.	LIC-601	50%	AUTO	Level on raffinate side of separator
17.	LIC-610	50%	AUTO	Tk-610 level control
18.	LIC-620	50%	AUTO	Tk-620 level control

If a system's complexity is measured by the number of items on the specification sheet, it would be easy to identify which system would be the most difficult to learn. Process technicians initially work on the simpler assignments and work their way up. Given proper time and training, a process technician can master very complex processes. In this text, the chapters work from simple to complex; however, by this point in the text a much higher level of understanding should be developing.

Pump Troubleshooting Scenario 1

Although a variety of pump problem scenarios could be included in this chapter, only one has been selected. This pattern will be followed throughout the following sections. In order for a technician to troubleshoot a process problem, it is necessary to see it as it truly appears. In order to simulate this feature, a unique problem will be introduced. Since a picture is worth a 1000 words, it will be worthwhile to look at how a specific problem causes variables to change. This is a small but important step. It will help you prepare for solving more difficult problems. The first step will be to identify each of the variables that can diverge from SP. This will appear in the process variable (PV) box. The next step will be to look at each instrument's typical value (TV) and compare it to the actual value (AV). This should have been made clear during the earlier sections in the text. It will then be necessary to identify each "secondary problem." Only one "primary problem" will exist; however, it will hide in the data.

In the following example, the primary problem is the failure of pump-202A. Review each of the variables carefully. The final exam associated with these chapters will not have the primary variable identified. In Figure 17-5, the pump fails. It is important for a new technician to see what this looks like before encountering it on the unit.

Compressor Troubleshooting Scenario 2

Figure 17-6 simulates what happens when the primary compressor fails. When the compressor fails, the pressure begins to drop slowly. A series of low-pressure alarms may sound, including an audible sound as well as flashing red lights. The pressure indicating controller has an SP of 100 psig. The process variable is at 50 psig and dropping. The valve has closed in response to the dropping pressure. It is important to note that PCV-100 opens to reduce pressure by allowing it to recirculate into the suction. The instrument air header has also dropped to 50 psig. The important thing in this simulation is to be able to quickly identify what has happened and get the air system back on line. If the air system completely fails, a wide variety of systems will be affected.

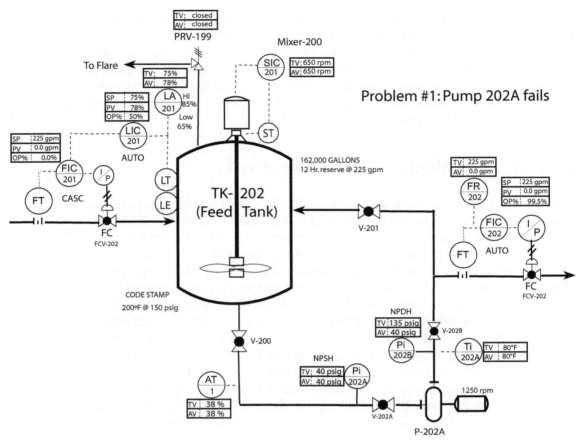

Figure 17-5 *Pump-202A Fails*

Heat Exchanger Troubleshooting Scenario 3

Heat exchangers are used in almost every chemical process in the chemical processing industry. These devices are designed to transfer heat energy between two streams through conductive and convective heat transfer processes. Typically, one of the streams is hotter and the other is cooler. Heat is transferred from areas of hot to cold. A special balance is established in a heat exchanger when both streams are flowing in and out of the heat exchanger. When one of the flows stops or is reduced, significant problems can occur. For example, if P-202 trips off, the shell-side flow to each exchanger stops. The tube-side flow to Ex-202 continues, and the temperature control valve 100 closes when the SP is exceeded. Residual heat from the tube side of Ex-202 continues to affect the instrumentation. When flow stops from P-202, it affects the operation of the entire system. Figure 17-7 illustrates what happens to process variables on a heat exchanger system when a pump trips.

Problem #2: Compressor fails

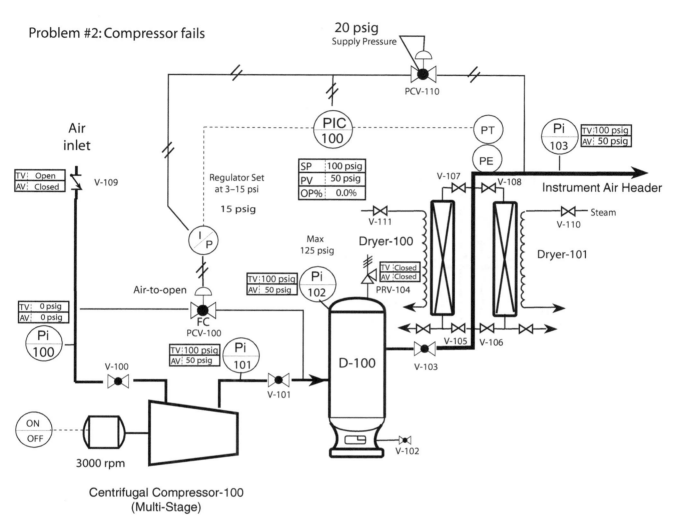

Figure 17-6 *Compressor-100 Fails*

Cooling Tower Troubleshooting Scenario 4

Tube fouling is a common problem and is characterized by a number of changes to the cooling tower's operational systems. A cooling tower is a device designed to cool water; however, the loss of cooling water flow to industrial devices will cause serious problems for the system it is associated with. As the tubes begin to foul, flow is reduced. Typically, this does not happen all at once and can be tracked by keeping a check sheet of inlet and outlet pressures on the condenser. Collecting data on the flow rate can also provide important information about potential plugging problems (Figure 17-8).

Problem #3: Pump fails

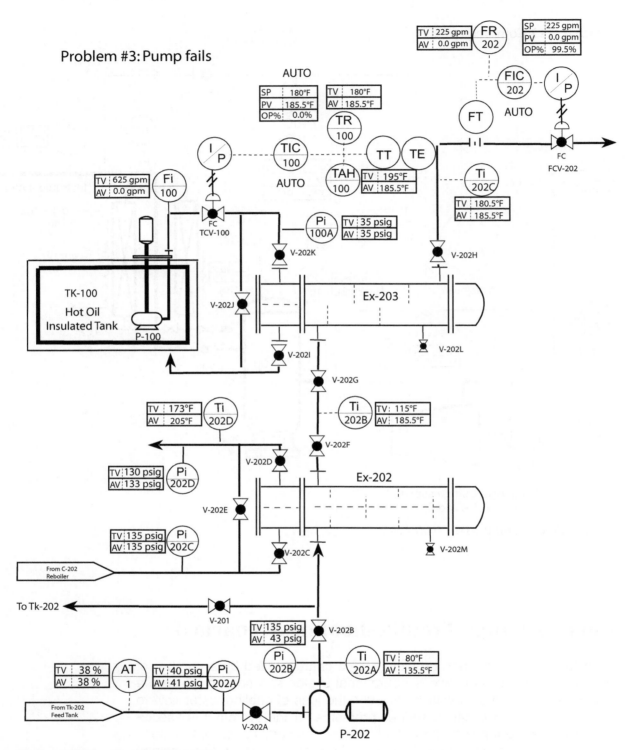

Figure 17-7 *Pump-202 Fails*

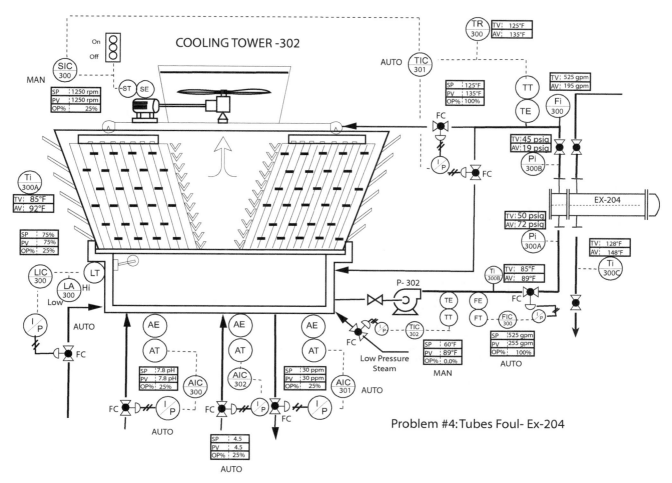

Figure 17-8 *Tubes Foul on Ex-204*

Boiler Troubleshooting Scenario 5

In Figure 17-9, the natural gas pressure drops below operational limits. This will have a series of cascading effects. Initially, FIC-402B falls from 50% to 30%, resulting in a 40°F decrease in the boiler steam discharge. The temperature near the burner drops from 600°F to 475°F. The stack discharge temperature drops from 450°F to 350°F.

Furnace Troubleshooting Scenario 6

Furnace-202 is classified as a small cabin furnace designed specifically to heat up and maintain the hot oil system. Pump-100A represents the feed charge device and is critical to the furnace operation. If the pump fails, the furnace will heat up quickly. This will cause the oil that is in the radiant and

Problem #5: Natural gas pressure drops

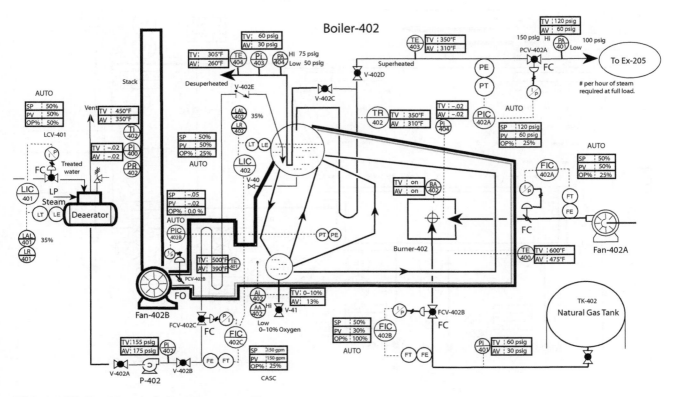

Figure 17-9 *Natural Gas Pressure Drops*

convective tubes to flash. This will result in erratic pressures and temperatures, and may result in damaging the furnace unless the burner is shut down (Figure 17-10).

Distillation Troubleshooting Scenario 7

Distillation column operation is dependent on a large variety of process variables working together at specific SPs. In Figure 17-11, one of these critical process variables fails, namely the cooling tower cooling water pump. When the cooling water pump fails, a cascading effect moves from one variable to the other, causing each to go out of specification. In order to better understand this process, it is important to look carefully at the original condition, or TVs and SPs. As the cooling water pump fails, the following variables are affected:

- FIC-300   525–55 gpm
- PIC-204   10.4–242 ft³/min   Flare lights up the night sky

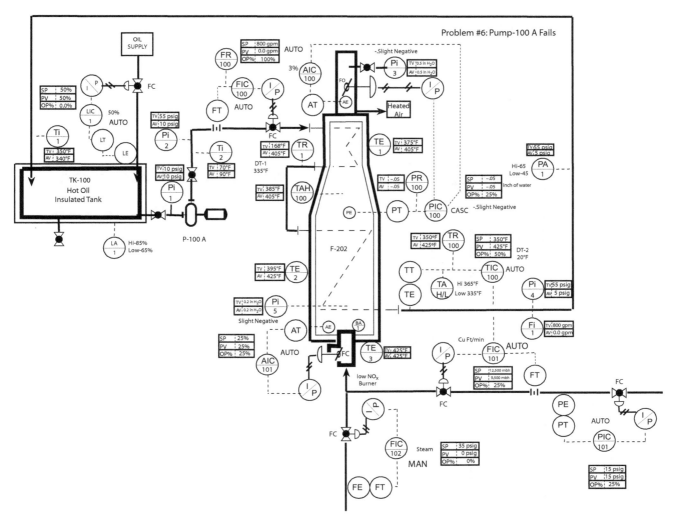

Figure 17-10 *Pump-100A Fails*

- LIC-204   50–10%
- FIC-203   Reflux increases to 100%
- TIC-203   158.7–187°F
- Pi-204A   105–150 psig
- Ti-204   128–187°F
- Pi-204C   160–180 psig
- DPT-202A   2.4–1.4 psig
- TE-202i   170.2–187°F
- AT-2   98.5–99.2%
- Fi-204A   60.5–0 gpm

This short list should illustrate how the loss of a single variable affects a wide assortment of other variables. As the temperature of the liquid butane

Problem #7: Cooling Tower Pump fails

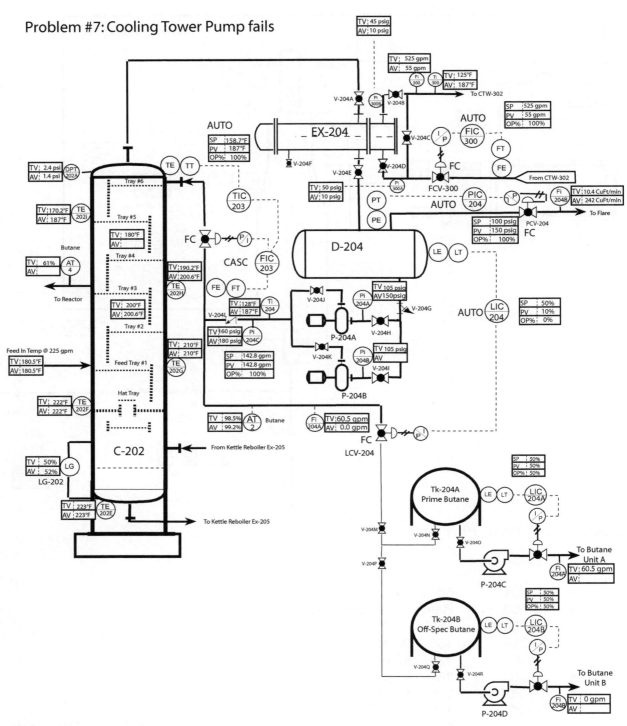

Figure 17-11 *Cooling Tower Pump Fails*

increases, it begins to boil and vaporize. This increases the pressure inside the drum and forces the pressure control valve to open up 100%.

Stirred Reactor Troubleshooting Scenario 7

The solvent feed introduced to the stirred reactor is designed to combine with the feed from the column to form a new product. In Figure 17-12, the solvent (toluene) control valve fails in the closed position. The result of this problem means that no reaction is occurring.

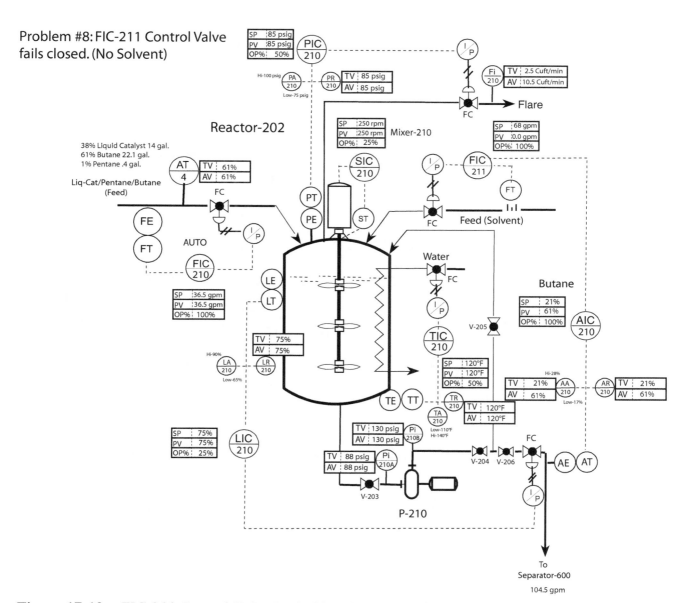

Problem #8: FIC-211 Control Valve fails closed. (No Solvent)

Figure 17-12 *FIC-211 Control Valve Fails Closed (No Solvent)*

415

Separator Troubleshooting Scenario 7

During operation, the stirred reactor and separator work together to produce and separate the products. Figure 17-13 is a continuation of the problem encountered in the previous section. This scenario illustrates the

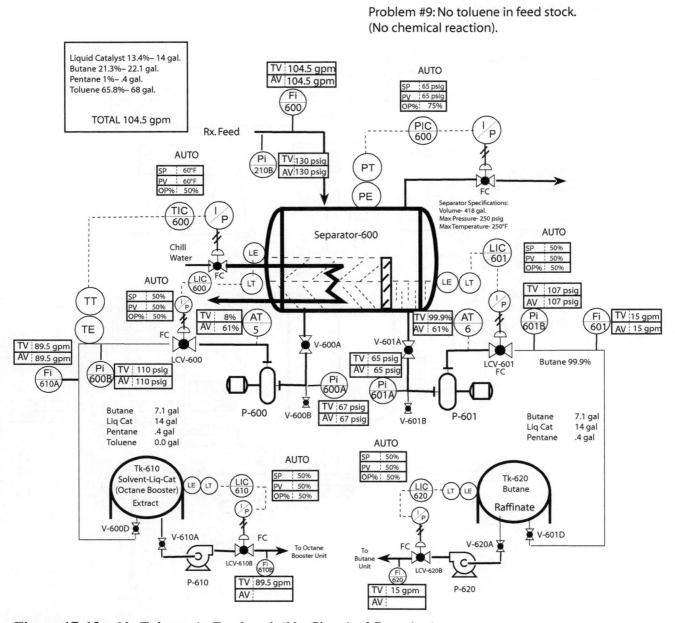

Figure 17-13 *No Toluene in Feedstock (No Chemical Reaction)*

interdependence of one system with another. Analyzers five and six indicate significant variation from the operating specifications.

Summary

Troubleshooting Methods
There are a number of troubleshooting methods that can be used with these models. Methods vary depending on individual educational faculty, consultants, and industry. The basic approach to most methods includes the development of a good educational foundation.

Method One: Educational
- Basic knowledge of the equipment and technology
- Understanding of the math, physics, and chemistry associated with the equipment
- Study equipment arrangements in systems
- Study process control instrumentation
- Operate equipment in complex arrangements
- Troubleshoot process problems.

Method Two: Instrumental
- Basic understanding of process control instrumentation
- Basic understanding of the unit process flow plan
- Advanced training in controller operation—PLC and DCS
- Troubleshoot process problems.

Method Three: Experiential
- Experience in operating specific equipment and system
- Familiarity with past problems and solutions
- Ability to think outside the box
- Ability to think critically—identify and challenge assumptions
- Ability to evaluate, monitor, measure, and test alternatives
- Ability to troubleshoot process problems.

Method Four: Scientific
- Grounded in principles of mathematics, physics, and chemistry
- Knowledge of theory-based operations
- Good understanding of equipment design and operation
- Ability to view the problem from the outside in
- Ability to utilize outside information and expertise, and reflective thinking
- Ability to generate alternatives, brainstorm, and rank alternatives
- Ability to troubleshoot process problems.

Troubleshooting Models

In this text, ten different troubleshooting models have been studied. These models include:

- pump and tank model
- compressor model
- heat exchanger model
- cooling tower model
- boiler model
- furnace model
- distillation model
- reactor model
- separator model, and
- combinations of models.

Review Questions

1. Identify the four methods of troubleshooting and list the most important feature of each.
2. Identify the ten troubleshooting models used in this text.
3. In Figure 17-5, explain why the FIC-202 is attempting to open the valve to 100%.
4. In Figure 17-5, explain why FIC-201 has closed the valve.
5. In Figure 17-6, explain why the pressure did not immediately drop to zero.
6. In Figure 17-7, explain why Ti-202D is reading 205°F.
7. In Figure 17-7, explain why TIC-100 has closed the valve.
8. In Figure 17-7, explain why Ti-202B is reading 185.5°F.
9. In Figure 17-8, explain why Pi-300A is reading 72 psig.
10. In Figure 17-8, explain why Pi-300B is reading 19 psig.
11. In Figure 17-8, explain why TIC-301 is opening the valve to 100%.
12. In Figure 17-9, explain why FIC-402B is opening the valve to 100%.
13. In Figure 17-10, explain why TR-1 is reading 405°F.
14. In Figure 17-11, explain why the level in D-204 is dropping while the controller is in auto.
15. In Figure 17-12, explain why AIC-210 is reading 61%.

Introduction to Quality Control

LEARNING OBJECTIVES

- Describe the history and philosophy of the quality movement.
- Describe the importance of quality control in the process industry.
- Define the term "quality control."
- Describe the significant contributions of Dr. Walter Shewhart.
- Explain the viewpoints of Taguchi and Ishikawa on quality.
- Describe how Dr. W. Edwards Deming and Dr. Joseph Juran were the intelligences behind the quality movement.
- Describe Phillip Crosby's entrepreneurial philosophies on quality.
- Explain the term "dimensions of quality."
- Explain the important role of Dr. John Dewey.
- Identify the contributions of George Box, Alex Osborne, and Larry Miles.
- Explain the connection among Frederick W. Taylor, scientific management, and the chemical processing industry.

Key Terms

Brainstorming—the unrestricted generation of ideas in a group process in response to a common or shared problem.

Orthogonal arrays and linear graphs—equations designed to isolate outside factors that affect quality (Dr. Taguchi).

Process—a set of interrelated activities designed to take specific products and produce marketable outputs or services.

Robustness—ability of a process to perform as intended (Dr. Taguchi).

Scientific management—a controversial theory that produced dramatic improvements in plant productivity.

Statistical quality control—an important branch of total quality management (TQM) that includes statistical process control and acceptance sampling. It is the collection, organization, analysis, and interpretation of data.

The loss function—quality equation that equates quality to cost (Dr. Taguchi).

The Shewhart learning and improvement cycle—the elements of plan, do, study, act (PDSA).

Total quality management (TQM)—a philosophy that represents the guiding principles of a continuously improving organization.

Quality—desired characteristics of a product or service (Thomas, 2004).

Zero defects—a term developed by Phillip Crosby that involves "getting it right first time."

Introduction to Quality Control

Statistical **process** control is an important practice and concept used by industrial manufacturers to control complex processes and provide their customers with products that exceed expectations. **Quality** control is grounded in mathematics, probability, and statistics. Modern computer technology has enhanced the ability of a process technician to monitor and control large processes. Video trends, analyzers, recorders, equipment checklists, instrumentation, modern process control, customer directives, and statistical process control combine to provide a stable and economically effective operation.

Key topics covered in this appendix include the following:
- introduction to quality control
- principles of **total quality management (TQM)**
- quality tools and techniques
- statistics
- control charts, and
- applied concepts.

In the chemical processing industry, a competitive and complex network exists that links products and services to the global economy. Many large companies have vast international holdings, which move beyond the control of any single country. A limited amount of resources and raw materials are available to these industrial giants. The relationship between suppliers and customers has developed into a complex system that responds to the successes and failures of other companies and countries. Competition among companies is intense and has recently forced many large organizations such as Chevron/Phillips, Exxon/Mobil, and BP Amoco to combine their vast resources in order to remain competitive. This combining process will continue into the first decade of the new century as industrial manufacturers compete for control of various markets.

Customer expectations have risen steadily over the past century in relation to the products purchased. When the expression "quality" is used, consumers and economists have a variety of definitions. Quality is easily defined as the desired characteristics of a product or service. Process technicians are directly responsible for a wide array of quality aspects associated with the complex processes they operate.

Quality control inside a chemical plant, a refinery, or a manufacturing complex is an integral part of the environment. Competition for customers is intense with all the various products produced. The economics of operating a large manufacturing complex is related to the quality of the product produced, the costs associated with the operation and maintenance of the facility, and the costs associated with marketing and shipping the products. Environmental regulations, safety guidelines, and new and innovative technologies are all part of a process technician's future.

Some quality experts believe that quality can be quantified using the following formula: Q (quality) $= P$ (performance) $\div E$ (expectations). When Q exceeds 1.0, customer expectations are satisfied. Chemical manufacturers determine performance characteristics in an operating unit with quality control techniques. The customer establishes product expectations. The key concepts associated with quality control will be discussed in Appendix II. Quality control involves the integration of five areas into the company's corporate culture: (1) product or service specifications, (2) design, (3) production, (4) inspection, and (5) evaluation or review. A number of quality gurus believe that there are ten objectives of quality. This basic framework can be used to develop the quality system for new or existing processes:

1. use of quality tools
2. product performance and characteristics: primary and secondary
3. conformance to product specifications
4. reliability or product performance
5. sensory characteristics or aesthetics
6. company reputation and long-term past performance

7. product life and durability
8. product support and service
9. personal response or person-to-person
10. equation of quality to cost using **the loss function.**

Another popular system used by the chemical processing industry is TQM, developed by W. Edwards Deming and Joseph M. Juran. TQM is defined as a philosophy that represents the guiding principles of a continuously improving organization. An important aspect of TQM is **statistical quality control,** which includes statistical process control and acceptance sampling.

The History and Philosophy of Quality Control

In the early 1900s, a young engineer, Frederick W. Taylor, introduced the principles of **scientific management.** This controversial theory produced dramatic improvements in plant productivity. According to some quality experts, the principles of scientific management included:

- Management should develop a scientific approach for each element of an individual's work to replace rule-of-thumb guidelines.
- Management should scientifically select, train, teach, and develop each worker so that the right person has the right job.
- Management should cooperate with workers to ensure that the job matches plans and principles.
- Management should ensure an equal division of work and responsibility among managers and workers.

Taylor believed that there was one and only one procedure to perform a job task. Under scientific management, managers would study the variety of methods used to accomplish a task. Elements from each of these methods were combined to provide one best way that had been time tested for efficiency. As a large number of non-English-speaking immigrant workers were found in the plants, Taylor did not believe they had the education to plan how work should be done. The remedy was to separate planning from execution. Taylor hired engineers to staff the planning department and gave them the responsibility to:

- develop scientific methods for performing job tasks
- establish goals for productivity
- establish reward systems for meeting goals
- train personnel.

The single most controversial element of scientific management was task allocation. As large tasks were broken into smaller and smaller parts, many workers believed they were being told by management what to do, when to do it, and the exact amount of time it would take to complete it. This created a dehumanizing rift between the management and the workforce.

Today, many of the elements found in Taylor's scientific method exist in the chemical processing industry. While the evolution of management systems included systematic, scientific, bureaucracy, administrative, human relations, quantitative, organizational behavior, systems theory, and contingency theory, the scientific management system has had the greatest impact on the process industry.

In 1922, a young engineer named Walter Shewhart became the first pioneer of the quality movement. Credited with being the grandfather of TQM, Shewhart proposed a variety of innovative methods, statistical process control, control charts, the learning and improvement cycle, and the application of statistics. A few years later, two associates of Dr. Shewhart, H.F. Dodge and H.G. Romig, pioneered "acceptance sampling" as a reasonable substitute for 100% inspection.

One of Dr. Shewhart's students was a young physicist named W. Edwards Deming. In 1940, Dr. Deming introduced statistical quality control to the U.S. Census Bureau and industrial operations. The teachings of Deming continued Shewhart's work and helped spark a quality revolution. The work of Deming was supported by Joseph M. Juran. Both Deming and Juran are credited with being the great intelligences behind the quality revolution. In 1926, a team of quality control pioneers from Bell Laboratories, Walter Shewhart, Donald Quarles, and Harold Dodge, introduced Hawthorne Works to the statistical approach and quality tools and techniques. Joseph Juran was a young electrical engineer who was selected to participate in this training program. In 1928, Juran authored the pamphlet *Statistical Methods Applied to Manufacturing Problems*. By 1937, Juran had risen to the position of a chief industrial engineer at Western Electric. Part of his job responsibilities included visiting other companies and teaching methods of quality management.

During World War II, the work of Juran and Deming was joined by a British chemist named George Box. In 1938, Box needed to develop a system to analyze experimental data. Unable to secure statistical advice, Box collected available data and taught himself. Over the next few years, George Box became one of the leading advocates for the use of statistical theory in experimentation. From the ashes of World War II, the Japanese nation attempted to rebuild its shattered economic structure. Although many U.S. manufacturers were neglecting the power of TQM, the Japanese were not. In 1949, Kaoru Ishikawa and Genichi Taguchi embraced the principles of quality control and initiated an effort which would later revolutionize the Japanese manufacturing process. During the 1950s, the Japanese were supported by Deming and Juran in their quality revolution. Another key contributor to the quality process during the 1950s was Larry Miles. Miles is credited with being the father of the value method, an effective problem-solving system developed in 1948. Many of the concepts taught by Shewhart, Juran, Deming, Taguchi, and Ishikawa were laced with advanced statistics and mathematics. In 1979, Phillip Crosby translated this terminology into language mere mortals could understand. Crosby's books *Quality Is Free* and *Quality without Tears* were easy to read and understand.

The impact of the quality movement is an integrated feature of our modern global economy. The legacy left by these quality pioneers is an important accomplishment in our history. Organizations such as the American Society for Quality (ASQ), the Registrar Accreditation Board (RAB), and the Quality Press continue the efforts of these pioneers.

Dr. Walter A. Shewhart

Walter A. Shewhart was born in New Canton, Illinois, on March 18, 1891. He earned a Ph.D. in physics at the University of California after completing his undergraduate work at the University of Illinois. In 1918, Shewhart accepted a position at Western Electric, a manufacturer of telephones for Bell Telephone. Shewhart was intrigued with the concept of reducing variation in their manufacturing process. Studies had shown that continual process adjustment in response to nonconformance actually increased variation and decreased product quality. On May 16, 1924, in a memorandum to his supervisors, Shewhart proposed "assignable-cause" and "chance-cause" variation as part of his control chart methodology. This process was designed to identify chance-cause variation while keeping the process in a state of statistical control.

Walter A. Shewhart is often referred to as the grandfather of TQM. Early in his career he believed that statistical theory could be applied to the needs of the industry. Most of his professional career was spent working for Western Electric, 1918–1924, and Bell Telephone, 1925–1956. Shewhart is characterized as having an intellectual ability that removed the dark clouds of ignorance, a generosity of pioneering spirit that led him to share his ideas with others, and the heart of an educator that endeared his students to him. Shewhart also had the ability to seek out talented individuals who could add to the body of scientific study and promote the quality movement.

In 1931, Shewhart authored *Economic Control of Quality of Manufactured Product.* He believed in the formulation of a scientific basis for securing economic control (Shewhart, 1931). In 1939, Shewhart authored *Statistical Method from the Viewpoint of Quality Control.* He was the first honorary member inducted into the American Society for Quality (ASQ). After retirement in 1956, Dr. Shewhart lectured at the University of London, at the Stevens Institute of Technology, and in India. Shewhart is credited with being the first to combine and apply the disciplines of engineering, statistics, and economics to manufacturing. He died on March 11, 1967, in Troy Hills, New Jersey. In the 1990s, the genius of Shewhart was rediscovered in the form of the "Six Sigma Approach."

Dr. Walter A. Shewhart's Key Contributions:

- Concept of reducing variation in the manufacturing process
- Application of statistical control methods to industry—control charts
- Grandfather of TQM
- Formulation of a scientific basis for securing economic control
- Combination of the disciplines of engineering, statistics, and economics
- The Six Sigma Approach
- Influenced Deming and Juran
- **The Shewhart learning and improvement cycle**—plan, do, study, act (PDSA).

Dr. W. Edwards Deming

Dr. W. Edwards Deming was born on January 1, 1900, in Iowa. In 1917, the family moved to Wyoming. While working on his B.S. degree at the University of Wyoming, Deming was employed as a janitor. He graduated in 1921 and transferred to the physics and mathematics M.S. program at the University of Colorado. Deming completed his doctoral studies in physics at Yale University. Soon after graduation, Dr. Deming was drawn to the field of statistics.

In 1940, the U.S. Census Bureau hired Deming to work on the conversion from total count to the new census sampling methodology. By 1941, Deming was teaching statistical quality control to engineers and inspectors working in industrial applications. In 1946, Deming left the Census Bureau and launched his own private consulting firm. The firm operated for more than 40 years with clients from all over the world. During this time, Deming spread the quality movement everywhere he went. Later in his career at New York University, professor emeritus Deming conducted classes on quality control and sampling. During a 10-year period, Deming averaged over 10,000 students per year.

In the 1950s, his ideas were used to usher in the export-led economic rise of the Japanese nation. American business, industry, government, and medical field officials noticed these changes overseas and quickly repented of their mistakes and lost opportunities. These quality concepts swept into the European nations and created a revolution. The concepts of suppliers, producers, and customers working together to improve quality linked the world together in a global economy. These quality concepts have opened the doors of communication among manufacturers, raw material providers, consumers, and competitors.

Dr. Deming Is Best Known for His Fourteen Points

1. Create and communicate to all employees a statement of the aims and purposes of the company.

2. Adapt to the new philosophy of the day; industries and economics are always changing.
3. Build quality into a product throughout production.
4. End the practice of awarding business on the basis of price tag alone; instead, try a long-term relationship based on established loyalty and trust.
5. Work to constantly improve quality and productivity.
6. Institute on-the-job training.
7. Teach and institute leadership to improve all job functions.
8. Drive out fear; create trust.
9. Strive to reduce interdepartmental conflicts.
10. Eliminate exhortations for the work force; instead, focus on the system and the morale.
11. (a) Eliminate work-standard quotas for production; substitute leadership methods for improvement. (b) Eliminate MBO; avoid numerical goals. Alternatively, learn the capabilities of processes and how to improve them.
12. Remove the barriers that rob people of the pride of workmanship.
13. Educate with self-improvement programs.
14. Include everyone in the company to accomplish the transformation.

By Dr. W. Edwards Deming, *Out of the Crisis*, 1986, The MIT Press, Cambridge, MA

Dr. W. Edwards Deming's Key Contributions
- Identification of system as a group of interrelated components
- Widespread use of SPC and quality techniques
- Widespread use of statistics
- 14 points from the text *Out of the Crisis*
- Meeting and exceeding customer expectations
- Continuous improvement
- Primary support for rebuilding of the Japanese economy after World War II.

In 1956, Deming received the Shewhart award from the American Quality Council. As Deming's seminars and lectures contributed to Japan's economic rising star, in 1960 he was awarded the Second Order Medal of the Sacred Treasure by the Emperor. The Japanese Science and Engineering group honored their mentor by establishing the Deming Prizes for significant achievement in product quality and dependability. In 1983, he was given the Samuel S. Wilks Award by the American Statistical Association. President Reagan gave Deming the National Medal of Technology in 1987. Dr. Deming was also honored with election to the Automotive Hall of Fame in 1991 and the Distinguished Career in Science award in 1988. Dr. Deming authored two books: *Out of the Crisis* and *The New Economics;* he also wrote 171 papers. Dr. W. Edwards Deming died in 1993 after a long and distinguished career.

Dr. Joseph M. Juran

Joseph Moses Juran was born in Braila, Romania, in December 1904. His family emigrated to Minnesota in 1912. At an early age, Joseph demonstrated a keen proficiency in math and science. This intellectual prowess allowed him to skip four grade levels. In 1920, Joseph became the first member of his family to enter college at the University of Minnesota. In 1925, he graduated with a B.S. in electrical engineering and in 1936 a J.D. in law at Loyola University. Shortly after graduating with his B.S., Western Electric's Hawthorne Works inspection office offered him a job. Over 40,000 employees were employed at the factory. In 1926, Walter Shewhart, Donald Quarles, and Harold Dodge from Bell Laboratories visited Hawthorne Works. The purpose of the visit was to teach Western Electric employees how to use and apply quality tools and techniques. Under the direction of Dr. Walter Bartky, a training program was established at the factory. Joseph M. Juran became one of twenty trainees selected to participate in this program. This single event set the internal compass of Joseph Juran on the quality path.

By 1928, Juran had authored a pamphlet called *Statistical Methods Applied to Manufacturing Problems.* In 1937, Juran was promoted to head of Industrial Engineering at Western Electric. His role, however, continued to move toward training and consulting inside and outside the company. During this time, Dr. Juran conceptualized the Pareto principle. In 1942, Juran requested a temporary leave of absence to serve as the assistant administrator of the Lend-Lease Administration in Washington. For the next 4 years, Juran managed the Lend-Lease Administration, which coordinated the transportation and shipment of products to friendly nations during the war. During this time, many of the experimental quality methods were tested and used to eliminate paper jams, reduce paperwork, redesign the shipment process, and cut costs. On September 1, 1945, Dr. Juran left Western Electric and Washington behind to enter private consulting.

In 1951, the *Quality Control Handbook* was published and quickly established Dr. Juran as a leading authority in the quality field. In 1954, Dr. Deming and Dr. Juran conducted a series of independent lectures for the Union of Japanese Scientists and Engineers and Keidanren. The message of these quality prophets was embraced by a nation still reeling from the devastation of World War II. Over the next 15 years, the Japanese completely redesigned their export system to fall in line with Shewhart's, Deming's, and Juran's philosophies of quality. By 1970, Japanese quality techniques were controlling the television and automobile industries. The prosperity of the postwar U.S. economy made the quality message fall on deft ears. By 1970, American manufacturers had lost considerable shares of the market to the Japanese. Significant efforts were made to recapture this market; however, only minor successes have been achieved. Dr. Juran does not believe that the American economy has been successful at implementing quality measures.

Dr. Joseph Juran was awarded the Order of the Sacred Treasure by Emperor Hirohito. He has also been honored with the National Medal of Technology by the president and the Deming Center Award by Dean Feldberg.

Dr. Joseph M. Juran's Key Contributions

- TQM—added human dimension
- Pareto principle
- Need for widespread quality training (teacher/lecturer)
- Top management involvement
- *Quality Control Handbook* (1951) and *Managerial Breakthrough* (1964)
- Quality improvement video series
- Quality Trilogy (1986, quality planning).

Dr. Kaoru Ishikawa

Kaoru Ishikawa is best known for the Ishikawa or fishbone diagram. As a leader in quality improvement, Ishikawa developed and enhanced the principles of cause and effect. According to Dr. Ishikawa, quality improvement is a continuous process that requires the company-wide use of seven quality tools: flowcharts, run charts, histograms, Pareto charts, scatter diagrams, fishbone diagrams, and control charts. As a champion of service after the sale, Ishikawa believed a customer should receive continued support and customer service even after the product is purchased. Dr. Ishikawa was the first to popularize the widespread use of quality circles. The fundamental structure of the cause-and-effect diagram was adopted by Dr. Deming to introduce total quality control. Like many other quality experts, Ishikawa believed in quality leadership flowing from the highest to the lowest levels of an organization. Ishikawa believed that standards should be constantly evaluated and compared with customer satisfaction. He also championed the application of quality throughout the life cycle of a product.

Dr. Kaoru Ishikawa was born in Japan in 1915. He was educated as a chemist in the Engineering Department at Tokyo University, where he graduated with a B.S. in 1939. In 1947, he was appointed as assistant professor at Tokyo University, and professor in 1960, shortly after completing his doctoral studies in engineering. He has authored *The Guide to Quality Control* in 1971 and *What Is Total Quality Control?* in 1985. Like many other quality experts, Ishikawa paid close attention to data collection, organization, and analysis, preferring to use Pareto diagrams, cause-and-effect diagrams, quality circles, group communication, control charts, and scatter diagrams.

Dr. Kaoru Ishikawa's Key Contributions
- Cause-and-effect diagrams
- Quality improvement is a continuous process
- Service after the sale
- Popularization of the widespread use of quality circles
- Application of quality throughout the life cycle of a product
- *The Guide to Quality Control* (1971) and *What Is Total Quality Control?* (1985).

Dr. Kaoru Ishikawa was given the following awards before passing away in 1989: the Deming Prize, the Nihon Keizai Press Prize, the Industrial Standardization Prize, and the Grant Award.

Dr. Genichi Taguchi

Dr. Genichi Taguchi was born in Japan on January 1, 1924. His early education was at Kiryu Technical College, where he earned a B.S. in textile engineering before the war claimed him. In 1962, he earned a Ph.D. from Kyushu University. From 1962 to 1982, he was a professor at the University of Tokyo. During World War II, he served in the Imperial Japanese Navy from 1942 to 1945. During the war, he was associated with the Astronomical Department of the Navigation Institute, the Ministry of Education, the Ministry of Public Health and Welfare, and the Institute of Statistical Mathematics. After the war, Dr. Taguchi was offered a position at the Electrical Communications Laboratory of the Nippon Telephone and Telegraph Company. From 1950 to 1962, Taguchi developed and applied his quality management approach. His understanding of quality was influenced by a chance meeting with the famous Dr. Walter Shewhart in 1954.

Many experts credit Taguchi with being the quality guru who revolutionized the Japanese manufacturing process. While Deming, Juran, and Shewhart were guest lecturers, consultants, and quality leaders, Dr. Taguchi was committed to the economic rebirth of his country. Taguchi's approach was to identify the key factors that had the most significant effect on product variability.

Dr. Genichi Taguchi's Key Contributions
- The loss function—equate quality to cost
- **Robustness**—ability of processing to perform as intended
- **Orthogonal arrays and linear graphs**—isolated outside factors
- Revolutionizing the Japanese manufacturing process.

Dr. Taguchi has been honored with the following awards: the Deming application prize, Deming awards for literature and quality 1951, 1953, 1984, and the Willard F. Rockwell Medal. He has authored *Orthogonal Arrays* (1951), *Design Experiments* (1958), *Management by Total Results* (1966), and two other books in the 1970s. Taguchi's primary principle is the optimization of the product and process before manufacturing, instead of optimizing before quality inspection. This prototyping method allows design engineers to troubleshoot, identify optimal settings, and produce a robust product.

In conventional quality control, loss occurs when the product shifts outside customer specifications. Dr. Genichi Taguchi believed that loss occurs as soon as the process deviates from the target value. Target values and product specifications are illustrated on the x-scale, while monetary loss is shown on the $f(x)$ scale. At the point where the curve intersects the specifications, the expense of discarding or fixing the product is illustrated at D, in dollars. The value at D can be used to plot the curve. Under this model, Taguchi has combined (1) target values, (2) specifications, (3) minimum variation, and (4) cost. Each of these variables is an important function under the TQM system.

Phillip B. Crosby

Phillip B. Crosby was born in Wheeling, West Virginia, on June 18, 1926. He graduated from Tridelphia High School in 1943 and entered military service during World War II and the Korean War. Between these two conflicts, Crosby briefly attended medical school. He entered the workforce in 1952 with a company called Crosley, where he worked on an assembly line and quickly realized that it was more cost effective to prevent problems than to fix them. Mr. Crosby worked for Crosley from 1952 to 1955, Martin-Marietta from 1957 to 1965, ITT from 1965 to 1979, and did independent consulting from 1979 to 2001.

From 1965 to 2001, Phillip Crosby published fourteen books on human relations, leadership, and quality management. Mr. Crosby's greatest ability was to take a complex topic and present it in common terms that were easy to understand. Crosby believed in teaching quality principles to all levels of an organization, creating a quality specialist department, emphasizing top-down quality management, getting it right first time—**zero defects,** and system modeling. He has received the Distinguished Civilian Service Medal in 1964, Edwards Medal of Honor in 1981, Marketeer of Year in 1985, Philanthropist of the Year in 1986, and was inducted as an Honorary Member of the American Society of Quality in 2001.

Phillip Crosby's Key Contributions
- Do it right the first time—zero defects
- Ability to simplify complex quality principles
- Top management responsibility for quality
- Four absolutes of quality management: (1) quality is defined as conformance to requirements, not as "goodness" or "elegance"; (2) the best system for causing quality is prevention, not appraisal; (3) performance standard must be zero defects, not "close enough"; and (4) measurement of quality is the price of nonconformance, not indices.

Other Leaders in the Quality Movement

A number of individuals could be added to the list of leaders in the quality movement, including John Dewey, Alex F. Osborne, George Box, and Larry Miles. John Dewey was a brilliant professional educator born in Burlington, Vermont, in 1859, who proposed a number of educational beliefs and techniques which were quite different from those around him. Central to his belief was the concept of experiential learning or learning through doing and experiencing. Dewey believed this process was life-long and allowed an individual to experience intellectual growth in a variety of social contexts. The concepts of thinking and reflecting also are connected to John Dewey. The idea or concept that people learned from their interactions with the environment had not been widely considered during this time frame. These philosophies were carried over into the manufacturing environment and influenced how on-the-job training was conducted. Dewey is considered to be the father of modern education by many people. He graduated from the University of Vermont in 1879 with a B.S. and later, John Hopkins University with a Ph.D.

Larry Miles is the father of the value method, a systematic approach used to analyze and improve value in a system, a facility, or a product. This proactive approach uses a concept called function analysis to convert a product or process into word-pairs labeled functions. The focus is now placed on these word-pairs as the analysis continues. In addition to function analysis, a variety of other elements were included function–cost, function–worth, create by function, implementation, the job plan, the team, and ownership. This system has been around since the 1950s and is a successful cost-saving alternative to trial and error.

Alex F. Osborne is the father of the brainstorm, who believed that "it is easier to tone down a wild idea than to think up a new one." This method of creating solutions to problems has been used successfully in the area of quality control. Creative thinking is enhanced when a group of individuals use the tools of **brainstorming** and affinity diagrams. Brainstorming is defined as the unrestricted generation of ideas in a group process in response to a common or shared problem. Rules of brainstorming include the following: creating an environment of free flowing ideas, not criticizing team members' ideas; using a panel format, including ten to twelve people, including a leader, a recorder, and the team members; maintaining a rapid flow of ideas; listing ideas when presented; and evaluating ideas after the brainstorming session.

George E.P. Box was a chemist born and raised in England. During World War II, Box served as a chemical weapons defense analyst and needed a new approach to analyze his research data. Unable to secure training in the statistical field, Box located a number of statistical textbooks and taught himself. From the classroom of World War II, Box became a leading authority on the statistical approach. After the war, George Box earned a Ph.D. in mathematics and statistics from University College in Great Britain. In 1960, he accepted a position at the University of Wisconsin and became the chair for the newly created statistics department. Box was awarded the British Empire Medal in 1946 and the Walter Shewhart Medal in 1968.

Summary

Quality is defined as the desired characteristics of a product or service. Process technicians are directly responsible for a wide array of quality aspects associated with the complex processes they operate. Chemical manufacturers determine performance characteristics in an operating unit with quality control techniques. The customer establishes product expectations.

In the early 1900s, a young engineer Frederick W. Taylor introduced the principles of scientific management. This controversial theory produced dramatic improvements in plant productivity. Taylor is primarily responsible for the operational structure of the modern chemical processing industry. Many experts believe that this system is antiquated and in need of change.

Dr. Walter A. Shewhart is often referred to as the grandfather of TQM. Key contributions included the invention of the control chart to reduce variation and the use of the statistical process control. Dr. Shewhart was responsible for launching the quality movement and sparking the keen intellects and abilities of Deming, Juran, Ishikawa, and Taguchi.

Dr. W. Edwards Deming was a key figure in the quality movement. In the 1950s, his ideas were rejected by American manufacturers, embraced by a devastated Japanese economy, and used to usher in the export-led economic rise of the Japanese nation. American business, industry, government, and medical officials took notice of these changes overseas; however, they were 20 years behind the new and improved Japanese nation. Key contributions include the widespread use of statistics, SPC, and quality techniques. He is well known in the industry for the concepts of meeting and exceeding customer expectations, and continuous improvement.

With Deming, Dr. Joseph Juran was the other great mind behind the quality movement. Juran was also responsible for the economic rise of the Japanese nation through the use of TQM. Key contributions include TQM—added human dimension, the Pareto principle, the need for widespread quality training, top management involvement, his books *Quality Control Handbook* (1951) and *Managerial Breakthrough* (1964), Quality Improvement video series, and Quality Trilogy, 1986 (quality planning).

Kaoru Ishikawa is best known for the Ishikawa or fishbone diagram. As a leader in quality improvement, Ishikawa developed and enhanced the principles of cause and effect. According to Dr. Ishikawa, quality improvement is a continuous process that requires the company-wide use of seven quality tools: flowcharts, run charts, histograms, Pareto charts, scatter diagrams, fishbone diagrams, and control charts. A champion of service after the sale, Ishikawa believed a customer should receive continued support and customer service even after the product is purchased. Dr. Ishikawa was the first to popularize the widespread use of quality circles.

Many experts credit Dr. Genichi Taguchi with being the quality guru who revolutionized the Japanese manufacturing process. While Deming, Juran, and Shewhart were guest lecturers, consultants, and quality leaders, Dr. Taguchi was committed to the economic rebirth of his country. Taguchi's approach was to identify the key factors, which had the most significant effect on product variability. Key contributions include the loss function—equate quality to cost, robustness—ability of process to perform as intended, and orthogonal arrays and linear graphs—isolated outside factors.

In 1979, another American quality guru appeared—Phillip Crosby. He published fourteen books on human relations, leadership, and quality management. Mr. Crosby's greatest ability was to take a complex topic and present it in common terms that were easy to understand. Crosby believed in teaching quality principles to all levels of an organization, creating a quality specialist department, emphasizing top-down quality management, getting it right the first time—zero defects, and system modeling.

There are a number of people who should be considered as honorary members of the quality movement. They include Dewey, Miles, Osborne, and Box. Dr. John Dewey believed that "education is not a preparation for life. Education is life itself." John Dewey taught the idea or concept that people learned from their interactions with the environment. A brilliant educator and philosopher, Dewey was considered to be the father of modern education. Larry Miles is the father of the value method, a systematic approach used to analyze and improve value in a system, a facility, or a product. Alex F. Osborne is the father of the brainstorm, who believed that "it is easier to tone down a wild idea than to think up a new one." From the classroom of World War II, Dr. George Box became a leading authority on the statistical approach.

Review Questions

1. Describe the history of the quality movement. List significant individuals.
2. Describe the importance of quality control in the process industry.
3. Write your definition for "quality."
4. List the significant contributions of Walter Shewhart.
5. List the significant contributions of W. Edwards Deming.
6. List the significant contributions of Joseph Juran.
7. List the significant contributions of Kaoru Ishikawa.
8. List the significant contributions of Genichi Taguchi.
9. List the significant contributions of Phillip Crosby.
10. List the significant contributions of George Box.
11. List the significant contributions of John Dewey.
12. Describe the principles of scientific management.

appendix 2

Principles of Total Quality Management

LEARNING OBJECTIVES

- Review the principles of total quality management.
- Define the term "quality management."
- Describe the term "conformance to specifications."
- Describe the term "maintenance of consistency."
- Describe the terms "internal failure and external failure."
- Compare and contrast the cost of quality.
- Describe the terms "appraisal and prevention."
- Describe the term "meet or exceed customer expectations."

Key Terms

Total quality management (TQM)—a philosophy that represents the guiding principles of a continuously improving organization.

Quality statements—include the quality policy statement, mission statement, core values, and vision statement.

Core values—include customer-driven excellence, visionary leadership, organizational and personal learning, valuing employees and partners, agility, management for innovation, focus on the future, management by fact, systems perspective, public responsibility and citizenship, and focus on results and creating value (Malcolm Baldrige Award).

Quality council—the heart of the TQM system that provides direction and builds a quality culture. It typically includes the chief executive officer (CEO), senior managers, and the quality coordinator.

Basic concepts of TQM—focus on customer, continuous improvement, top to bottom commitment, involvement of work force, treatment of suppliers as partners, and establishing performance measures for processes.

Introduction to Total Quality Management

Total quality management (TQM) has been defined as a philosophy that represents the guiding principles of a continuously improving organization. This process involves the organizational use of quantitative tools and techniques, management principles and practices, and human resource development (HRD). The primary purpose of TQM is to provide a quality product to the customer, which will result in improved sales and lower production costs. The philosophy of TQM must be embraced by the corporate culture from the top to the bottom. Product quality is the responsibility of the entire organization.

The first hard lessons of the quality war were learned by American business people as the Japanese captured significant market shares in the automotive and electronics industries. Consumers preferred the reliability and cost of Japanese products. In a period of less than 20 years, the Japanese erased their poor reputation for exported goods. As many business and industry leaders flew to Japan in the late 1970s and 1980s to learn the secrets of their success, they were amazed at what the Japanese were able to do with limited resources. Old equipment and technology were being used to produce superior products. The TQM system was integrated into the entire culture. Complex operations were broken down into smaller processes by this system, and better ways of doing business was identified. What these business people found was very distressing. The quality movement led by a number of Americans had revitalized Japanese manufacturing. It is only after a company loses significant shares of the market that they realize the need to implement TQM. According to several quality experts, quality should be first among equals of cost and service (Figure A2-1 and Figure A2-2) (Table A2-1).

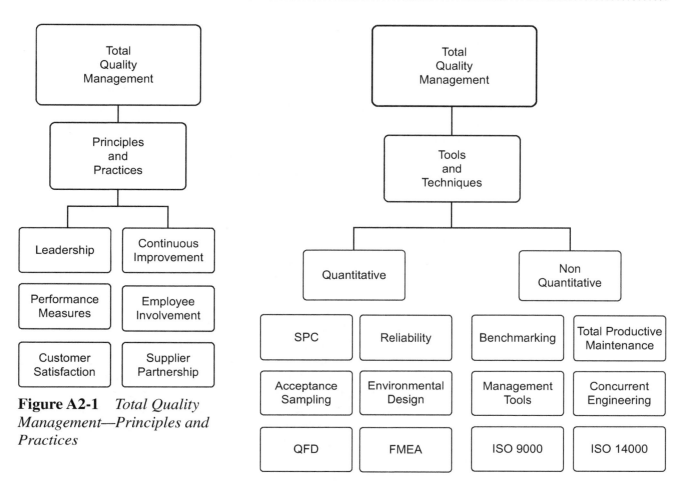

Figure A2-1 *Total Quality Management—Principles and Practices*

Figure A2-2 *Total Quality Management—Tools and Techniques*

New	Old
Manager teaches, facilitates, and delegates	Manager controls, plans, and assigns
Customer-oriented	Product-oriented
Long-term goals	Short-term goals
Prevention	Detection
Managers handle problems	Teams handle problems
Quality #1	Cost and service #1

Table A2-1
The TQM Change Process

Principles and Practices of TQM

Continuous Process Improvement

In the chemical processing industry, the focus is on continuous improvement and perfection. Although perfection is difficult to achieve, improvement is not. Under the continuous improvement model, each operating unit is broken into individual processes. A company can continually improve by:

- making our processes adaptable, effective, and efficient
- planning for the changing needs of our customers
- visualizing all work as a process
- evaluating in-process performance: control charts, scrap reduction, and paper-work reduction
- eliminating waste and rework
- evaluating value-added and nonvalue-added work activities
- using statistical process control (SPC) tools and techniques
- benchmarking.

Process technicians have a number of steps that can be used during the quality problem-solving process. These steps include (1) identify and prioritize improvement opportunities, (2) analyze current process, (3) develop and implement solutions, (4) study and evaluate results, and (5) standardize work practices and plan for the future. This section will include discussions on the technology that provides the foundation for quality improvement. Process technicians use this technology, which is a valuable component of the continuous quality improvement team. The principles of continuous quality improvement include:

- innovation and improvement of services and products
- innovation and improvement of processes
- commitment to quality
- integration of suppliers and customers into the quality process
- using quality tools
  - SPC
  - flowcharts
  - cause-and-effect diagrams (fishbones)
  - Pareto charts
  - run charts
  - control charts
  - planned experimentation
  - histograms or frequency plots
  - forms for collecting data
  - scatter plots
- auditing and evaluation
- providing continuous quality improvement training to all employees

- unrelenting commitment and involvement of all levels in the organization
- documentation of what you do and doing what you say.

Quality Improvement Cycle

Phase One—Plan
The first step in the improvement cycle is to increase current knowledge of the process. The more the team knows about the process, the more likely the changes submitted by the team will impact on quality. At the conclusion of phase one, a plan should be developed that will (1) address specific questions and (2) consider methods, resources, schedules, and people. Phase one will take a significant amount of time for the team to complete. The planning phase should address:

- specific objectives and questions
- predictions, and
- plan for test.

Phase Two—Observe and Analyze
Phase two implements the data collection process. The data collected will be used to address the questions from phase one.

Phase Three—Learn
This phase combines phase one and phase two activities. The results of the data analysis can be compared with current knowledge to see if contradictions exist.

Phase Four—Act
The results from phase three are used to decide whether a change to the process is required or not. If a change is required, a modified brainstorming session should be conducted to determine what changes to the process would result in improvement. These changes should be clearly stated (Figure A2-3).

Supplier–Customer Relationship

Industrial manufacturers buy raw materials from suppliers to produce products for their customers. Companies depend on suppliers to provide them with quality raw materials. Customers depend on companies to provide them with quality products. In today's global economy, a new relationship exists among suppliers, companies, and customers. Each is dependent upon the other for financial success. Companies are becoming more and more involved with customers and suppliers. Products and raw materials are tracked from inception. Documentations, quality charts, and external

Figure A2-3
Improvement Cycle

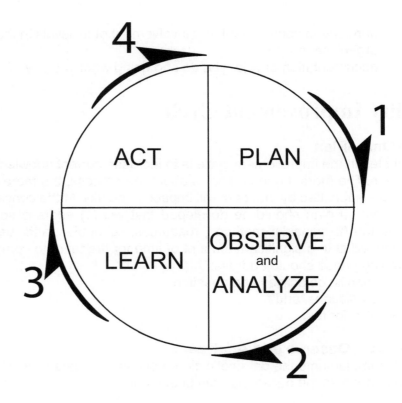

audits follow products and raw materials from cradle to grave. Customers are providing more information about their needs to companies.

Leadership

Quality control is the responsibility of every person in the organization; however, it must take a top-down approach. True leadership is involved in the quality process at each step of the analysis, design, development, implementation, and evaluation. TQM must be infused into the corporate culture.

Employee Involvement

Each of the quality gurus mentioned in Chapter 1 believed the use of quality tools and techniques that should be taught at all levels of the organization. The most valuable resource of an organization is its employees. In many cases, quality problems are attributed to operating personnel. Dr. W. Edwards Deming believed that only 10–20% of the quality problems could be attached to operations. This leaves 80–90% of all quality problems firmly attached to management. Employee involvement requires an open communication among all levels of an organization. This requires an open and receptive environment that allows every employee to participate. Other common techniques include project teams, education and training, and suggestion systems.

TQM Tools and Techniques

In the TQM system, a number of quality tools are traditionally used. These tools are designed to improve the quality of the products or services offered by the company. A short list of these tools include:
- SPC—Pareto diagrams
- SPC—Ishikawa cause and effect
- SPC—checksheets
- SPC—process flow diagrams
- SPC—scatter diagrams
- SPC—histograms
- SPC—control charts
- acceptance sampling
- reliability
- experimental design.

The Quality Council

Quality councils are established to create an atmosphere inside an organization where the principles of quality can take root. Some quality gurus believe that the quality council is the heart of a TQM system. The council is composed of the CEO, senior managers of functional areas, and the quality coordinator. The responsibilities of the quality council include improving communications, developing trust with customers and employees, training and coaching rather than directing and supervising, emphasizing improvement rather than maintaining, empowering rather than controlling, encouraging cooperation among groups, learning from problems, choosing quality rather than price, developing quality support systems, recognizing team efforts, and demonstrating a total commitment to quality (Figure A2-4).

Core Values of Total Quality Management

TQM has a number of core concepts. The Malcolm Baldrige Award has identified a number of **core values** that can be used to develop a quality system. Some of these concepts include:
- top-down management support—visionary leadership
- customer focus—customer-driven excellence
- organizational and personal learning
- valuing employees and partners
- agility
- management for innovation

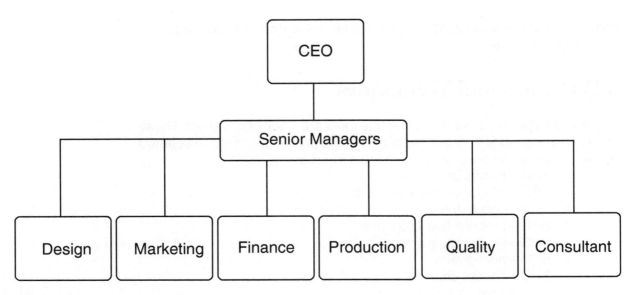

Figure A2-4 *Quality Council*

- focus on the future
- management by fact
- systems perspective
- public responsibility and citizenship
- focus on results and creating value.

Quality Statements

The core values listed in the previous section are typically included in the **quality statements** of the organization. These statements also include core values, a mission statement, a vision statement, and a quality policy statement. A company's quality statements help define the organization and give direction and purpose. Each level of the organization should have input into the development of these four statements. Once developed, quality statements should not be changed or modified.

Summary

In TQM there is an old saying "the customer is always right." Company individuals who deal closely with customers should be aware of the critical role they play. Diplomacy, communications, and willingness to continually improve services and products are important variables when working with customers. This same role can also work with company vendors and suppliers. In the quality circle, each side is equally important to the process. According to a number of experts, "quality should be considered first among equals of cost and service."

TQM has been defined as a philosophy that represents the guiding principles of a continuously improving organization. This process involves the organizational use of quantitative tools and techniques, management principles and practices, and HRD. The primary purpose of TQM is to provide a quality product to the customer, which will result in improved sales and lower production costs. Under the continuous improvement model, each operating unit is broken into individual processes. Process technicians have a number of steps that can be used during the quality problem-solving process. These steps include (1) identify and prioritize improvement opportunities, (2) analyze current process, (3) develop and implement solutions, (4) study and evaluate results, and (5) standardize work practices and plan for future.

Quality councils are established to create an atmosphere inside an organization where the principles of quality can take root. Some quality gurus believe that the quality council is the heart of a TQM system. Quality statements include (1) core values, (2) a mission statement, (3) a vision statement, and (4) a quality policy statement. A company's quality statements help define the organization and give it direction and purpose.

Review Questions

1. Describe the purpose and role of quality councils.

2. Describe the individual elements of the quality statements of a company.

3. Explain how continuous improvement helps a chemical company.

4. What is total quality management?

5. List and describe the quality tools used by process technicians.

6. What are the five steps listed in the quality problem-solving process?

7. What is the heart of total quality management?

8. List the core values of the Malcolm Baldrige Award.

9. A customer has a serious problem with one of your products. Describe what you would say and do.

10. What is percentage of quality problems management responsible for according to Dr. W. Edwards Deming?

11. What are the listed responsibilities of the quality council?

12. Describe the statement "quality is first among equals of cost and service."

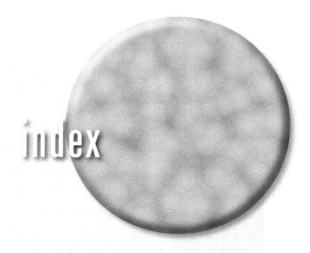

index

CPSIA information can be obtained
at www.ICGtesting.com
Printed in the USA
BVHW060338060422
633520BV00007B/55